A handbook to the Common Intertidal Invertebrates of the Gulf of California

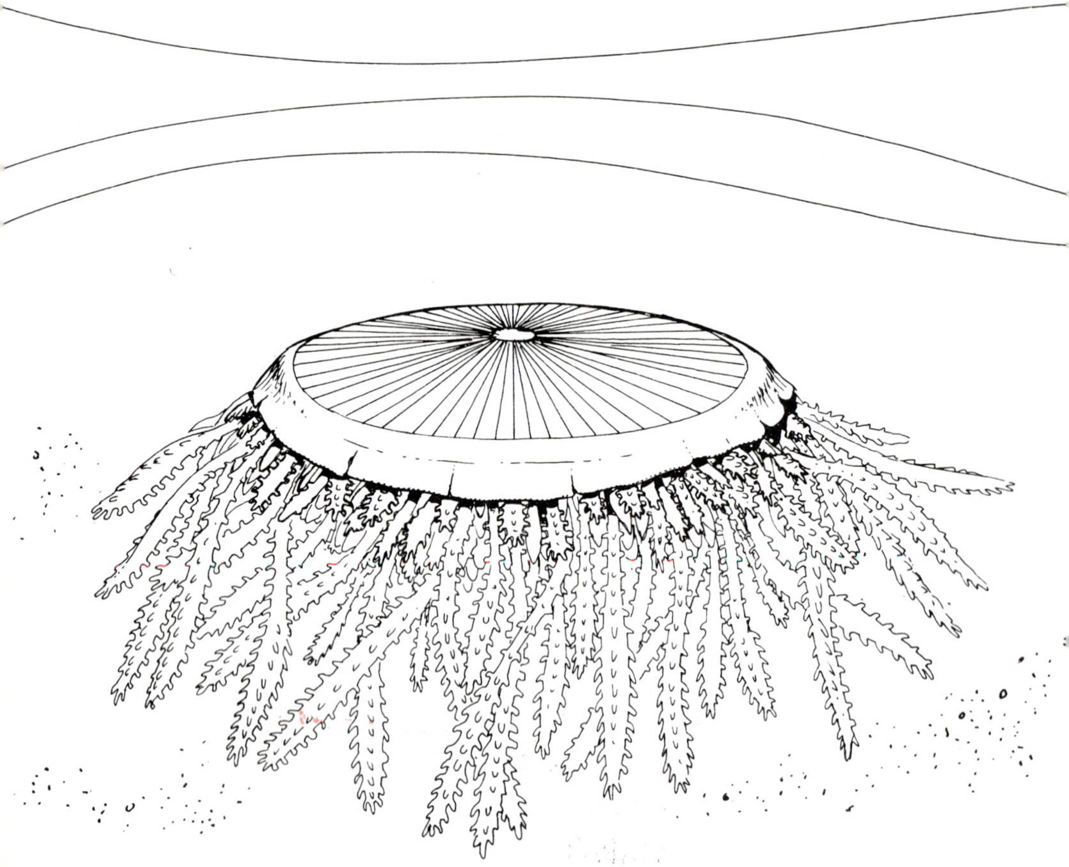

A handbook to the Common Intertidal Invertebrates of the Gulf of California

RICHARD C. BRUSCA

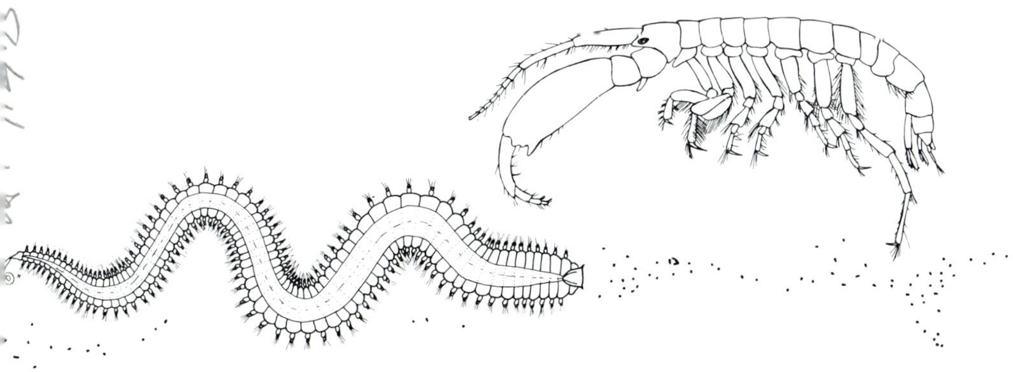

THE UNIVERSITY OF ARIZONA PRESS • Tucson, Arizona

About the Author . . .

RICHARD C. BRUSCA researched and completed this project as Resident Marine Biologist and Station Manager at the University of Arizona-University of Sonora Cooperative Marine Station at Puerto Peñasco (Rocky Point) on the upper Sonoran coast of the Gulf of California, while he was working toward a Doctor of Philosophy degree at the University of Arizona. Born and raised in California, he has studied Pacific Coast marine life for more than 10 years. His B.S. degree is from California State Polytechnic University, his M.S. degree from California State University at Los Angeles, and he is a member of several scientific research societies, including the honorary Sigma Xi.

Illustrations in the section "Mollusca (Snails, shellfish, etc.)" (pages 132 to 180) reprinted from *Sea Shells of Tropical West America: Marine Mollusks From Baja California to Peru, Second Edition* by A. Myra Keen with the permission of the publishers, Stanford University Press. ©1958, 1971 by the Board of Trustees of the Leland Stanford Junior University.

THE UNIVERSITY OF ARIZONA PRESS

Manufactured in the U.S.A.

I. S. B. N.-0-8165-0356-7
L. C. No. 72-76901

to my
Mother and Father
and to
Waren Stauls
(poet, scientist, friend)

Contents

ILLUSTRATIONS

Platyhelminthes

Nemertea

Annelida (Polychaetes)

Mollusca

Mollusca (cont'd.)

Arthropoda

Preface

This handbook is the first publication of comprehensive taxonomic keys and species characterizations for the intertidal fauna of the Gulf of California. It is designed for use by students, teachers, and interested amateurs, and is intended to be an easily read source of rapid identification and ready information on the fauna of this region. Perhaps as important as the keys, most of which are original, is the author's attempt to bring together a great deal of miscellaneous information that is scattered through the literature and so lost to the eyes of most amateur zoologists and students of the seashore. This accounts for the larger lists of references than one usually expects to find in a handbook.

The value of taxonomic keys in identifying animals of a specific region cannot be underestimated; yet, odd as it may seem, there are few such publications available. For the west coast of North America, this volume joins two earlier principal works of this type: Light, et al. (1961), and Allen (1969). The first of these, S. F. Light's classic *Intertidal Invertebrates of the Central California Coast,* deals mainly with the animals found between Monterey and Bodega Head, in California. R. K. Allen's *Common Intertidal Invertebrates of Southern California* deals principally with the animals found south of Point Conception, many of which occur as far south as the outside tip of the Baja California peninsula. This handbook, the third source of comprehensive keys, deals with the invertebrates of the Gulf of California, many of which occur as far south as Ecuador and the Galapagos Islands.

In addition to these three, there are a handful of well-known volumes dealing principally with the animals of the west coast of North America north of Mexico, but lacking in keys for their identification. These include MacGinitie and MacGinitie (1968), Johnson and Snook (1927), Ricketts, Calvin, and Hedgpeth (1969), Hedgpeth (1964), Hedgpeth and Hinton (1961), and Reish (1968). Only two prior popular books dealt specifically with the fauna of the Gulf of California: Wesley Farmer's *Tidepool Animals from the Gulf of California* (1968), and Steinbeck and Rickett's *Sea of Cortez* (1941). The former is quite limited in scope, although possessing an excellent section on the opisthobranch molluscs; the latter is out of print.

This handbook was prepared during the two-year period 1970 to 1972, while the author was stationed in residence at the University of Arizona-University of Sonora Cooperative Marine Station at Puerto Peñasco, Sonora, Mexico. During this time, numerous collections were made throughout the Gulf. Collections from the University of Arizona and the University of Sonora invertebrate museums, as well as several private collections, also were available for study.

This first edition of *Handbook to the Common Intertidal Invertebrates of the Gulf of California* is expected to be only a beginning. Data and specimens from the Gulf of California have been accumulating at the University of Arizona almost faster than they can be processed. I hope that anyone — student, professor, or amateur — who has new information that would be useful, or who has found any information in this handbook to be in error, will forward his comments, so that such information can be used in future revisions.

RICHARD C. BRUSCA

Acknowledgments

The author's most sincere gratitude and appreciation are extended to the scientists listed below, for without their cooperation and efforts this handbook could not have been prepared.

Dr. Donald P. Abbott, Hopkins Marine Station, for identification of compound ascidians; Dr. Richard K. Allen, California State College at Los Angeles, for reviewing the entire manuscript, donating several illustrations, and preparing the author for his own taxonomic niche; Dr. Gerald J. Bakus, Allan Hancock Foundation, for a great deal of assistance with the poriferan section of the manuscript; Dr. William C. Banta, American University, for identification of bryozoans; Dr. J. Laurens Barnard, Smithsonian Institution, for writing the amphipod key, for reviewing the polychaete section, for helpful advice, and for good companionship; Dr. Thomas Biffar, University of Miami, for identification of callianassid Crustacea; Dr. James A. Blake, Pacific Marine Station, for reviewing the polychaete section of the manuscript; Dr. Gary J. Brusca, Director, Trinidad Marine Laboratory of Humboldt State College, Arcata, California, for reviewing the entire manuscript, setting goals to strive for, serving good wine, and publishing the *Stomatopod;* Dr. T. H. Bullock, University of California at San Diego, for identification of hemichordates, and his review of that section of the handbook; Mr. Martin Burkenroad, San Diego Museum of Natural History, for identification of thalassinid Crustacea, suggestions for improvement of the manuscript, and stimulating conversations on the philosophy of taxonomy; Dr. Peter Castro, University of Puerto Rico, for advice on natantiate Crustacea; Dr. Pérez Farfante, Smithsonian Institution, for help with the penaeid shrimp section; Dr. Fenner A. Chace, Jr., Smithsonian Institution, for identification of shrimp, and his review of that section of the manuscript; Dr. Kristian Fauchald, Allan Hancock Foundation, for reviewing the polychaete section; Dr. John S. Garth, Allan Hancock Foundation, for a tremendous amount of help in identification of brachyuran Crustacea, and his review of that portion of the handbook; Miss Janet Haig, Allan Hancock Foundation, for the investment of a great deal of time and effort in identifications, and for her review of the porcellanid section; Dr. Cadet Hand, Bodega

Marine Laboratory, for a review of the cnidarian section; Dr. Olga Hartman, Allan Hancock Foundation, for looking over the polychaete section; Dr. Joel W. Hedgpeth, Oregon State University Marine Science Center, for reviewing the introductory sections of the manuscript; Roy S. Houston, University of Arizona, for preparation of the mollusc section; Jerry D. Kudenov, University of Arizona, for preparation of the polychaete section; Dr. Raymond B. Manning, Smithsonian Institution, for identification of stomatopod crustaceans; Dr. John McKay, Interlibrary Loans Librarian, University of Arizona Library, and his staff, for obtaining hard-to-find journals; Doctors A. R. Mead and W. B. Miller, University of Arizona, for reviewing the polychaete chapter; David H. Montgomery, California State Polytechnic College at San Luis Obispo, for reviewing the introductory sections of the manuscript, introducing me to the sea, and for putting up with me, periodically, for many years; Dr. David L. Pawson, Smithsonian Institution, for identification of holothuroids; Dr. Donald J. Reish, California State College at Long Beach, for identification of polychaete annelids; Dr. Mary E. Rice, Smithsonian Institution, for identification of peanut and spoon worms; Dr. Arnold Ross, San Diego Museum of Natural History, for advice on the barnacle section; Mr. Charles R. Smith, California State Department of Public Health, for collaboration and partnership on the echinoderm section; and Dr. John S. Sumner, Department of Geosciences, University of Arizona, for review of the section on the physical evolution of the Gulf of California and for his many helpful suggestions.

Sincere thanks are also extended to: Mr. Richard Inglis, Mrs. Dorothy Inglis, Miss Kathy Jepma, and Mr. George Staley, for help in collecting specimens, gathering data, etc.; Dr. Agustín Ayala Castañares, Director del Instituto de Biología, Universidad Nacional Autónoma de México, for advice regarding commercial shrimp of the Gulf; Dr. Donald A. Thomson and his crew in the fish museum at the University of Arizona; Mr. Alex Spear; Juan Pistola and Edith in La Paz; Sr. Ramón Durazo in Puerto Peñasco; and Mr. and Mrs. Richard Dwyer, for transporting Dr. Barnard and me to islands in the southern Gulf.

Special thanks go to Mr. Jerry Battagliotti, artist for most of the illustrations in this work. Mr. Battagliotti's illustrations are unquestionably one of the book's most valuable assets. His skill in lifelike free rendering, while maintaining the taxonomic validity of his drawings, is both impressive and appreciated.

Acknowledgments also are given to Drs. A. R. Mead, W. B. Miller, and P. E. Pickens, for preliminary work in assembling the University of Arizona invertebrate collection, and, in Dr. Pickens' case, for special encouragement to the author.

Appreciation is extended to the initiators of this project, the Marine Sciences Committee of the University of Arizona, whose director, Dr. John R. Hendrickson, gave the author his unflagging support and assistance throughout the writing of the handbook.

The research and writing of this manual would not have been possible were it not for the full cooperation of the University of Sonora, and especially that of Dr. Carlos Enrique Peña, Director of C.I.C.T.U.S. (Centro de Investigaciones Científicas y Tecnológicas de la Universidad de Sonora), and Sr. Agustín Cortés-Chávez, Coordinador Administrativo Unidad Experimental Peñasco, C.I.C.T.U.S.

I am deeply grateful to Marshall Townsend, Director, the University of Arizona Press, and other members of his helpful staff, for all of their efforts in transforming my manuscript into published form, with special thanks to Gale Monson for his excellent editorial work.

Lastly and most importantly I want to thank my wife Caren, my son Alec, and my daughter Carlene, for making the time spent in Mexico not only academically rich but a beautiful and fulfilling experience.

RICHARD C. BRUSCA

introduction

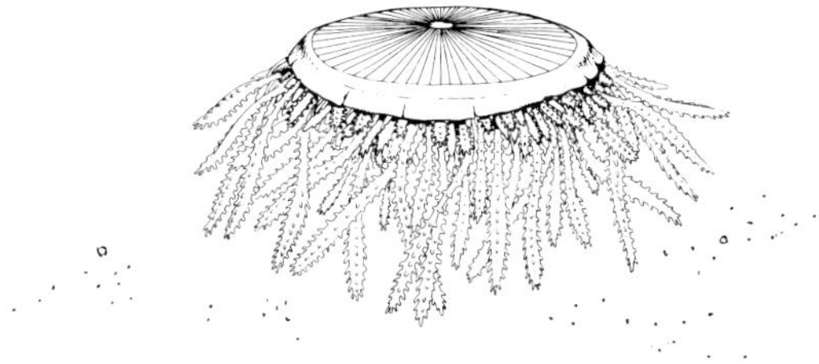

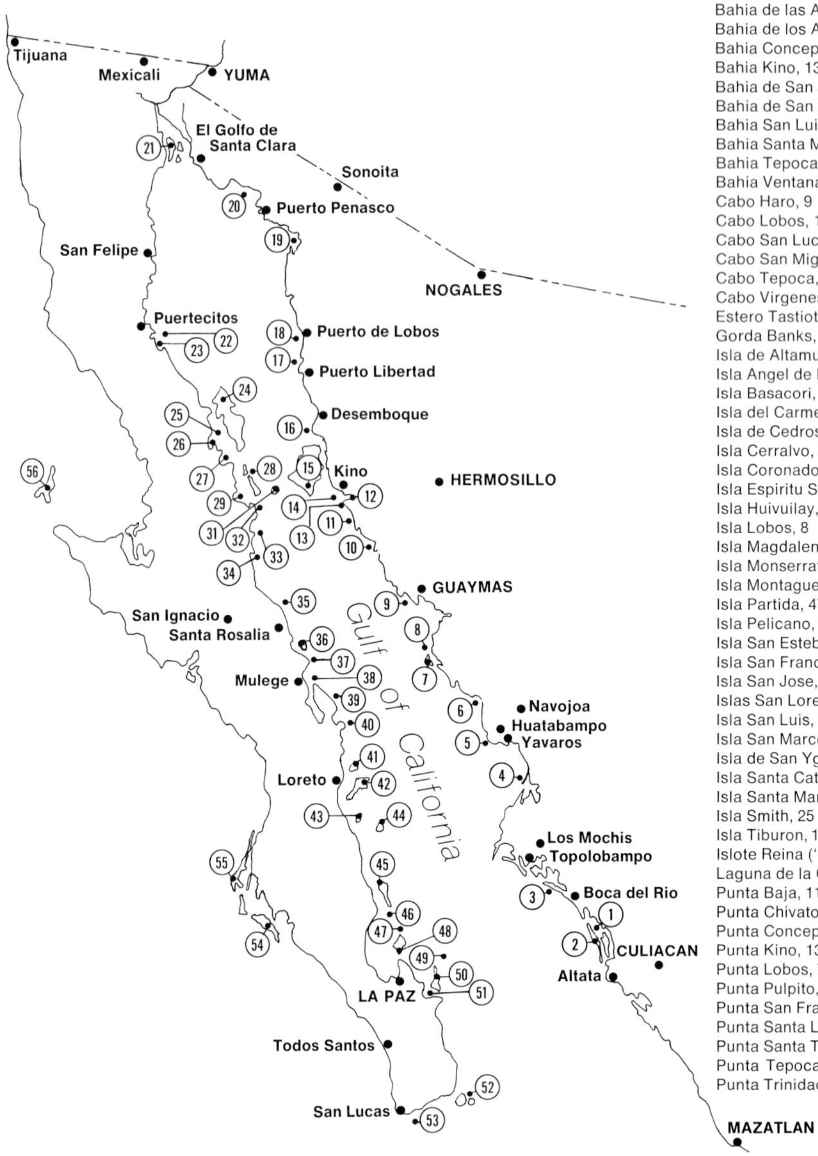

Bahia Adair, 20
Bahia de las Animas, 27
Bahia de los Angeles, 26
Bahia Concepcion, 38
Bahia Kino, 13
Bahia de San Jorge, 19
Bahia de San Rafael, 29
Bahia San Luis Gonzago, 23
Bahia Santa Maria, 1
Bahia Tepoca, 18
Bahia Ventana, 51
Cabo Haro, 9
Cabo Lobos, 17
Cabo San Lucas, 53
Cabo San Miguel, 33
Cabo Tepoca, 16
Cabo Virgenes, 35
Estero Tastiota, 10
Gorda Banks, 52
Isla de Altamura, 2
Isla Angel de la Guarda, 24
Isla Basacori, 4
Isla del Carmen, 42
Isla de Cedros, 56
Isla Cerralvo, 50
Isla Coronado, 41
Isla Espiritu Santo, 48
Isla Huivuilay, 6
Isla Lobos, 8
Isla Magdalena, 55
Isla Monserrat, 43
Isla Montague, 21
Isla Partida, 47
Isla Pelicano, 14
Isla San Esteban, 31
Isla San Francisco, 46
Isla San Jose, 45
Islas San Lorenzo, 28
Isla San Luis, 22
Isla San Marcos, 36
Isla de San Ygnacio, 3
Isla Santa Catalina, 44
Isla Santa Margarita, 54
Isla Smith, 25
Isla Tiburon, 15
Islote Reina ("Seal Rock"), 49
Laguna de la Cruz, 12
Punta Baja, 11
Punta Chivato, 37
Punta Concepcion, 38
Punta Kino, 13
Punta Lobos, 7
Punta Pulpito, 40
Punta San Francisquito, 32
Punta Santa Lugarda, 5
Punta Santa Teresa, 39
Punta Tepoca, 18
Punta Trinidad, 34

Key to this map is an alphabetical list of pertinent sites with locations numbered in consecutive order northward from the southernmost point on the mainland coast, down the Baja California coastline and on around the Baja California tip.

Gulf of California

Introduction

1. The Physical Evolution of the Gulf of California.

Data and theories regarding the physical history of the Gulf of California and the peninsula of Baja California have improved and advanced considerably in the last few years. What follows is a brief summary of some of the most significant recent literature.

The Gulf of California is a unique body of water that so far has had a short but dynamic life. Unlike most other marine basins of similar size, the Gulf was comparatively shallow until relatively recent times. The Gulf originated in the Tertiary period of the Cenozoic era, most likely during the Miocene or Pliocene epoch. Larson et al. (1968) placed the date of the opening of the deeper southern Gulf at 4 to 4.5 million years ago, while Rusnak and Fisher (1964) placed the initial opening spasms throughout the length of the Gulf closer to 25 million years ago. The former authors based their conclusions on recently obtained magnetic data, the latter on paleontology.

Durham and Allison (1960) stated, "Throughout the Cenozoic period the boundaries of the Gulf expanded and contracted in harmony with submersion and emersion of the peninsula and adjacent areas on the eastern margin of the Pacific and with sedimentation at the mouth of the Colorado River. The peninsula of Baja California at times was restricted to a small northern area with its southern extent expressed in a series of islands. During times of emergence the peninsula expanded slightly to include areas beyond its present coasts." These authors also presented a series of interesting maps, illustrating the theoretical shape and location of the Gulf and Baja California through the Paleocene, Eocene, Oligocene, Miocene, and Pliocene.

Schuchert (1935) stated that the present elevation of mainland México is the result of block faulting that also occurred in the late Pliocene and Pleistocene; however, the geologic origins of the Gulf are probably unrelated to the geologic phenomena that produced the topography of present-day mainland México.

The present shape of the Gulf is most likely the result of vertical and horizontal movements that took place along lines of structural weakness along the East Pacific Rise, tearing the Baja California peninsula away from the mainland as the result of great blocks of the earth's crust being pulled away from the crest of the Rise. The crest of the East Pacific Rise intersects the North American mainland very near the mouth of the Gulf, extending south from this region and continuing as a globe-circling oceanic ridge system. The Rise disappears as such from the Gulf northward, becoming the San Andreas Fault, and finally reappears again in the north Pacific off Oregon in the form of the Gorda Rise and the Juan de Fuca Ridge (Menard, 1964). So, the Gulf formed as the peninsula of Baja California split off from the west coast of México, this split probably being due to the northwestward movement of elevated fractured plates of crustal material that separated from the mainland to leave a deep basin. Accompanying this movement was a complex variety of gravitational slidings and volcanic eruptions, many of the latter due to a weakening of the earth's crust, and the production of fault zones caused by the northwest movements of the land mass. Rusnak and Fisher (1964) stated the source of this uplift and tilting may be ascribed to the development of the East Pacific Rise, and the Gulf would therefore remain as an extension of this Rise. Larson et al. (1968) extended this concept, and reported the Gulf as a product of simple ocean-floor spreading from the crest of the Rise.

According to paleontologic data, the northern Gulf is older than the southern portion. Evidence indicates it was in existence (in some form) during the late Miocene (Larson et al., 1968). It has been postulated that the Gulf initially consisted of a narrow channel or very shallow southern Gulf leading up to the larger northern basin. The southern Gulf then could have arisen as a series of 'offsets' or spreadings, initially moving from north to south.

The lower two-thirds of the Gulf appears to be a rift between two continental blocks, separated by a crust of oceanic character, while the upper third of the Gulf is more strongly affected by the deltaic deposits of the Colorado River. There is an extensive deep water region in the lower Gulf, with a few of the numerous basins plunging to depths of more than 9,000 feet. The upper Gulf is much flatter and more shallow, averaging around 600 feet in depth. The ocean floor of the Gulf is also characterized by a high heat flow, which is probably related to the presence of the East Pacific Rise in this region.

Anderson (1971) placed the origin of the Gulf at about 25 to 30 million years ago. He postulated that mainland México had been decoupled from the Baja California strip before this time, the Baja region being part of the Pacific Plate of crustal material, west of the East Pacific Rise. As a segment of oceanic rise moved eastward to approach the southern half of Baja California, the peninsula was pulled away from the mainland. Further, as magma from the upper mantle welled up into the rift, it formed a new rise that worked its way north into the widening Gulf. This tearing off of Baja California from the mainland therefore formed the southern part of the San Andreas Fault system.

The San Andreas Fault appears to form a boundary between the North American Plate of crustal material and the North Pacific Plate, separating the southwestern part of California from the rest of North America. The Pacific Ocean and that part of California west of the San Andreas Fault seem to be moving northwest in respect to the rest of the continent. Anderson (1971) pointed out that the theory of moving plates is fundamental to the theory of continental drift, which was long disputed but now generally is accepted due to the voluminous evidence from geological, geophysical, biological, and geochemical data.

The Gulf is bordered on the west by the high mountains and pre- to mid-Cretaceous igneous and metamorphic rocks of the Baja California peninsula, consisting largely of granitic blocks. On the east, the Gulf is bounded by the sedimentary and volcanic blankets of the coastal plains of Sonora and Sinaloa, which are largely covered by broad alluvial fans and deltas of sand and gravel. These are in turn covered, in places, by high sand dunes. The head of the Gulf is bordered by the Colorado River delta, which itself rests on the floor of the old Colorado Desert and now serves to keep the waters of the Gulf from entering the low Imperial Valley and Salton Sea area of California. Much of the region around the northern Gulf consists of low-lying coastal plains, whose mud and salt flats bear sediment derived from the Colorado River.

The Imperial Valley is a structural continuation of the Gulf. It forms a trough whose land surface rests more than 250 feet below sea level, and in places now contains more than 19,000 feet of recent deltaic and marine deposits. Even in the northern Gulf, these deposits often exceed 15,000 feet in depth. Most of the deposits that have been examined are silt and clay of recent origin, although sediments of Miocene and Pliocene age have also been reported.

The recent history of the Salton Sea and the northern Gulf of California is an intriguing story. Briefly, it is as follows. In the recent geologic past, the Gulf of California extended northward over the present Imperial and Coachella valleys to near San Gorgonio Pass, north of Palm Springs. The Colorado River was, at this time, emptying into the Gulf somewhere near the present site of Yuma, Arizona. As silt at the river mouth built up, it began forming a massive delta that eventually extended the entire width of the river's valley, closing it off to isolate the northern part as an ancient salt water lake, covering an area that would now include the cities of Mexicali, Brawley, and Indio. This lake had no outlet, and, like the Great Salt Lake of Utah, its evaporation rate exceeded its influent and it eventually dried up. Later the Colorado River changed its course to begin dumping water into the dry lake bed instead of the Gulf, creating prehistoric Lake Cahuilla (or Lake LeConte).

During this period, the lake was drained into the Gulf by the now extinct Rio Hardy. Finally the Colorado River again changed its course, to resume emptying directly into the Gulf. Lake Cahuilla dried up, and remained dry until 1905, when the river course was accidentally diverted toward the west to once again begin pouring water into the bed of Lake Cahuilla. This was the result of the construction of irrigation headworks to divert part of the Colorado's flow into the Imperial Valley by means of a long level canal. The headworks were destroyed by floodwaters and the entire river began to flow down the canal instead of following its own channels to the Gulf (winter 1904–1905 and summer 1905). For two years the river emptied into the Imperial and Coachella valleys, finally being rediverted to its former course in 1907, when the break in the irrigation headworks was repaired. From this came the no-outlet Salton Sea.

Sykes (1937) reported that in 1906 he observed the level of the Salton Sea rising at the rate of 4 inches per day, with a rapid advance of the water across the nearly level floor of the basin. After 1907 evaporation lowered the Salton Sea level, with a corresponding increase in salinity. More recently (since about 1948), the Sea has risen from irrigation drainage to decrease the salinity. The surface elevation of the Salton Sea fell from −195 feet in 1907 to −250 in 1925, and from 1925 to 1935 the level fluctuated between −250 and −243 feet (Walker, 1961). Recent estimates of future stabilization levels vary from −220 feet (Blaney, 1954) to −235 feet (Walker, 1961). It is of interest to note that the salinity of the Colorado River water at the head of the Alamo channel irrigation canal, prior to its destruction in 1905, was only 0.7% (Ross, 1914) and yet the Salton Sea had, by 1907, attained a salinity of 3.6%. The greater salinity of the sea in respect to the river was due to salts leached from the bottom deposits left by evaporation on previous occasions. The salinity of the Salton Sea remained below 1.5% until as recently as 1930, reached 4.04% in 1948 and fell to 3.07% in 1955. The present water source for the Salton Sea is largely the All-American Canal, an irrigation tap off the Colorado River at Imperial Dam above Yuma.

Before the building of Hoover Dam (1935) and the advent of large scale irrigation projects in the Imperial Valley, the effects of the Colorado River drainage could be seen in the Gulf as far south as La Paz. The average annual flow, in the vicinity of Yuma, for the years 1902 to 1934 was over 15 million acre-feet. In recent years, however, the flow of the river into the Gulf has been drastically and continually reduced. Thompson (1965) stated that little river detritus has reached the Gulf of California in the last 55 to 60 years. The average annual river flow at the mouth for the years 1935 to 1965 was a little over 4 million acre-feet. In the early 1960s the annual discharge recorded at a station in El Maritimo, México, was only 85,000 to 503,000 acre-feet.

The origins of the numerous islands in the Gulf of California are complex, and often are based on independent structural events, not necessarily dependent upon each other, or the massive movements of the peninsula itself. Most of the islands broke from the peninsula in the late Miocene. The Islas Angel de la Guarda and San Lorenzo may have been cut off along the same fault line that developed up to the Imperial Valley and gave rise to the infamous San Andreas Fault. The Islas Las Tres Marias, off the coast below Mazatlán and Isla San Francisco, near La Paz, were probably formed by volcanic upwellings, as were some of the other small islands of the Gulf.

At times during the past there existed a water connection between the Atlantic and Pacific oceans, across Central America, probably in the region of Panamá (although a few workers place this extinct water channel at southern México, where the Isthmus of Tehuantepec now exists). Throughout most of the Tertiary, the North Equatorial Current of the Atlantic, aided by the ever present westerly trade winds, funneled water through this island-studded land gap in middle America. This connection permitted the immigration of West Indies fauna to the Pacific coast of México, and a much lesser amount of easterly immigration. The evolutionary remnants of these migrations are seen among the animal assemblages of the Gulf of California and the West Indies today. Some of the California fauna has even been said to be remnants of an original Caribbean stock that has moved north from its point of immigration 10 to 50 million years ago.

2. Water Temperatures and Salinities in the Gulf of California.

There is a large seasonal temperature fluctuation in the northern Gulf (above lat. 28°N). At Puerto Peñasco this annual fluctuation may exceed 16° C, while at Bahía de Los Angeles it may exceed 11° C. Onshore water temperature at Puerto Peñasco may reach 30 to 32°C (88 to 91°F) in the summer, and may drop to 10 to 11°C (50 to 52°F) in the winter, although the usual temperature is around 15 or 16° C. Offshore sea surface temperatures, however, remain relatively constant throughout the northern Gulf, usually falling only 2 to 5°C midway between the eastern and western shores. In early summer a tongue of cold water, probably an upwelling, usually projects north from the vicinity of Isla Angel de la Guarda.

Sea surface temperatures in the southern Gulf (below lat. 28° N) exhibit less seasonal fluctuation than the waters of the northern Gulf. Onshore water temperatures drop to 17 to 18°C (63 to 65°F) in the winter, and rise to around 32 to 35°C (90 to 95°F) in the summer. Because of its warmer winter waters, the southern Gulf is generally considered a tropical or subtropical habitat, as opposed to the northern Gulf, whose cold winter water temperatures and cooler summer temperatures in respect to the southern Gulf give it a more temperate climate.

The salinity of the northern Gulf generally ranges from 36 to 37 parts per thousand, although exceptional months do occur — for instance, in April of 1954, a salinity of 39 parts per thousand was recorded at Puerto Peñasco (Thomson, 1969). The slightly higher salinities common to the northern Gulf are partly a result of increased evaporation in this semi-isolated region. Roden (1964) stated that evaporation exceeds precipitation by approximately 250 cm/year in the northern Gulf. In the southern Gulf the salinity is nearer that of the open sea, approximately 34.5 to 36 parts per thousand.

3. Faunal Provinces and the Fauna of the Gulf of California.

A faunal province has been defined by Schenck and Keen (1936) as a region populated by a distinctive assemblage of species. The uniqueness of this assemblage varies with the province and the specific animal groups used. Transition zones always exist between such general faunal assemblages, the width of these zones being different for each animal group examined. In other words, if the zoogeography of various groups of animals were to be studied, each being mapped individually, it may be found that each group would be represented by faunal zones with slightly different boundaries.

Johnson and Snook (1927) considered the west coast of North America, north of Mexico, to consist of three faunal regions: the northern region, extending from Alaska to Oregon; the central region, ranging from Oregon to Point Conception; and the southern region, from Point Conception to Mexico. McLean (1969) divided the west coast of North America into three regions also: the Oregonian, from southern Alaska to Point Conception; the Californian, from Point Conception to the southern tip of Baja California; and the Panamic, from the tip of Baja California southward. Abbott (1968) divided the west coast into five provinces (or four provinces and one subprovince): the Boreal, that portion of Alaska north of the Aleutian Islands; the Aleutian, from the

Islands south to northern British Columbia; the Oregonian, from British Columbia to Point Conception; the Californian, from Point Conception to the Mexican border, and continuing offshore to the tip of Baja California and into the Gulf of California; and the Panamic, ranging intertidally from the Mexican border and the head of the Gulf of California to South America.

Steinbeck and Ricketts (1941) spoke of the Panamic Province, identifying its boundaries within rather narrow limits. They defined the southern limit as lat. 4°30′ S, and the northern boundary as about lat. 23° N, near Cabo San Lucas, although they gave evidence of its northward extension to approximately lat. 28° N, near Isla de Cedros. Their collections, and other collections, indicate a certain percentage of the Panamic fauna may be found as far north as Point Conception, California.

Dall (1909) described the Panamic marine province as extending from Lower California, including the Gulf of California, south to the Bay of Guayaquil, Ecuador.

One most commonly encounters references to faunal provinces when reviewing the literature pertaining to molluscs, and much of the designation of these regions is based on molluscan distribution records. This is what should be expected, since the molluscs are probably better known than any other group of intertidal animals.

Certain generalizations can be made about the Pacific coast of North America, without becoming involved in faunal province definitions. These are: a. The fauna north of Point Conception, California, is largely different from the fauna south of this region. However, a great deal of overlap does occur, and much more in some groups than in others; for instance, the author has found that among the opisthobranch molluscs there is a 32 percent overlap of species at Point Conception. b.The common intertidal forms found below Point Conception tend to range farther offshore as one moves south from the Mexican border, becoming almost entirely absent near the southern tip of Baja California. These cold water forms must move into deeper water in the lower latitudes to maintain a suitable environmental temperature. c. The west coast of Baja California appears to be an overlap zone, containing numerous representatives of both the California fauna and the tropical Panamic fauna. However, the northern limit of *most* of the Panamic fauna appears to be in the area between Bahía Magdalena and the U.S.-Mexico border. d. The Panamic fauna shares its affinities principally with the Indo-Pacific and West Indies regions, the latter relationship being a result of the ocean bridge across Mexico in the Cenozoic era (see preceding section).

The distribution of onshore water temperatures along the outer coast of Baja California exhibits anomalies which are reflected in the distribution of the marine life. Alternating areas of warm and cold water cause a discontinuous distribution of warm and cold water faunas along the coast, rather than the gradual transition seen along many coastlines. This makes defining the boundary between the tropical and temperate zones along this coast difficult and arbitrary. The boundary might be established at Bahía de Sebastián Vizcaíno, which is the northernmost stronghold of much of the tropical eastern Pacific fauna, or as far south as Cabo San Lucas, where the California abalones finally cease to exist. Even these arbitrary boundaries do not accommodate the species of tropical invertebrates that occur in southern California (such as the heart urchin, *Lovenia cordiformis;* the pale urchin, *Lytechinus pictus;* the fiddler crab, *Uca crenulata;* and others) or the temperate species that occur in the Gulf (the lined shore crab, *Pachygrapsus crassipes;* the long-fingered shrimp, *Betaeus longidactylus;* etc.).

As for the faunal affinities of the intertidal animals of the Gulf, much work remains to be done. This region is largely unknown, compared to many areas of equal size, such as California or the east coast of the United States. Some of the best known groups of animals from the Gulf are those that have been studied through the efforts of the Allan Hancock Foundation and its Pacific Expeditions. The majority of these expeditions were, however, involved with subtidal and pelagic collecting.

The zoogeographical affinities of each group of animals are discussed in the corresponding chapters. Generally speaking, the Gulf exhibits an interesting assemblage of animal types, consisting of species that have close affinities with: (1) the Panamic fauna, (2) the West Indies fauna, and (3) the California fauna. Also, there is a surprisingly large number of species that appear to be restricted to the Gulf entirely. The origins of these endemic species present a number of theoretical problems, whose answers could shed light on important questions in the fields of zoogeography and animal speciation.

The Panamic fauna of the Gulf represents those animals originating in Central and South America. Panamic animals from the Gulf with which California students may be familiar include: the sea porcupine, *Lovenia cordiformis;* the pale urchin, *Lytechinus pictus;* the brittle stars, *Ophiothrix spiculata* and *Ophioderma panamense;* the sand star, *Astropecten armatus;* the fiddler crab, *Uca crenulata* (probably the only fiddler occurring in the Pacific United States); and the bland nemertine, *Lineus flavescens.* A number of the Panamic animals had their origins in the Indo-Pacific region.

The West Indian Gulf fauna that immigrated to the west coast of America from the Caribbean came via the extinct water connection that existed between the Atlantic and Pacific oceans, probably where Panamá is today (see preceding section). Much of the Gulf sponge and hydroid fauna had its origin in the Caribbean. Most of the original West Indies stock that immigrated to the Pacific evolved into new, but similar, species or subspecies.

The Californian fauna probably immigrated to the Gulf by one or both of two possible ways. The first of these could have been a past water connection across the Baja California peninsula. Garth (1961) stated that there are three areas where this extinct water bridge can be shown to have possibly existed in the geological past: at the level of Llano de Magdalena, at the midpeninsular level below the Sierra Vizcaíno, or above the Sierra Vizcaíno, perhaps at the level of Laguna Ojo de Liebre (Scammons Lagoon). The second pathway the Californian fauna may have taken to the Gulf, and the most probable, would be by a southward displacement of present day isotherms to move the colder temperate sea below the level of Cabo San Lucas, permitting immigration around the tip of Baja California and back up into the Gulf. Some of the Californian animals living in the Gulf are: *Pachygrapsus crassipes,* the lined shore crab; the barnacles *Chthamalus fissus* and *Balanus tintinnabulum;* the polyclad *Pseudostylochus burchami;* the crabs *Randallia ornata* and *Lophopanopeus frontalis;* the shrimp, *Betaeus longidactylus,* and the peanut worm, *Phascolosoma agassizii.*

Another factor adding uniqueness to the Gulf of California-Baja California region is the phenomenon of insular species, encountered in the region of Cabo San Lucas. This area is largely cut off and isolated from the mainland by deep water to the east and cold winter temperatures to the north. This means the Cabo is, in regards to its specialized fauna, somewhat of an island. Because of this, the area shares many tropical affinities, not with mainland Mexico, but with offshore islands, such as Socorro, Clarion, and the Galapagos.

Garth (1955) stated that the intertidal regions of the Gulf of California, from Bahía Agua Verde on the west to Puerto San Carlos on the east, constitute in effect a Pacific Mediterranean region, of which the present communication with the ocean, unlike the Strait of Gibraltar, now lies in the tropics (referring to the region of Cabo San Lucas and Mazatlán). Garth (1961), in his excellent discussion of brachyuran affinities and distributions in the Gulf, also stated, "The west coast of Baja California, at least in its northern portion, provides a continuum of habitat for north-temperate species found in the littoral zone of southern California south of Point Conception. This favorable environment for temperate species exists at least as far south as Punta Eugenio, opposite Isla de Cedros, and locally as far south as Punta Entrada, outside Magdalena Bay." Garth further notes, "It is also noticeable that when the Gulf of California range of these species is compared with their west coast of Baja California range, the southern limit in the Gulf of California is almost always farther north than on the open Pacific coast. This corresponds with what is known of temperatures in the region, as the outer coast is under a northerly current regime, with frequent upwelling, while the Gulf of California is cut off from northerly currents, with upwelling and associated phenomena modified by local conditions."

Walker (1960) distinguished four faunal areas in the Gulf, which he labeled as: (1) the *upper Gulf,* extending north from Bahía San Francisco past the southern tip of Isla Tiburon; (2) the *central Gulf,* which includes the short length of shoreline from Bahía Kino to Guaymas on the eastern shore of the Gulf, and the much longer shoreline between Bahía San Francisquito and La Paz on the west side; (3) the *Cabo San Lucas area,* extending north from Cabo San Lucas to La Paz; and (4) the *southeastern Gulf,* from Guaymas south on the eastern shore of the Gulf. He includes a good discussion of each of these areas, particularly in regards to the fish populations of the Gulf.

The interesting phenomenon of ecological equivalence is strikingly seen in the Gulf. In fact, a student of the California intertidal zone can expect to find many of the same types of animals, in the same habitats, in the tidepools of the Gulf that he is familiar with along the Pacific shores of the United States. For instance: the common California rocky intertidal sea urchin, *Strongylocentrotus purpuratus,* is replaced in the Gulf rocky intertidal by the nearly identical *Echinometra vanbrunti;* the California mudflat crabs of the genus *Hemigrapsus* are replaced in the Gulf by similar-appearing mud crabs of the genus *Eurytium;* the ubiquitous California sea star *Pisaster* is represented in the Gulf by the equally ubiquitous sea star *Heliaster;* the mud snail, *Cerithidea californica,* by its near twin, *Cerithidea mazatlanica;* the sand dollar, *Dendraster excentricus,* by *Encope* and *Micropora,* and so on. Suffice it to say, the more one learns to know the animals of the seashore, the more one may predict a certain type of animal to be living in a certain habitat. The most conspicuous differences between these two regions are the lack of giant kelp, goose barnacles, and large mussel beds in the Gulf, all three of which are familiar scenes to students of the west coast of the United States.

4. Habitats of the Gulf Intertidal.

Hedgpeth (1969) stated that there are three principal factors influencing the distribution of intertidal invertebrates: (1) amount of wave shock; (2) bottom type; and (3) tidal exposure. By varying the combinations and degree of these three factors one can come up with just about every conceivable type of shoreline habitat, regardless of where you are, and expect to find approximately similar animal types in similar situations in any temperate or tropical sea of the world.

The Gulf of California possesses a great diversity of shore types. The most conspicuous of these could be loosely grouped into: rocky shores, sandy beaches, and mud or sand-mud flats (of bays and estuaries). The most productive of these areas, in terms of numbers of species, is the rocky shore. The high number of species present in this habitat is largely due to the great number of available niches. For discussions of the general ecology of shoreline habitats on the Pacific Coast of America, the reader is referred to those references cited in the Preface, particularly Steinbeck and Ricketts; and Ricketts, Calvin, and Hedgpeth.

A brief description of a rocky shore fauna in the Gulf of California would go something like this: On the exposed surface of the rocks are vast numbers of barnacles (*Balanus, Chthamalus,* and *Tetraclita*), coiled worm tubes *(Hydroides* and *Spirorbis),* and snails *(Littorina, Nerita, Collisella* [formerly *Acmaea*], *Columbella* [formerly *Pyrene*], and *Acanthina*). If the tide is low enough, we may also find the beautiful green coral (*Porites*), sea anemones (*Bunodactis, Anthopleura, Bundosoma,* and *Phyllactis*), and sea stars (*Othilia tenuispina, Phataria unifacialis,* and *Pharia pyramidata).* In the crevices of the rocks are hidden the large crabs (*Eriphia* and *Grapsus*) and sea urchins (*Echinometra* and *Eucidaris*). If we turn over one of the large rocks, vast numbers of small crustaceans scurry off it and into the water below. These are mostly amphipods, isopods, hermit crabs, shrimp (*Lysmata, Alpheus,* and *Synalpheus*), and smaller porcelain crabs *(Petrolisthes* and *Pachycheles).* Clinging to the bottom of the rock we can expect to find a multitude of animal types, from sponges (*Verongia, Geodia,* and *Haliclona*) to those distant relatives of man, the ascidians, who curiously are often difficult to tell apart from the lowly sponges. Also found clinging to the bottom of the rocks are sea cucumbers (*Neothyone, Pentamera,* and *Selenkothura*), sea stars (*Heliaster* and *Pharia*), small brittle stars, (*Ophiothrix spiculata, Ophionereis annulata,* and *Ophiactis),* flatworms *(Pseudoceros, Prosthiostomum,* and *Stylochoplana*), the beautiful sea slugs or opisthobranchs (*Casella, Chromodoris, Pleurobranchus,* and *Aplysia),* polychaete worms, porcelain crabs, a variety of gastropods and chitons, and some of the smaller true crabs *(Ala cornuta, Pilumnus,* and *Leptodius).* In the pool left behind by our overturned stone we can find the large brittle stars (*Ophioderma* and *Ophiocoma*); the ubiquitous fireworms (*Eurythoe* and *Notopygos),* and other polychaetes; many of the small to medium sized crabs (*Pilumnus, Ala, Xanthodius, Geotice, Glyptoxanthus,* and *Leptodius*); shrimp, including the pistol shrimp (*Alpheus);* and small tidepool fish (mainly Gobiidae). If we were to pick up a handful of the sand underneath where the rock was, we would find a large number of small crustaceans (amphipods and isopods), many different types of polychaete worms, probably a few peanut worms (sipunculids), and perhaps a burrowing anemone or two (*Metapeachia* or *Pachycerianthus*). Living in the interstitial material between and around the rocks is an interesting assemblage of animals, consisting largely of polychaete worms, including the beautiful fan worms *Sabella* and *Megalomma,* sea cucumbers *(Brandtothuria* and *Fossothuria),* amphipods, and sipunculids.

At first glance, we may think that this exhausts our rocky shore habitat. But several interesting areas still remain to be explored. One of these is the rock itself. If the stone is hard, such as basalt, we won't worry about it. But let's assume the rocks on this beach are of sedimentary nature, perhaps sandstone or a conglomerate (as much of the rocky coastline of the Gulf is). By splitting the rock open, either with a hammer or by dropping it on another rock, we will find it infested with a myriad of tubes and tunnels, whose principal inhabitants are sipunculids (*Themiste lissum),* boring clams *(Lithophaga),* polychaete worms, and holothurians. Another easily overlooked habitat is the mat of algae covering most of the rocks below the high tide zone. By careful examination of this material, or by rinsing it in a *mild* formalin solution to chase out the inhabitants, we will find large numbers of amphipods, isopods, worms, and small crabs. One last habitat still remains to be explored — those large chunks of sponge underneath the rock that we just placed carefully back in its original location. When the sponges are broken open, we find their interior resembles that of the soft rock we just cracked apart, a vast storehouse for small crabs, amphipods, brittle stars, and worms.

The vast expanses of large mudflats, perhaps more appropriately called *tidal flats,* of the Gulf are often overlooked by the student of the seashore. At first glance, these bleak, quiet, occasionally foul-smelling areas appear quite barren when one is used to the colorful and active habitats of the rocky coast. Upon close examination, however, these regions will reveal themselves as worthwhile collecting areas. Walking across the bay or estuary at low tide, we will almost invariably encounter the mud snails *Cerithidea mazatlanica, Natica chemnitzii,* and species of *Nassarius.* If the substrate has enough sand in it, we will find the olive shells (*Oliva* and *Olivella*). Clumps of compound ascidians and large chunks of sponge are often found on the sand-mud surface. Protruding from the ground or near the surface will often be the tubes of various polychaete worms (*Diopatra, Chaetopterus,* and *Arenicola*).

Just below the surface of the mud (or sandy mud), we may find a variety of polychaetes, the odd little acorn worms (enteropneusts), the ghost shrimps *Callianassa* and *Upogebia,* and, if luck is with us, the beautiful peanut worm *Sipunculus nudus*. Usually a number of small water channels will cut through a mud flat or bay, draining its upper reaches and often carrying water seaward even after the tide has begun moving in. These channels are good areas to look for fish, and swimming crabs (*Portunus* and *Callinectes*) and other invertebrates. If the tide is a very low one, and we were to walk all the way out to the water's edge, or even wade in to our waists, we may be fortunate enough to come across a bed of sand dollars (*Encope, Mellita,* or *Clypeaster*), some gorgonian coral, and an occasional sea pen. On the way back in, if we were to stop and browse along the upper margin of the bay, where there are often large beds of algae, *Salicornia,* or marsh grass, we may find great colonies of fiddler crabs (*Uca*) and probably some of their predators, the larger crabs of the genus *Eurytium.* If we had taken a shovel along with us, we would have dug up large numbers of clams (*Chione, Protothaca, Tagelus,* and others) and would be well on our way to a delicious clam chowder.

This still leaves the sandy beach unexplored. Hunting for animals in this habitat usually leads to disappointment, and we may prefer to collect on the pure sand beaches while involving ourselves with some other activity, such as swimming or picnicking. If our luck holds out we should find some ghost crabs (*Ocypode occidentalis*). If we brought our shovel along, and the tide is out, we might find mole crabs (*Emerita* or *Hippa*) and surely a number of amphipods and isopods; and, if the tide is very low indeed, we could probably come upon some sand dollars, heart urchins, some of the large sea stars (*Astropecten, Luidia, Astrometis,* and *Oreaster*), and a few swimming crabs.

The foregoing account has mentioned only the most common inhabitants of the most common types of shorelines. The Gulf offers a number of other habitats, including wharf pilings, freshwater estuaries, and mangrove swamps. In each habitat, the animals can be seen to distribute themselves in relation to their particular needs. Those animals that can cling tightly to the substrate will be found along wave-swept shores, those animals that cannot will be forced to live in quieter waters, and so on. One of the principal factors affecting the distribution of the intertidal animals is the degree of daily or monthly exposure during the low tides. This has led to the concept of intertidal zonation, which is discussed in the following section.

5. The Tides and Intertidal Zonation in the Gulf of California.

The upper Gulf of California has some of the largest tides to be found anywhere in the world (20 to 30 feet of vertical displacement). Matthews (in Thomson et al., 1969) stated the earliest known tide records from the northern Gulf and the Colorado Delta region are those from Francisco de Ulloa, made in 1539. At the present time, there are five primary tide stations in the Gulf. These are at Puerto Peñasco, Guaymas, Topolobampo, La Paz, and Mazatlán. There are a total of 15 tide stations on the west coast of Mexico.

No attempt will be made here to go into the physics and astrophysics of tides. The dynamics of the tides have been amply explained in a great number of texts, and the reader is referred to the references following this section for information beyond what is given in this handbook. One of the most lucid accounts on the basics of this subject is Hedgpeth's excellent description in *Between Pacific Tides,* pp. 383–395.

The tidal flow of the Gulf of California, and indeed that of nearly the entire Pacific coast of North America, is generally described as *semidiurnal* or *semidaily*. These terms describe a tidal pattern of two high tides and two low tides a day. This pattern of highs and lows is associated with, and largely dependent upon, the phases of the moon and the height of the moon in relation to the celestial equator. The tides are also influenced to a lesser degree by the relative position of the sun in respect to the moon and earth, the centrifugal force of the earth-moon system around the sun, the ocean basin topography, and that phenomenon known as *sieche,* or the inherent oscillations of large bodies of water. During the new and full moons the earth, moon, and sun are in a straight line, and the tidal range is at its greatest. This is due to the multiplying effect of the gravitational pull of these other two celestial bodies on the earth's seas. These periods of extreme high tides and extreme low tides are referred to as *spring tides*. During the quarters of the moon, when the sun and moon are at right angles to the earth, the tidal range is least, and this is the period known as *neap tides*.

At Guaymas the tides are of fairly small amplitude, but at Puerto Peñasco the amplitude of all the tidal constituents is increased markedly over their values at Guaymas to produce a well-marked semimonthly inequality. The degree of inequality and the existence of semidiurnal tides is partially dependent also upon what is known as the natural period of oscillation of the body of water in question. Defant (1961) stated the natural period for the entire Gulf is 7.66 hours.

Unlike the northeastern coast of North America, whose two daily highs are usually approximately equal and whose two daily lows are usually approximately equal, much of the Gulf of California, due to its geography, experiences *irregular* or *mixed semidiurnal* tides, in which one daily high is much higher than the other high and one daily low is much lower than the other low. This is the same phenomenon seen on the west coast of the United States. This irregularity is quite pronounced in the northern Gulf, but decreases in magnitude as one moves south. In the southern Gulf, the amplitude is small enough that it is often overcome by a given combination of the many interacting forces to produce periods of simple diurnal tides (one high and one low daily), these periods lasting a few days to a few weeks.

So, we have two low tides and two high tides occurring each day, and twice a month (during the full and new moons) developing into very high highs and very low lows. The breadth of the beach that extends from the highest point the high tides cover, to the lowest point the low tides expose, is known as the *intertidal zone*. It is with this area that this handbook is primarily concerned. Below the low tide line lies the *subtidal* zone, accessible only by wading, skin diving, or trawling. Above the high tide line, in the narrow band where the winds and waves toss their salty spray, is the *splash* or *spray zone*. Above the splash zone is the vast expanse of the north Mexican deserts.

For the sake of convenience, the intertidal zone, or more correctly the true tidal fluctuation, is classified quantitatively into a number of arbitrary divisions. In North America the unit of measurement used to divide the zone is feet. For example, in Puerto Peñasco the lowest low tide of the year is separated from the highest high tide of the year by about 24 feet of vertical displacement. In other words, if a pole were fixed in the water and observed daily, it would be found that the water level on the pole moves up and down about 24 feet throughout the year, between its highest point and its lowest point. In order to allow for long-term fluctuations in the level of the sea, we don't start measuring with 1 at the lowest point on the pole, or the lowest low tide, but rather compute a 0 point at a defined level between the extremes and call everything above this level a plus tide (e.g., +2 feet) and everything below it a minus tide (e.g., −2 feet). Thus the actual, physical point of 0 tide may fluctuate from decade to decade as the level of the seas rise or fall and new calculations are made. The 0 tide level or "0 datum" is set at the mean (average) height of the low water of the spring tides, which is the conventional datum for Mexico, Central America, and most other countries. This means that the average of the lowest lows of the year is the 0 tide level. Thomson (1971) states, "Mean low water springs is defined as the mean of the low waters of the spring tides which occur a day or two after the moon is new or full."

The vertical displacement of water along California shores is 7 or 8 feet, while in the northern Gulf of California it is around 24 feet (in the Bay of Fundy the vertical displacement occasionally exceeds 50 feet, and on the east coast of the United States it is around 2 to 3 feet). The southern

Gulf exhibits much less tidal fluctuation, as the range of spring tides is only about 4 feet. The tidal range increases rapidly from Isla Tiburon north, reaching a maximum at the mouth of the Colorado, where spring ranges of 30 feet are encountered. It must be remembered that these figures are for vertical displacement, up and down the pole we fixed in the sea bottom. When these vertical displacements are translated to horizontal displacements the figure increases as a factor of the slope. On a rocky shore with a good slope the tide may go out several hundred feet, while in a gently sloping bay or tidal flat, several miles may be uncovered as the tides retreat.

Probably since the first naturalist took the time to carefully observe the seashore, it has been apparent to man that the intertidal zone could be divided not only quantitatively by increments on a measuring stick, but qualitatively, by observing the organisms living in this area. E. F. Ricketts' original studies of intertidal zonation on the Pacific Coast of California recognized four zones or regions in the intertidal. Most students of the west coast have become so familiar with this scheme that its concept and practical application is almost subconsciously applied and rarely questioned. An excellent, firsthand discussion of this and other zonational schemes is given by Hedgpeth in *Between Pacific Tides.*

I have found Ricketts' time-proven scheme of intertidal zonation to work as well in the Gulf as it does elsewhere. The faunal and floral communities will of course vary in species names, but not in concept. Unlike *Between Pacific Tides,* the approach of this handbook is systematic, and no extensive discussion of the zonation of the Gulf is included. However, with the descriptions of most of the species included in the text I have indicated whether the animal is commonly found in the high, mid-, or low intertidal regions in general. Peter Pickens has spent some time and effort collecting and correlating data in the northern Gulf and the following passage is from a section of his field guide (1970):

"Zones are designated as I, II, III, and IV and correspond somewhat to Ricketts' four zones for the California coast. There are differences, however. For example, Zone I in California has littorinid snails and extends from the splash zone about the highest tides to the level of the mean of the high tides. At Rocky Point, on the other hand, Zone I, which also has *Littorina,* extends from the splash zone to a line which is about half way between the mean of the high tides and mean sea level. Thus, Zone I extends from +17 feet to +12 feet, Zone II from +12 feet to +8 feet (approximately mean sea level), Zone III from mean sea level to 0 feet (mean of the low water spring tides), and Zone IV from the Zero Datum to −6 feet (the lowest of the low tides). The intertidal zone is, therefore, approximately 24 feet high, but may vary in width . . ."

As more data are accumulated, especially from an ecological approach, the zonational factors of the Gulf will come into better focus.

Tide calendars for the Gulf of California are available from the United States Government Printing Office, Washington, D.C., or the Mexican Government. In the latter case, write to: Anales del Instituto de Geofisica, U.N.A.M., Mexico, D.F. for their *Tablas de predicción de mareas*. The Department of Biological Sciences of the University of Arizona has, for several years, been publishing a tide calendar for the northern Gulf of California. Copies of this calendar should be ordered, well in advance, by writing: Tide Calendar, Bureau of Mimeographing and Multilithing, The University of Arizona, Tucson, Arizona 85721.

6. Synopsis of Taxa Treated in the Handbook.

Phylum Porifera

The sponges are a large, abundant group, well-represented in the Gulf intertidal and subtidal. They are the simplest of all multicelled animals and usually occur as colorful encrustations on the substrate. Sponges may range from minute encrustations on rocks to massive shelves several meters in length. Seventeen species are included.

Phylum Cnidaria (=Coelenterata)

The cnidarians are a highly diversified, abundant group, found at all depths, in all seas, and are especially common in the intertidal region. This phylum includes such familiar forms as sea anemones, sea fans, sea pens, corals, hydroids, and jellyfish. Forty-seven species of Gulf cnidarians are discussed.

Phylum Platyhelminthes

The flatworms are a small group of delicate animals, common in the Gulf intertidal but often overlooked due to their secretive habits. They are most commonly found as thin brown or black ovals, a few centimeters to several inches long, clinging to the undersides of rocks. Fourteen species of intertidal flatworms are discussed.

Phylum Nemertea

The ribbonworms or proboscis worms are a small group of invertebrates, rarely encountered, but on occasion found in large numbers. They are, like the flatworms, often overlooked because of their secretive habits. Most ribbonworms are 10 to 20 inches long but may contract to less than half that length. Ten species of Gulf ribbonworms are discussed in the handbook.

Phylum Annelida

The polychaete annelids (Class Polychaeta) make up an extremely large, abundant, and common group of invertebrates, found in the intertidal zone and at all depths the world over. Their extreme diversity, and often small size, make them a somewhat difficult group to work with. Fifty-four species are discussed.

Phylum Mollusca

The mollusca are probably the best known group of invertebrates in the world. Their diversity and numbers of species are exceeded only by the Arthropoda. Mollusca are common to all marine environments and, along with the polychaetes, usually make up the dominant fauna of sandy beaches, mud flats, and tidal shallows. Included in this phylum are the snails (gastropods), clams (pelecypods), chitons (amphineurans), tooth shells (scaphopods), and squids and octopi (cephalopods). Eighty-two species of molluscs are included.

Phylum Arthropoda

An extremely large and diverse group of invertebrates, represented in the intertidal environment by a number of distinct groups, as follows (all arthropods included in this handbook belong to the class Crustacea):

Subclass Cirripedia: The barnacles. A large group of common crustacea, usually found attached to rocks in the high and mid-intertidal zones. Eight species are treated.

Order Amphipoda: The amphipods. A large and abundant group of small, shrimplike crustaceans, extremely common in the water, under rocks, in algal masses, etc., throughout the intertidal zone and the open sea. About 15 genera are discussed in the handbook.

Order Isopoda: The isopods. Small, shrimplike crustaceans, common but not abundant in the intertidal zone. Included in this group is the rock louse *Ligia,* so often seen scurrying over the rocks of the splash zone or running about the wharves. Eleven genera of Gulf isopods are included.

Order Stomatopoda: The mantis shrimp. Dorsoventrally flattened, shrimplike creatures inhabiting the shallow subtidal zone. The second thoracic appendages are greatly enlarged, with the terminal segment being turned back on itself and bearing a sharp toothed edge. Mantis shrimp are common but not abundant in the Gulf. They are rarely found in the intertidal zone but are often brought up in shrimp trawls.

Suborder Natantia: The true shrimp represent a large group of crustaceans, common in the intertidal region as well as in the open sea and on the sea floor in waters of all depths. It is to this taxon that the prized commercial shrimp belong. Intertidal representatives are smaller than the commercial shrimp, *Penaeus,* but usually bear more distinctive and colorful markings. Thirty species of true shrimp are discussed.

Section Macrura: As defined by the taxonomy used in this handbook, the section Macrura contains the spiny lobster *Panuliris* and the flat lobster *Evibacus,* but not the ghost shrimp of the superfamily Thalassinoidae. Other authors prefer to put the ghost shrimp into section Macrura; however, in this handbook they are placed in section Anomura. The spiny and flat lobsters are principally bottom dwellers in subtidal depths.

Superfamily Thalassinoidae: This taxon includes the ghost shrimp (*Callianassa* and *Upogebia*) and the spiny red thalassinid *Axius* (*Neaxius*). The ghost shrimp characteristically live on muddy tidal flats or mud-rock substrates in the mid- and high intertidal, and are often abundant. *Axius* lives in burrows in the rocky low intertidal, and is rarely encountered. Six species of thalassinids are discussed.

Suborder Brachyura: The true crabs. Brachyuran crabs are one of the most abundant and characteristic inhabitants of the intertidal region (as well as of the shallow subtidal). Their abundance and diversity allows them to be used as zonational markers and biological indicators for a number of natural phenomena. Species range in size from minute to 6 inches or more in width. Fifty-six species of true crabs are discussed.

Family Porcellanidae: The porcelain crabs. These small crabs resemble the brachyuran crabs, except that the last walking leg is greatly reduced and folded dorsally to give them the appearance of having only three pairs of walking legs, plus the chelipeds. Porcelain crabs are common and abundant in all habitats of the rocky intertidal. Thirty-eight species are included in the handbook.

Superfamily Paguroidae: The hermit crabs. These are anomuran crabs that have lost the abdominal carapace and inhabit mollusc shells, usually shells of gastropods. Hermit crabs are the most familiar animals on the seashore and are well-represented in nearly all intertidal and shallow water habitats. Although a good number of species occur in the Gulf intertidal, the taxonomy of several groups is very poorly known and for this reason only seven species are discussed.

Superfamily Hippoidae: The mole crabs or sand crabs. A small group of anomuran crustacea, typically found burrowing in the sand of the intertidal zone. These curious crabs are rarely found without a great deal of diligent searching. Three species are discussed.

Phylum Sipunculida

The peanut worms represent a small phylum of unsegmented, rather seclusive worms, found in the intertidal region throughout the Gulf. Most species are only an inch or so long, tan, and resemble a miniature peanut when the eversible proboscis is drawn into the body, hence the common name. One exception is *Sipunculus nudus,* a large iridescent form. Seven species are included.

Phylum Echiuroidea

The spoon worms. A small phylum of unsegmented, somewhat fleshy marine worms, found principally in burrows in sand and mud, or in rocky crevices of the intertidal. The sausage-shaped body has a flattened, spoon-shaped proboscis. Three species are treated in the handbook.

Phylum Bryozoa

This group actually contains two distinct phyla, the Ectoprocta and the Entoprocta. However, for convenience they are often studied together because of their similarity in appearance and habitat. The bryozoans may form thin incrustations on rocks or other hard surfaces, or they may appear as short, bushlike branching colonies. A single colony is made up of many individual members, called zooids. Entoprocts are rarely encountered and none are included in the handbook. The ectoprocts, however, are often found, and six genera are discussed. This is a very large phylum, but a poorly studied one, and for this reason it is a difficult and often confusing group to work with.

Phylum Brachiopoda

The lampshells. Small, bivalved animals resembling pelecypods but having the shell valves arranged dorsal-ventrally (as opposed to right and left as in the clams). They are nearly always subtidal and found attached by a short, fleshy peduncle. The Brachiopoda are a small group, rarely encountered, and only three species are included.

Phylum Echinodermata

A large, abundant, and extremely diverse group of marine invertebrates. Four classes of echinoderms are treated in the handbook, as follows: *Asteroidea,* the seastars; *Ophiuroidea,* the brittle or serpentstars; *Echinoidea,* the sand dollars and sea urchins; and *Holothuroidea,* the sea cucumbers. The echinoderms constitute one of the dominant groups of animals in the intertidal and subtidal zones of the Gulf and all marine environments of the world. Fifty species are included.

Phylum Hemichordata

A small group of wormlike marine animals that until recently had been considered a subphylum of the chordates. Only one of the two classes of hemichordates are considered in this handbook, the *Enteropneusta* or acorn worms. These worms inhabit shallow water, usually burrowing into the sand or mud substrate. Exposed tidal flats are often dotted with the coiled castings of these animals. This is a small taxon, its members only infrequently found. Three species are included.

Phylum Chordata

Class Ascidiacea: The ascidians or sea-squirts. These are soft-bodied, sessile animals, usually found attached to rocks, shells, pilings, etc., in the intertidal and shallow water zones. Solitary forms ranging from minute to several inches in length occur, as do colonial forms that may appear as colorful encrustations, greatly resembling encrusting sponges, or as pulpy masses fixed in sand or mud. This is a large group of very common animals that are, however, difficult to work with, requiring a specialist for accurate species determinations. The ascidians of the Gulf are little studied and poorly known. Six genera are included in the handbook.

using the handbook

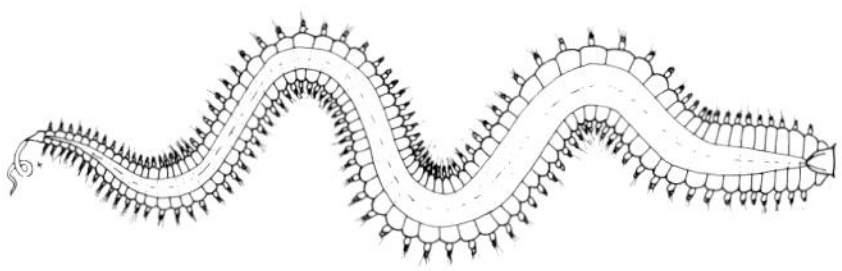

Using the Handbook

Each phylum of animals is treated as a chapter. No key to the phyla has been included, although the phyla Arthropoda and Echinodermata have keys to the classes (or other subdivisions) and these taxa have been treated individually.

For each group of animals treated, the following sections are presented: preservation, taxonomy, occurrence in the Gulf, dichotomous key, illustrations, species descriptions, and literature references. References, however, are not with the respective chapter but are found at the back of the handbook, presented separately by groups.

The *taxonomic scheme* used for each group is taken from what the author hopes is one of the most up-to-date and authoritative sources for that particular taxon. In many groups of animals, the systematics below the class level are unstable, and different workers often utilize different schemes of classification. In such cases, I have tried to use the classification of the last published monograph dealing with the animals of this particular region, or what I believe is a generally agreed upon taxonomy.

To use the *keys,* begin at couplet number one, with the animal to be identified in front of you. After reading each choice of the first couplet, decide which set of characters applies to the specimen you are identifying. Look at the number following the description fitting your animal, and go to the couplet with that number. Continue going from couplet to couplet, each time making the choice of which of the two sets of characters applies to your animal. Eventually you will reach a name instead of a number, and that is the name of your animal. You can now look up that name in the descriptions following the key. By comparing your animal to the species description you can double-check the accuracy of your initial identification based on the key.

The principal function of a key is utility, and for this reason the keys are primarily artificial and not phyletic. That is, they are largely based on species characters alone, as opposed to characters of the higher taxa. In nearly all cases, key couplets apply equally well to all individuals of a population, regardless of sex or age. The choices are to be considered absolute and descriptive. The keys may be used either forwards or backwards (by using the numbers in parentheses that indicate the source of the couplet).

Definitions of terms used in the keys and in other sections of the handbook may be found either in the taxonomy section of the particular chapter or in the glossary. The keys have been, whenever possible, written from observations of the living animal. However, enough characters are given in almost all instances to identify an animal after it has been properly preserved. It is important to remember that an *improperly* preserved animal will often be impossible to identify, thus rendering it totally useless.

A great deal of effort has been made to present accurate *distribution records* for each species. Many of the ranges included in this handbook represent significant extensions of previously known distributions. These extensions are based on the author's own collecting data.

Most of the *species descriptions* are based on a combination of the literature and observations of the live animal. The person's name following each species name indicates the author of that species (the person who first gave the animal that specific name). Parentheses around the author's name indicates the species is now in a different genus than when originally described. When recent name changes have occurred or when an animal is commonly known by two or more names I have included these synonomies in the species descriptions. Complete synonomies, however, have not been attempted for any groups.

The *literature references* for each group are not meant to be complete bibliographies, but are designed to present the major works relating to the Gulf fauna, mainly those with descriptions or keys for two or more species. Occasionally a reference is included that deals with animals of a different region. The reason for this inclusion is that reference has been found to be a key taxonomic source for the entire group, or the paper contains an extensive bibliography, thereby making it a valuable source. Publications mentioned in a specific section of the handbook, but not cited in the references for that section, will be found in the introductory reference section.

Certain limitations had to be placed on groups containing vast numbers of species (such as the arthropods, molluscs, bryozoa, polychaetes, and Porifera). In these cases the only species included in the keys are those I have actually collected, plus any additional species that the literature has reported to be common. In the other groups I have tried to include all of the common intertidal forms of the Gulf, whether collected by myself or not. This means, of course, that some of the characters in the keys are reliant upon descriptions published in the literature. In these cases, attempts were made to utilize original descriptions or other reliable sources.

The animals included in the keys are primarily intertidal. I have, however, included the more common forms of shallow water subtidal invertebrates to make the handbook useful to skin and SCUBA divers.

References to Steinbeck and Ricketts throughout the text refer to John Steinbeck and E. F. Ricketts, *Sea of Cortez,* Viking Press, New York, 1941. Anyone sincerely interested in the Gulf should have a copy of this remarkable book, a "leisurely journal of travel and research, with a scientific appendix on the marine animals of the Panamic Faunal Province." Unfortunately it is no longer in print, although most university libraries have copies. References to Knudsen throughout the text refer to J. W. Knudsen, *Biological Techniques,* Harper and Row, New York, 1966. This is a book that belongs on the shelf of every student of the life sciences. In it may be found formulas for preparing the preservatives, relaxants, fixatives, etc., mentioned in this handbook, as well as the techniques with which to apply them.

When discussing ranges of the various species, I have used the phrases "northern Gulf" and "southern Gulf." For this handbook, "northern Gulf" includes any and all of the area above Islas Angel de la Guarda and Tiburón, and "southern Gulf" includes any and all of the Gulf below these islands, to lat. 23° 27′ N (Tropic of Cancer).

The name "Bahía Cholla" used throughout the handbook is one not usually found on maps of Sonora. It refers to a shoreline indentation above Puerto Peñasco that is actually a part of the much larger Bahía López Collada. The latter is often shown on maps as Bahía Adair, or as Adair Bay.

Generally, whenever "Baja California" is mentioned, it refers to the Lower California peninsula and not to the Mexican State of the same name. The State actually includes only about the northern half of the peninsula, the remainder being in a Territory of Mexico, Baja California Territorio Sur.

In an attempt to add a touch of familiarity to the text, where applicable, I have briefly compared or mentioned certain animals from the shores of California, Oregon, and Washington. This will perhaps allow students to relate more directly to the overwhelming barrage of new animals they will encounter when visiting the Sea of Cortez.

Lastly, before progressing to the collecting and keying of the marine life of the Gulf, it becomes again necessary to remind everyone, students and teachers alike, of the growing need for conservation and ecology-conscious explorations of the seashore. Over the years, as interest in the sea has increased, so has an increased demand been placed on the intertidal habitat. The bus loads and vast car caravans of students that have been descending on the shores of California over the last 10 years have literally wiped out areas that once were nearly as productive as most of the Gulf of California still is. Thank Nature that the Gulf is protected by a factor of distance. In California places like Hazard Beach, Newport Bay, Dana Point, and the rocky coastline of the Santa Barbara region now harbor only the most common, drab, and fertile forms of marine life. Just 3 years ago the beautiful chestnut cowry (*Cypraea spadicea*), the only member of this tropical genus of shells known to occur north of Mexico on our coast, was so common at Dana Point that it could be picked up by the handfuls. Apparently it was. However, the Point's open, rocky coastline habitat was also largely altered by a series of marinas. My last visit to Dana Point turned up but a single specimen of *Cypraea*. But, reminiscing does little good, for if there is one thing man learns from experience, it is the fact that man doesn't learn from experience. Needless to say, when you are exploring the vast virgin coastlines of the Sea of Cortez, take only what you need, return what you don't want (alive), and replace your overturned stones. And right now might be a good time to remind you that to legally collect animals anywhere in Mexico you need an official Mexican collecting permit, which may be obtained, in due time, from the Instituto Nacional de Pesca e Industrias Conexas, Departamento de Estudios Biologico Pesqueros, Mexico, D.F. Also, to transport living or dead echinoderms, molluscs, or crustaceans across the border into the United States, you will need a special permit from the United States Department of the Interior, Fish and Wildlife Service.

The Common Intertidal Invertebrates of the Gulf of California

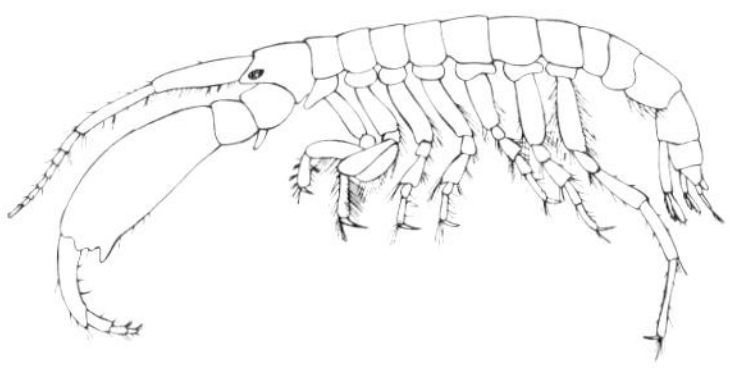

A SHORT LESSON ON ANIMAL PHYLOGENY

Long ago in some primeval slime,
When the protists were already here.
A metazoan appeared for the very first time
But its origin was never quite clear.

For Hadzi and Hanson have chosen their view
From ciliates to acoels, that's best.
But Hyman, supported by more than a few,
Said, "Gentlemen, surely you jest."

For what then becomes of the poor radiates,
Are they degenerate from some higher form?
Or is there a polyphyletic mandate;
These theories must really be scorned.

Hyman goes on to say flagellates were
The metazoan ancestor, hollow,
And most of us have decided to listen to her,
Then the planula — acoel thought follow.

The bilateria now have appeared,
With mesoderm and all sorts of things,
It really doesn't matter how you got there,
Just believe in one of the kings.

From the flatworms radiation must have flared,
'Cause variety looms up ahead.
But few have remembered, and fewer have cared,
That most of the ancestors are dead.

No matter, trudge on, through the pseudocoels plight,
(Thanks to Shapeero the priapulids dropped),
And their precursers are certainly nowhere in sight,
So why worry from where they cropped.

On to the schizocoels boldly we flail
With the spiralling cells of the flatworms.
Through segmented worms and torsioning snails,
Evolving, evolving, and making up terms.

Let Lempke and Saunders study specifics,
We don't give the small groups a nod.
Our encompassing schemes are really terrific,
We're working with *phyla,* by God!

Back to the flatworms we gracefully strut
To find where the deuterostomes start.
How about pouching an anemone's gut?
That Jäegersten really was smart.

Forget spiral cleavage, that's the protostome's using
Get on with a new coelom type.
And throw out the lophophorates, they're too confusing,
They mess up a theory that's otherwise right.

With chaetognaths and echinoderms branching off well,
We near the chordate's haunt.
Their three (count 'em) magic traits all evolve swell,
Choose anyone's theory you want.

Onward and oneward to the top of the pile,
Past amphibians, reptiles and kin.
Where man sits alone, but the lower forms smile,
Shake their heads, and say, "Listen to him."

— WAREN STAULS

Porifera (Sponges)

Preservation: Sponges should be preserved in 70 percent alcohol. Formalin should not be used for preservation, as it will eventually deteriorate the sponge spicules, which are required for proper identification. To prepare calcareous sponges for histological studies, specimens should first be fixed in a seawater Bouin's solution or other similar fixative. To prepare siliceous sponges and demospongia for histological observations, the sponges should be fixed in 10 percent buffered seawater formalin for 1 to 2 days prior to preservation. Knudsen describes techniques for preparation of slides for examination of sponge spicules and fibers. Hartman (1961) discussed a technique for freeing sponge spicules from their surrounding tissue for subsequent microscopic examination. He suggested treating a small part of the specimen with sodium hypochlorite (or a chlorine bleach) until the spicules are free. The sample is then washed several times with water, centrifuging between washings to make certain that the minute microscleres settle to the bottom. The water is finally replaced with 95 percent alcohol, and the suspension of spicules poured onto a slide and allowed to dry. Before examining under a microscope, a drop of xylene, glycerine, or other suitable mounting medium, and a cover slip, are placed on the smear.

Taxonomy: Field identification of sponges is difficult. Proper identification of species usually requires one or more laboratory procedures, such as: determining if the sponge is calcareous or siliceous (which is done by putting a drop of dilute acid on the sponge surface, a great deal of foaming and bubbling indicating a calcareous sponge, as the calcium carbonate spicules are dissolved by the acid); microspcopic observation of spicules to determine type and size; and determination of body plan. Caution must be used when observing the action of an acid on a sponge, because if the specimen is attached to (or has embedded in it) substrate of a calcareous nature, this will also bubble when touched by the acid. Be careful to place the drop of acid directly (and only) on the sponge surface.

The taxonomy of this group is based largely on the spicule and sponge fiber anatomy, which is impractical for field key usage. Accurate species determinations are often difficult (even for a specialist) because the fauna of this region is so poorly known. For these reasons, only the most common and distinctive sponges of the Gulf are included in the following key. It is occasionally difficult to tell an encrusting sponge from some of the encrusting ascidians. Proper identification can always be made with the use of a microscope, but in the field the best way to be certain you are observing a sponge is to look for multiple excurrent openings (two or more oscules). Other less reliable, but *usually* satisfactory, methods include gently running your finger across the surface, feeling for the minute spicules that many sponges have protruding above the epidermis, giving a sponge a rough texture as opposed to the smooth slimy surface of an ascidian. Also, the compound ascidians usually have the individual members of the colony visible as a more or less petaloid pattern across the surface.

The sponges of the Gulf of California are poorly known. The systematics of this phylum in general have presented problems for students of the group for many years, and numerous taxonomic schemes have been presented. Significant work in sponge taxonomy has been published by Vosmaer (1882–1886), Minchin (1900), Hentschel (1923), de Laubenfels (1936a), and Dickinson (1945). A recent paper by Bakus (1966) presented one of the first significant contributions to Pacific coast sponge taxonomy since de Laubenfels. The following classification is based on Hartman (1958) and Lévi (1957).

Sponges of the Gulf: There are probably a hundred or more species of sponge in the Gulf of California, many as yet undescribed. Steinbeck and Ricketts recorded 14 species from the Gulf intertidal, and concluded that the poriferan affinities of this region were closest to the Californian Province, less so to the West Indies fauna, and nearly lacking with the Panamic Province. Dickinson (1945) stated that the Demospongia are by far the dominant type of sponge in the Gulf (as is usually true wherever sponges are found), and that most of the sponge fauna of the Gulf probably arose as an "off-branching" from West Indies stock. This probably occurred, he said, in the late Miocene, when the sea channel across Central America was open. Dickinson further stated that this Gulf sponge fauna probably gave rise to the majority of the Californian sponge species, which, if true, could account for the Steinbeck and Ricketts findings of a close affinity between these two areas. I have, however, seen no other references in the literature to this relationship. Bakus (1971) stated, regarding the northern Gulf, that 58 percent of the sponges have tropical affinities and 28 percent have cool temperate affinities.

Within the intertidal zone, the majority of the sponges are found in the mid- and low intertidal region, many extending into the subtidal.

Some of the best sponge collecting can be done by skin-diving beyond the lowest low tide line, where one often finds massive colonies or whole reefs of a species of sponge that may be represented in the intertidal as a mere patch encrusting on a rock ledge. In the intertidal zone, sponges may be found in permanent tidepools, crevices, and ledges, or under rocks. The interior of a sponge usually serves as a haven for vast numbers of smaller animals, such as amphipods, isopods, young crabs, polychaetes, shrimp, and brittle stars.

KEY TO THE COMMON INTERTIDAL SPONGES OF THE GULF OF CALIFORNIA

1 Sponge forms a thin encrusting colony on substrate (to 15 mm thick) **2**
-Sponge upright, a thick encrustation (20 or more mm thick), or cup-shaped, or finger-like **10**
2(1) External ostia visible to the naked eye **3**
-External ostia not visible to the naked eye **9**
3(2) Sponge yellow; skeleton composed entirely of spongin fiber, no spicules present ...... ***Verongia thiona***
-Sponge not yellow (color variable); skeleton composed of spicules, usually with spongin fiber also **4**
4(3) Sponge asconoid or leuconoid; spicules calcareous; spicules of one type only (triaxons) **17**
-Sponge leuconoid; spicules siliceous; spicules of two or more types **5**
5(4) Sponge bright red ...... ***Tedania nigrescens***
-Sponge not red (lavender, orange, or white) **6**
6(5) Sponge lavender **7**
-Sponge orange or white **8**
7(6) Ostia surrounded by collars ...... ***Adocia gellindra***
-Ostia mounted in tall, fingerlike projections, or in volcanolike elevations ...... ***Haliclona permollis***
8(6) Sponge rough and irregular, with ostia in large volcanolike elevations that are often paired; sponge orange to white ...... ***Hymeniacidon sinapium***
-Sponge smooth, with ostia in small, unpaired elevations; sponge white ...... ***Hymeniacidon rugosus***
9(2) Sponge smooth and even; sponge slate colored; sponge lacking spicules ...... ***Chondrosia reniformis***
-Sponge rough and irregular; sponge white; spicules present ...... ***Hymeniacidon sinapium***
10(1) Sponge brick red **11**
-Sponge not red **12**
11(10) Majority of sponge surface smooth to the touch; no forked or branched spicules; spicules 150 to 200μ long ...... ***Tedania nigrescens***
-Majority of sponge surface rough to the touch; some spicules forked or branched; spicules 185 to 400μ long ...... ***Acarnus erithacus***
12(10) Sponge resembles a great mass of upright, fused fingers; sponge bright yellow, but usually with a green hue due to a commensal algae ...... ***Terpias zeteki***
-Sponge form not as above; sponge not bright yellow **13**
13(12) Sponge dark brown, purple, or black externally, but pure white internally; subspherical when young, massive when mature (this may be the commonest species in the Gulf intertidal) ...... ***Geodia mesotriaena***
-Sponge not colored as above; body form variable **14**

14(13) Sponge without monaxon spicules (tetraxon only); sponge with long, sharp spicules that easily break off in the skin .... ***Stellata estrella***
-Sponge with tetraxon, triaxon, or monaxon spicules; spicules not as above **15**
15(14) Sponge chocolate-brown to orange; ostia in pits or cones, evenly spaced all over the colony (fig. 1.1) .... ***Pseudosuberites pseudos***
-Sponge white or yellow; ostia not as above **16**
16(15) Sponge forms yellow spots or raised yellow areas in calcareous material, often overgrowing the material completely; spicules siliceous .... ***Cliona celata***
-Sponge not as above, usually white in color; spicules calcareous **17**
17(4,16) Sponge forms a mass of convoluted folds and ridges; spicules vary greatly in size; sponge always white (fig. 1.3) ...... ***Leucetta losangelensis***
-Sponge forms erect, fingerlike processes; spicules all about the same size; sponge white to pink (fig. 1.2) .... ***Leucosolenia sp.***

SYSTEMATICS AND SPECIES DESCRIPTIONS
(Porifera)

CLASS DEMOSPONGIAE

SUBCLASS TETRACTINOMORPHA

Order Tetractinellida

Family Ancorinidae

Stelletta estrella de Laubenfels

A massive, subspherical, cartilaginous sponge with a tough cortex. This species has long, sharp spicules that easily break off in the skin and can irritate. Its range is from southern California to the Gulf of California.

Family Geodidae

Geodia mesotriaena von Lendenfeld

When young this species is subspherical, but as it grows it spreads laterally into a massive cake, often forming very large colonies in the low intertidal. There is a thick, tall spicule plush which is easily felt and easily rubbed off. The oscula are restricted to special pore areas. The colorful cortex may often be thick and tough, difficult to cut through, even with a sharp knife. This may well be the commonest sponge of the Gulf intertidal. It is light or dark brown to purple externally (in extreme cases dirty white, or even pitch black), and white or pale cream internally. The range is from southern Alaska to the Gulf of California. The internal chambers of this and many other sponges house a great number of small invertebrates. One piece of *Geodia,* approximately 15 by 15 cm, dissected by Mr. Dick Inglis at Puerto Peñasco, was found to have living in it over 100 species of animals, mostly amphipods, polychaete worms, and small crabs.

Order Carnosa

Family Chondrosiidae

***Chondrosia reniformis* Schulze**

A smooth sponge, entirely lacking spicules. The skeletal support is derived from a very thick mesogleal layer. The dark encrustation it forms makes it almost indistinguishable from compound tunicates. This sponge is generally slate colored, and external ostia are rarely seen with the naked eye. The distribution is given by Steinbeck and Ricketts as cosmopolitan.

SUBCLASS CERACTINOMORPHA

Order Haplosclerida

Family Haliclonidae

***Haliclona permollis* Bowerbank**

This sponge is pale lavender to a vivid purple, soft and spongy to the touch, not at all slimy. Most Gulf specimens have the regularly spaced oscula surrounded by volcano-shaped cones, 2 to 3 mm wide. This encrusting sponge is about 1 cm thick, and forms growths of flat sheets or meandering strands across the under surface of rocks. It is cosmopolitan in distribution. This species also is one of the more common encrusting sponges of the California coast.

Order Poecilosclerida

Family Tedaniidae

***Tedania nigrescens* (Schmidt)**

This common sponge varies from encrusting to massive, and may or may not have visible oscula. The surface is generally smooth, unless damaged. It is characteristically bright red or orange. Another cosmopolitan species, ranging on our coast from the Gulf to the Galapagos Islands.

***Acarnus erithacus* de Laubenfels**

A massive, brilliant red sponge, with very large spicules, making it rough to the touch. This sponge occurs from Washington state to central and southern California, the outer coast of Baja California, and the Gulf of California.

Family Adociidae

Adocia gellindra de Laubenfels

A thick encrusting form, 2 to 4 cm thick. This sponge is pale lavender with few oscula, all irregular in shape, often having raised collars. The surface is smooth. It ranges from central Calfornia to the Gulf of California.

Order Halichrondrida

Family Hymeniacidonidae

Hymeniacidon sinapium de Laubenfels

A dull orange, yellowish green, or white sponge that forms a flat, thin encrustation on rock surfaces. The oscula are in numerous volcanolike elevations, often paired, giving the sponge surface a papillated appearance. Often encrusts small stones and empty shells in the midintertidal zone. It ranges from southern California to the Gulf of California.

Hymeniacidon rugosus (Schmidt)

This sponge is a thin encrusting form, usually about 1 cm thick and 6 cm square. The surface is fairly smooth and has numerous cones about 5 cm high by 1 cm at the base. This is a cosmopolitan species, occurring in the lower Gulf.

Order Hadromerida

Family Suberitidae

Terpias zeteki (de Laubenfels)

A massive sponge, resembling a mass of large, fused fingers. The surface is lumpy and bright yellow to cinnamon brown, but often has a distinct green hue due to commensal algae. It is reported from both sides of the Panama Canal and the Gulf of California.

Family Pseudosuberitidae

Pseudosuberites pseudos Dickinson

A chocolate-brown, orange, or reddish sponge, with craters, pits, and cones evenly placed across the surface. The oscula are few in number, but large and obvious. Most colonies are about 4 by 5 inches across and 2 to 3 inches thick, are oval, and bear 3 to 4 large oscula, although I have found some colonies up to 15 inches in length and 6 inches in width. The species is easily recognized by its distinct cortex and large rounded growth form. It is usually found during minus tides, sitting free on the sand of an exposed bay or protected tidal flat, where it may have washed up from greater depths. It may be restricted to the Gulf of California, where it is especially common at Bahía Cholla and in all of the numerous tidal inlets just north and south of Puerto Peñasco.

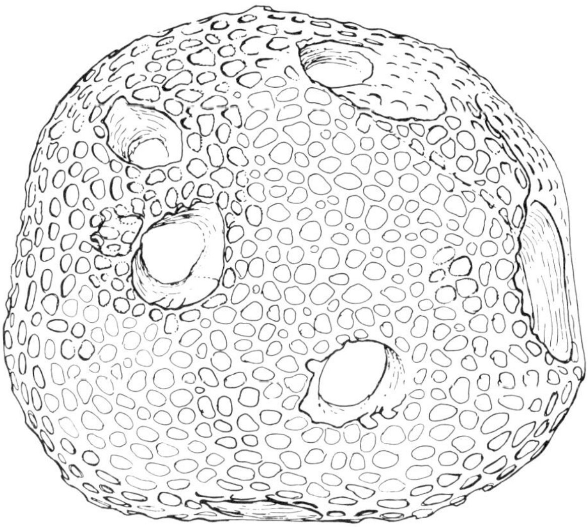

Fig. 1.1 *Pseudosuberites pseudos*

Family Clionidae

Cliona celata Grant

This sponge may appear as a massive, vaselike structure, white or yellow in color, or more commonly as yellow patches in some calcareous structure it is boring and living in. Apparently the older the sponge becomes, the more it loses its tendency to bore, and instead becomes free-living. Steinbeck and Ricketts reported it as an obvious and important feature of the lowest intertidal at Bahía de los Angeles. I have recorded it regularly from the Guaymas area. It is cosmopolitan.

Order Dictyoceratida

Family Spongiidae

Verongia thiona de Laubenfels

Verongia thiona forms a coarse-textured, spongy, yellow encrustation that turns purple or blue when dried out. This sponge has no spicules, the skeleton being composed entirely of spongin fiber. It is a fairly common sponge throughout the Gulf, and ranges from southern California to Guaymas and possibly further south. It is also a common species in the intertidal of southern California, where it is called the sulphur sponge.

CLASS CALCAREA

Order Leucosoleniida

Family Leucosoleniidae

Leucosolenia sp.

Members of this genus of sponges are typically small and white, with a fingerlike growth pattern. They are rather fragile and break easily. The development of the body form is on the primitive asconoid plan, and the spicules are always calcareous. Dickinson (1945) mentions only one species of this genus, *Leucosolenia irregularis* Jenkins, in the Gulf. Steinbeck and Ricketts also mention only one from the Gulf, *Leucosolenia*

Fig. 1.2 *Leucosolenia sp.*

coriacea Montagu. Both of these species are cosmopolitan. Dickinson (1945) stated of *L. irregularis* ". . . a small white sponge rather brittle to fragile in consistency. The shape of this sponge is not very symmetrical; in fact, it tends to be leafy." Specimens collected by the author, that have been assigned to this genus, range from dull or bright white to pale pink. There are so many species of this genus known throughout the world, and surely so many yet to be described, that it may often be better to leave an identification at the generic level than to make even a good guess at its specific name.

Order Leucettida

Family Leucascidae

Leucetta losangelensis de Laubenfels

Specimens of this calcareous sponge are usually quite amorphous. They form highly folded colonies that are nearly pure white. This is the most common calcareous sponge of the Gulf, and it ranks closely behind *Geodia* in abundance among all the intertidal sponges of the Gulf. It ranges from southern California into the Gulf, and usually occurs as an upright mass, several inches across, in the mid- and low intertidal. I have found it to be most common in narrow crevices between large boulders, but it is also quite common on and under stones.

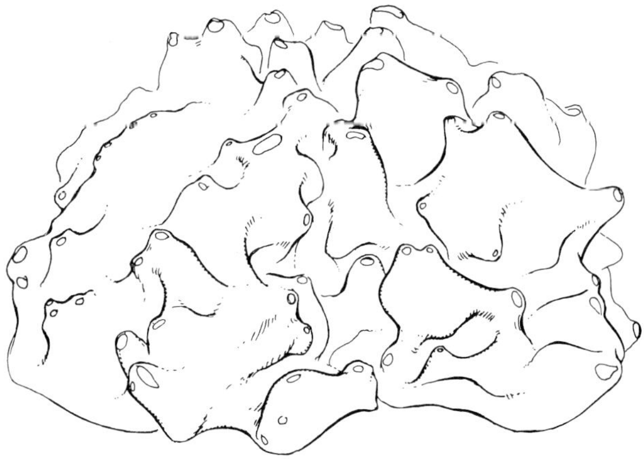

Fig. 1.3 *Leucetta losangelensis*

Cnidaria (Sea anemones, coral, etc.)

Preservation: All of the cnidarians can be relaxed in magnesium chloride prior to preservation. The hydroids also will relax with menthol, the sea anemones with clove oil or chloretone, and the jellyfish with ethyl urethane or chloretone. Knudsen described a method for relaxing and fixing hydroids, in which hot Bouin's solution (50°C) is poured over fully expanded specimens, allowed to sit a few minutes, then poured off, to be replaced by a fresh solution of Bouin's for 30 more minutes. The hydroids are then subjected to a series of alcohol washes, beginning with a 30 percent solution and ending up with 70 percent, in which they are preserved. Dr. Cadet Hand informs me he has had good success relaxing cnidarians with an isotonic solution of 7.5 percent $MgCl_2$ and seawater. Steinbeck and Ricketts (1941) recommended using cocaine or novocaine for relaxing sea anemones. Final preservation of the hydroids, sea anemones, and jellyfish can be in 5 percent formalin or 70 percent alcohol. The soft and stony corals should be preserved only in 70 percent alcohol.

Taxonomy: Classification of the Hydrozoa is based on a large number of morphological features of the zooids and the colony itself. The few morphological terms used in the key for this group can be found in the Glossary. Unfortunately, the class Hydrozoa, especially the order Hydroida, has become plagued with the phenomenon of a double taxonomy. Many workers who have restricted their studies to the sessile, benthic stages of these creatures have placed species names on these polypoid stages, while other workers interested in the planktonic stages have placed entirely different names on the medusae of many of the same species. Only recently has extensive work begun to try and associate medusa with polyp, and give the animal a single species name. At first glance this double system of naming may appear cumbersome, but one should remember that it is quite often far better to have a name on an animal now, so that it can be identified and studied, even though you know that that name may well be changed some time in the future.

Classification of the sea anemones is based largely on body morphology, tentacle arrangement, nematocyst data, histology, and internal anatomy. The body of a typical sea anemone may be divided into three regions: the oral disk, the column, and the pedal disk. An effort to simplify the key as much as possible has resulted in the elimination of all specific anemone terminology beyond this.

Classification of the corals is based on growth form and skeletal anatomy. The classification of corals is in a state of constant revision. Hyman (1940) divides the class Anthozoa into two subclasses, putting the stony corals into subclass Zoantharia, order Madreporaria. Durham (1947) puts them into several orders, although the Gulf corals herein treated all fall into the order Scleractiniae. The skeleton of the coral colony as a whole is termed the *corallum,* of each polyp a *corallite.* The corallite consists of a cup with internal, vertical *septa* that rest on the *basal plate.* The wall of the cup is termed the *theca.* The inner ends of the septa are often fused to a central column called the *columella.*

The cnidarians form a large phylum of animals, displaying a great variety of body forms, but all possessing the same basic, radial-biradial symmetry. The entire phylum is treated in a single key to maximize utility. To facilitate rapid identification of specific groups, the various major taxa (class, order, etc.), are marked in the key in bold type. The conspicuous lack of published data on the Cnidaria in general makes this an extremely difficult group with which to work. A glance at the bibliography will reveal a singular lack of monographic material dealing with almost all of the group in this region, except the stony corals and hydroids.

The systematics used in this section are based on Fisher (1938–1948) for the hydroids, Durham (1947) and Durham and Barnard (1952) for the stony corals, Hyman (1940) and Carlgren (1951) for the sea anemones, and Hyman (1940) for the hydromedusae and gorgonians.

Cnidaria of the Gulf: The Cnidaria are well represented in the Gulf, especially in the subtidal, where an enormous number of species of hydroids, true corals, and gorgonians exists. Fraser (1948) reported 91 species of hydroids from the Gulf, from 23 different genera. Of these, 28 species in 11 genera are known to occur intertidally. The hydroid genera most often encountered in the Gulf intertidal are *Aglaophenia, Obelia, Plumularia, Sertularia,* and *Eudendrium,* although none of the hydroids could be considered abundant in the Gulf intertidal except perhaps *Aglaophenia.* In the intertidal zone, the phylum Cnidaria is represented principally by the sea anemones, and only occasionally will one encounter a sea pen, gorgonian, or hydroid.

The most often encountered anemones of the Gulf intertidal are probably *Bundosoma* (fig. 2.6), *Aiptasia, Phyllactis* (fig. 2.7), and *Palythoa* (fig. 2.8) on rocky coasts; and *Calamactis* (fig. 2.5) and *Metapeachia* on the tidal flats. The rocky coastlines also harbor what is probably the only species of truly intertidal stony coral of the Gulf, *Porites californica*. During the lowest low tides a drained tidal flat may often expose living gorgonians and sea pens, although these and the majority of the corals are best collected with trawling gear or by SCUBA diving.

The faunal affinities of the Gulf Cnidaria are probably largely tropical, although much more work needs to be done with this group throughout the eastern Pacific before the American fauna can be considered well-known, especially the sea anemones and Alcyonaria. Many of the deep-water hydroids of the Gulf extend their ranges to both the temperate and tropical seas of America. Fraser (1938) stated, "The hydroid fauna of the Gulf seems to be rather a mixture. While that of the west coast of Lower California is definitely northeast Pacific in type, that of the Gulf, if this collection is representative, is a mixture of species from the north and from the south, the latter related to those from the West Indian region," and (1939), "When the Caribbean Sea was wide open to the Pacific, before the Panama land connection appeared, the hydroid fauna of the West Indian region must have been much similar to that of the contiguous part of the Pacific, since 85 species (out of the 119, 71 percent) in this list have been reported in the North Atlantic, mostly from the West Indian region or along the course of the Gulf Stream. These must all be old established species." The true corals, on the other hand, are distinctly tropical in origin and affinity. Of the 11 genera of true corals reported from the Gulf by Durham (1947), four are cosmopolitan and seven are Indo-Pacific. Of the 21 species reported from the Gulf, 12 are unique to this region, while nine extend into the Panamic region. Durham states the Gulf coral probably originated in the Indo-Pacific, reaching the Gulf via the Panamic fauna. For a discussion of zoogeographical affinities of the sea anemones, see Carlgren (1951).

The true jellyfish (Scyphozoa) typically inhabit the open sea and are not intertidal. These animals are therefore not included in the handbook, although two hydrozoan forms of floating coelenterates are included (*Physalia* and *Porpita*), as they are commonly blown into the intertidal by high winds at sea.

KEY TO THE COMMON INTERTIDAL CNIDARIA OF THE GULF OF CALIFORNIA

1 Animal free floating in sea **(Hydromedusae)** **2**
-Animal sessile or sedentary **3**

Hydromedusae

2(1) Animal flattened and disk-shaped, without a conspicuous float (fig. 2.3) .... ***Porpita***
-Animal not flattened and disk-shaped, with a large oval float (fig. 2.4) .... ***Physalia***

3(1) Animal colonial, in the form of featherlike plumes, or branching bushlike colonies, or erect stalks with secondary stems bearing zooids **(Hydroids)** 4
-Animal solitary or colonial, but without above forms, or, if featherlike, main stem is much broader than secondary branches **(Corals, Sea Anemones, Alcyonaria)**.......... 8

Hydroids

4(3) Hydranths (gastrozooids) athecate, not encased in a chitinous outer covering .... ***Eudendrium***
-Hydranths thecate, encased in a protective opaque outer covering 5
5(4) Hydrotheca (perisarc) with annulations or rings below polyp; colony unusually bushlike, several inches tall .... ***Obelia***
-Hydrotheca without annulations; colony not bushlike, but in the form of a feather or small stalked rows of zooids 6
6(5) Polyps on both sides of branches; nematophores never present; usually found growing on blades of algae or other plants .... ***Sertularia***
-Polyps on one side of branches only; nematophores often present; rarely found growing on plants 7
7(6) Colony formed from clusters of featherlike branches; main stem stiff but flexible; hydrotheca close-set; colony brown to greenish-brown (figs. 2.1 to 2.2) .... ***Aglaophenia***
-Colony slender and not resembling a cluster of feathers; main stem flaccid, never stiff; hydrotheca set far apart; colony largely colorless .... ***Plumularia***

8(3) Animal with 8 pinnately branching tentacles around mouth of each polyp; all colonial, but not encased in rigid, calcareous exoskeleton, rather without a skeleton or with a flexible skeleton **(Alcyonaria,** the sea pens, sea fans, sea pansies, and gorgonians) 9
-Animal with more than 8 tentacles surrounding the mouth of each polyp; tentacles not pinnately branched; animal solitary and without a rigid skeleton (sea anemones), or colonial with polyps embedded in hard, calcareous mass (true corals) 14

Alcyonarians

9(8) Polyps not embedded in a chitinous matrix; body a single, flaccid stalk with leaflike folds bearing polyps **(Sea Pens)** (Fig. 2.10) .... **order Pennatulacea**
-Polyps embedded in a chitinous matrix; body a highly branched, rigid structure **(Gorgonians)** 10

Gorgonians

10(9) Branches of colony intertwined, anastamosed and fused to form a single, large fanlike structure (sea fans) .... ***Gorgonia* spp.**
-Branches of colony do not anastomose and fuse as above; body resembles a tree or bush 11
11(10) Gorgonian not red, usually bright orange .... ***Eugorgia ampla***
-Gorgonian red, or red with yellow 12
12(11) Gorgonian red and yellow .... ***Eugorgia aurantica***
-Gorgonian pure red 13
13(12) Surface smooth, but with numerous small holes; ends of branches blunt, not tapering to narrow point .... ***Psammogorgia arbuscula***
-Surface very bumpy; ends of branches taper to thin, narrow point (fig. 2.9) .... ***Muricea californica***
14(8) Animals solitary, or if colonial *not* encased in a calcareous matrix **(Sea Anemones)** 15
-Animal always colonial and encased in a calcareous matrix **(True or Stony Corals)** 30

Sea Anemones

15(14) Animal with a collar or serrated ruff below tentacular ring (ruff resembles a circle of short, fused tentacles 16
-Animal without a collar or serrated ruff as above 17

16(15) Column pink or white; tentacles without longitudinal brown stripes (fig. 2.7) .... ***Phyllactis cocinnata***
-Column orange; tentacles with brown, longitudinal stripes .... ***Phyllactis bradleyi***

17(15) Animal long and thin, without a basal disk; usually found burrowing in sand or mud; entire column often in a mucus-lined tube 18
-Animal short and stout, with a basal disk; usually found attached to rocks or other solid object; column exposed, or covered with adhering sand grains but not in a mucus-lined tube 22

18(17) Column white 19
-Column gray, brown, or some other color, but not white 21

19(18) 16 tentacles; siphonoglyph tubular and elevated, distinct from rest of oral disk and actinopharynx .... ***Metapeachia* sp.**
-More than 16 tentacles; siphonoglyph low and indistinct, not as above 20

20(19) Animal extremely thin and threadlike or wormlike .... ***Isarachanthus panamensis***
-Animal elongate but stout, not threadlike or wormlike (fig. 2.5) .... ***Calamactis praelongus***

21(18) Tentacles with brown longitudinal stripes; column dark gray, long .... ***Pachycerianthus insignis***
-Tentacles without longitudinal stripes, but occasionally with brown blotches; column not dark gray; column short and stubby .... ***Pachycerianthus aestuari***

22(17) Anemones colonial, individuals fused at bases; tentacles very short and stout, leaving entire oral disk exposed 23
-Anemones solitary, not fused at bases; tentacles usually long, partially or entirely obscuring oral disk 24

23(22) Tentacles brown; polyps usually over 20 mm long; mouth region brown or faintly green; oral disk brown or faintly green (fig. 2.8) .... ***Palythoa ignota***
-Tentacles green; polyps less than 20 mm long; mouth region yellow; oral disk green .... ***Zoanthus danae***

24(22) Anemone small, column less than 0.5 inch in diameter 25
-Anemone medium to large, column greater than 0.5 inch in diameter 27

25(24) Tentacles very thin and fine, banded; column not translucent .... ***Epizoanthus gabrieli***
-Tentacles stout, not banded; column often translucent 26

26(25) Column differentiated into two regions, an upper, delicate thin-walled region, and a lower, thick-walled region; column yellowish brown .... ***Diadumene leucolena***
-Column not differentiated as above; column white to greenish brown .... ***Aiptasia californica***

27(24) Column with numerous longitudinal rows of small warts (verrucae) between septal attachments 28
-Column without longitudinal rows of warts .... ***Anthothoe carcinophila***

28(27) Warts not adhesive (so sand and bits of shell do not adhere to column); warts arranged in rows only near the pedal disk, becoming random over majority of column; tentacles red, brown, or green, often with yellow hashmarks around the mouth (fig. 2.6) .... ***Bunodosoma californica***

-Warts adhesive (sand and bits of shell usually adhering to column); warts in distinct rows entire length of column; tentacles green, or transparent with white patches; without yellow hashmarks around the mouth **29**

29(28) Tentacles green, without white patches; polyps usually over 40 mm tall; rows of warts on column close together; often a white, fusiform mark at the base of each inner tentacle .... ***Bunodactis mexicana***

-Tentacles transparent with white patches; polyps usually less than 30 mm tall; rows of warts widely spaced; without a white mark at the base of the inner tentacles .... ***Anthopleura dowii***

Stony Corals

30(14) Coral polyps solitary or only loosely joined at base; 30 or more septa .... ***Astrangia pedersenii***

-Coral polyps all joined into a rigid calcareous structure; 25 or fewer septa **31**

31(30) Colony encrusting; colony brilliant green .... ***Porites californica***

-Colony not encrusting, but erect and often branching; colony not brilliant green **32**

32(31) Orifice of corallite highly irregular and spinulose; septa highly eroded; colony broadly lobed, not distinctly branched ... ***Porites nodulosa***

-Orifice of corallite rounded and relatively smooth; septa not highly eroded; colony distinctly branched **33**

33(32) Columella well developed; 6 or 12 septa; not known to form large reefs .... ***Porcillopora robusta***

-Columella minute or absent; 12 septa; may form large reefs .... ***Porcillopora capitata***

SYSTEMATICS AND SPECIES DESCRIPTIONS

Cnidaria

CLASS HYDROZOA

Order Hydroida

Suborder Calyptoblastea (Leptomedusae)

Family Campanulariidae

Obelia spp.

Members of this genus grow in bushlike clusters, and are usually more easily seen and more robust appearing than the other hydroids, even though they are often mistaken for clumps of algae or even overlooked entirely when covered with adhering mud and sediment, as they often are. The hydroid forms a branched colony growing from a creeping rootstalk. The feeding polyps have bell-shaped hydrotheca, and are usually borne on short stalks. Members of this genus release free medusae. Fraser (1937, 1938), reported three species of *Obelia* from the Gulf intertidal: *O. hyalina* Clarke, northern Gulf to Panamá; *O. tenuis* Fraser, lower Gulf to

Ecuador; and *O. plicata* Hincks, Alaska to Ecuador. Steinbeck and Ricketts reported *O. plicata* as a profusely branching form with tufts several inches long. Their record was from the Estero de la Luna, Sonora. They also reported *O. dichotoma* (Linneaus) from the back of a turtle in the same locale. I have not collected this genus in the Gulf.

Family Sertulariidae

Sertularia spp.

In this genus the feeding polyps are sessile, sitting right on the stalk, not held up on pedicles as in *Obelia*. The polyps are arranged in a definite pattern along the stem, usually occurring in opposite pairs. Fraser (1937, 1938) reported three species from the Gulf intertidal: *S. desmoides* Torrey, northern California (at least) to the west coast of Mexico and southern Gulf of California; *S. exigua* Allman, southern California and the southern Gulf to Central America; and *S. stookeyi* Nutting, Baja California and the Gulf, south to Ecuador. Steinbeck and Ricketts reported *S. versluysi* Nutting from Cabo San Lucas. This identification may be questionable.

Family Plumulariidae

Plumularia spp.

Members of this genus consist of colonies of pinnately arranged branches, the main stems being solitary stalks, not themselves branching. The hydrothecae have smooth margins, and the nematophores are movable. The gonangia are sac-shaped or bottle-shaped. Fraser (1937, 1938) listed two species of *Plumularia* as occurring in the Gulf intertidal, *P. lagenifera* Allman, Alaska to Peru, a large, robust hydroid common to the low intertidal of California and Canada, and *P. sinuosa* Fraser, southern California to Panama and throughout the Gulf. Although Fraser (1938, 1949) gives no records of *P. setacea* (Ellis) from the Gulf intertidal, and I have not collected it there, it very likely ranges into the exposed littoral, as it does on the Pacific coast, and it is very common subtidally in the Gulf. This is the common California plumularian. It is nearly transparent, and when viewed under water it seems to disappear. The delicate glasslike stalks are 2 to 7 cm tall, unbranched beyond the lateral stems of zooids. The gonangia are bottle-shaped. This species occurs as far north as Vancouver Island and possibly Alaska.

Aglaophenia spp.

This genus of "ostrich plume" hydroids is characterized by its featherlike growth form, sessile nematophores, hydrothecae-bearing toothed margins, and gonangia enclosed in a corbula formed by overlapping leaflike structures. Fraser (1937, 1938) reported four species of *Aglaophenia* from the Gulf intertidal: 1. *A. diegensis* Torrey, ranging from Alaska to Ecuador and throughout the Gulf. A delicate elongate form, solitary or occurring in clumps of a few stems. This may be the commonest intertidal hydroid of the Gulf. I have found large areas of the rocks in Bahía San Carlos covered with this hydroid, from the subtidal up to about the 0 tide level, the stems attaining heights of 10 to 15 cm. 2. *A. inconspicua* Torrey, from north of San Francisco to Isla de Cedros and in the northern Gulf. This is a common California ostrich-plume, small (usually about 3 to 5 cm tall), and often found growing on algae or broken shells. The polyp cup (hydrothecum) bears nine teeth on its margin. The plumes are yellowish tan to light brown. 3. *A. pinguis* Fraser, southern California, Pacific coast of Baja California, and the lower Gulf. The polyp cups of this species also bear nine teeth, but the median tooth is greatly reduced so as to give the appearance of only eight. Fraser (1948) stated the other teeth may occasionally be reduced also, giving the appearance of a mere wavy line for a margin. 4. *A. propingua* Fraser, known only from Bahía Agua Verde in the Gulf and Ensenada on the outer side of Baja California. In this species the colonies stand to 12 cm high and may be slightly branched. The stem is dark brown, and distinctly divided into internodes that are completely occupied by the hydrotheca, so that the margin of one hydrothecum is practically touching the base of the next. The hydrotheca bear nine very low, rounded teeth, the median tooth being the most acute. Members of this genus often have large numbers of caprellid amphipods clinging to the stems.

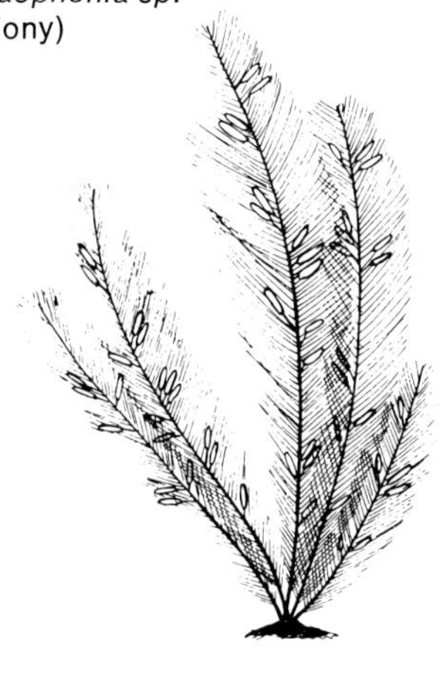

Fig. 2.1
Aglaophenia sp.
(colony)

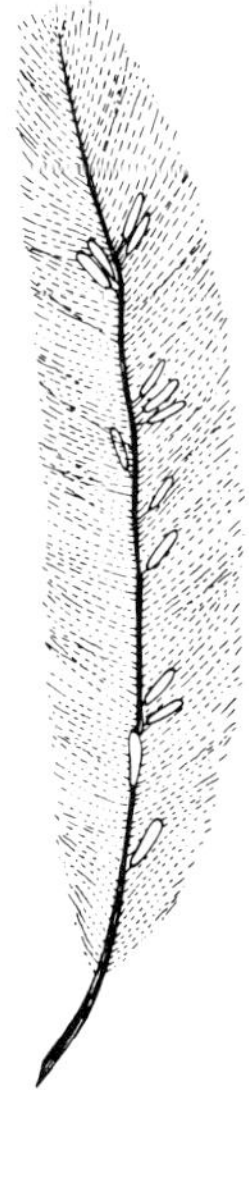

Fig. 2.2
Aglaophenia sp.
(individual stalk)

Suborder Gymnoblastea (Anthomedusae)
Family Eudenriidae

Eudendrium spp.

Members of this genus form small colonies resembling miniature flowering bushes. The branches of the colony take off in all directions, intertwining, as compared to the more upright growth form of *Obelia.* Even so, a thick growth of these hydroids can be easily mistaken for an alga. Close examination will reveal the flowerlike hydranths, usually borne at the tips of the secondary branches. The main stalks of many members of this genus are annulated. Fraser (1937, 1938) reported five species of *Eudendrium* from the Gulf intertidal: *E. cochleatum* Allman, Baja California and the lower Gulf; *E. eximium* Allman, northern Gulf; *E. ramosum* (Linneaus), Alaska to Peru; *E. tenellum* Allman, northeast Pacific shores to Ecuador; and *E. tenue* A. Agassiz, southern California to Colombia. Members of this genus are seldom collected due to their small size, fragility, and scarcity.

Order Chondrophora

Porpita sp.

A small, disk-shaped hydrozoan medusa, blue to lemon colored, cosmopolitan in tropical and subtropical seas. The margin of the disk of *Porpita* is fringed with short, hollow, tentaclelike dactylozooids. The gonozooids possess a true mouth. The medusoid gonophore is set free to discharge the sex cells (as in *Velella,* the north Pacific "by-the-wind sailor," or "purple sailor"), but the medusae soon die because they cannot feed. I have collected this hydrozoan washed ashore at El Golfo de Santa Clara, Puerto Peñasco, and Bahía Kino, all in the northern Gulf. The ontogeny and systematics of *Porpita* (and *Velella)* have long been debated. These animals have been, at various times, regarded as modified medusae, hydroid colonies related to siphonophores, and hydroid individuals. Most modern workers admit a close relationship with the sessile hydroids, and many place them in the separate order Chondrophora. A recent paper by Fields and Mackie (1971) describes them as being a single, large tubularian hydranth, floating upside down at the surface instead of being attached by a stalk to the bottom. Their classification of these two genera is as follows: order Anthomedusae; suborder Capitata; superfamily Tubularoidea; family Velellidae.

Fig. 2.3 *Porpita* sp.

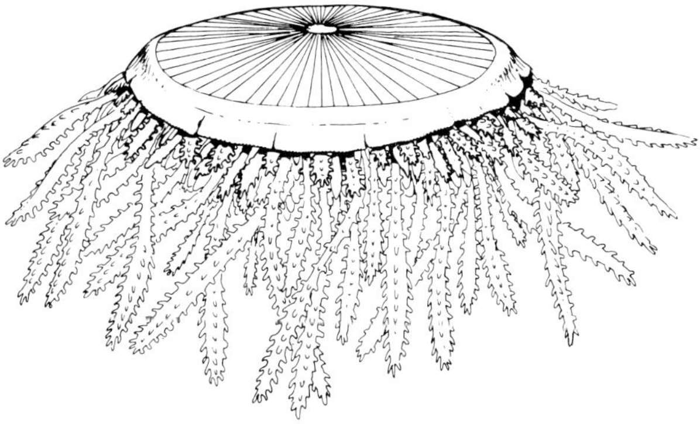

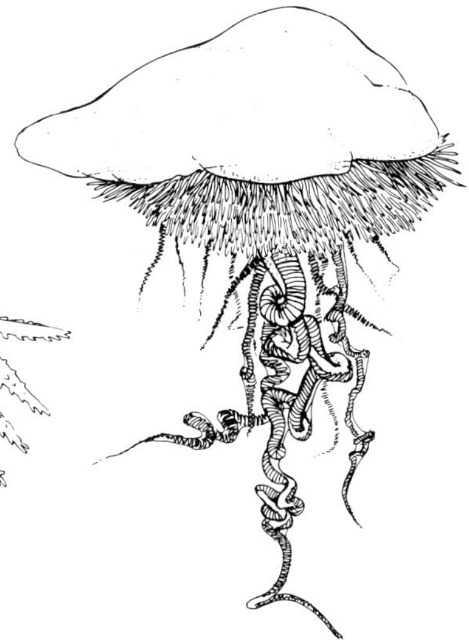

Fig. 2.4
Physalia sp.

Order Physophorida

Physalia sp.

The infamous Portuguese man-of-war. This siphonophore, unlike most of its relatives, has no swimming bell with which to propel itself, but instead has an oval, saclike float that allows it to remain on the surface of the ocean to be blown about by the ocean breezes. (Hyman gives the composition of the gas in the float of *Physalia* as 85 to 91 percent nitrogen, 1.5 percent argon, and 7.5 to 13.5 percent oxygen). *Physalia* cannot deflate its float (although the crest of it is quite contractile), and thus is at the mercy of the ocean winds, which is the reason it is often found cast ashore after a stormy sea. A number of specialized polyps trail into the water from the underside of the float. These are several centimeters to several meters in length, depending on the type and the size of the individual. The specialized structures arise from a very short and inconspicuous "stem," which is in reality a budding coenosarc on the ventral surface of the float. Close examination (a microscope is not needed) will reveal clusters of attached medusae, tangles of long stinging tentacles (the dactylozooids) armed with nematocyst batteries, as well as the gastrozooids.

The beautiful bluish purple float of this dangerous jellyfish is a familiar but carefully avoided sight in tropical and subtropical seas throughout the world. The sting from this creature is one of the strongest known among the cnidarians, rivaled only by that of the sea wasps (*Chironex* and other Cubomedusae). MacGinitie and MacGinitie (1968) state that the exact nature of the toxin (called hypnotoxin by Hyman) is not known, but it is a protein and is not formic acid as some assumptive authors have labeled it in the past. The MacGinities described an interesting and rather simple method of observing the discharge of cnidarian nematocysts (MacGinitie and MacGinitie, 1968, p. 121), and Knudsen (p. 122) describes several other methods of discharging and examining them. The study of nematocysts and nematocyst toxins is an interesting area of research, yet very little modern biochemically oriented work has been done in this field.

The early development of *Physalia* is apparently passed in deep water, as it has not yet been observed (Hyman, 1940). Members of this genus attain float lengths of 5 to 30 cm (2 to 12 inches), although I have never collected any from Gulf waters longer than 2.5 inches (6 cm). During the months of July, August, and September, these creatures are quite often cast ashore by summer winds throughout the Gulf.

CLASS ANTHOZOA
SUBCLASS ZOANTHARIA
Order Actiniaria (the true anemones)
Tribe Nynantheae
Subtribe Athenaria
Family Haloclavidae

Metapeachia sp.

An elongate white anemone without a pedal disk and its corresponding basilar muscles. This anemone burrows into mud and mud-sand with its rounded foot, called a physa. It has 16 short, banded tentacles, and a single, tubular siphonglyph that is elevated above the level of the oral disk and actinapharynx. The tentacles are arranged in only two circles around the mouth.

Family Halcampoididae

Calamactis praelongus Carlgren

Another elongate white anemone without a pedal disk (see *Metapeachia*), but possessing a well-developed physa. This species has around 24 tentacles, and its smooth column reaches lengths in excess of 50 mm. Members of the genus have only one siphonoglyph. *Calamactis praelongus* is found burrowing in mud and sand-mud bays. Pickens (1970) reported it as the most common anemone in Bahía Cholla.

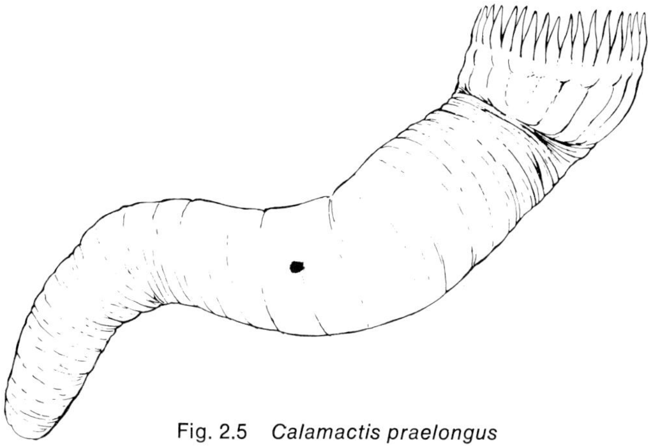

Fig. 2.5 *Calamactis praelongus*

Subtribe Endomyaria

Family Actiniidae

Bunodactis mexicana Carlgren

This and the following two species of the family Actiniidae have the column covered with small round warts or vesicles. In *Bunodactis* and *Anthopleura* these warts, or verrucae, are arranged in longitudinal rows between the septal attachments, and are actually hollow projections of the body wall. They are provided with a special musculature and lined with a glandular epidermis, making them adhesive to allow the anemones to cover themselves with bits of shell and sand, probably an adaptation for physical protection or camouflage. The tentacles of *Bunodactis* are olive green, fairly long, and often number over 100. The column color is variable, but is usually red, brown, or light green. This species often has a white fusiform mark around the base of each inner tentacle. The polyps are large, the column reaching 50 to 60 mm in length. The tentacles are equally long.

Anthopleura dowii Verrill

This species resembles *Bunodactis* (above) but may be differentiated by the rows of verrucae, which are very close together in *Bunodactis* but widely spaced in *Anthopleura,* leaving portions of the column exposed between the rows of vesicles. *A. dowii* is also smaller than *Bunodactis,* and dark green or brown, the tentacles being a translucent green, spotted with white patches. The average specimen will have a column about 30 to 50 mm long, and tentacles 15 to 20 mm long. In large individuals, two broad siphonoglyphs are visible.

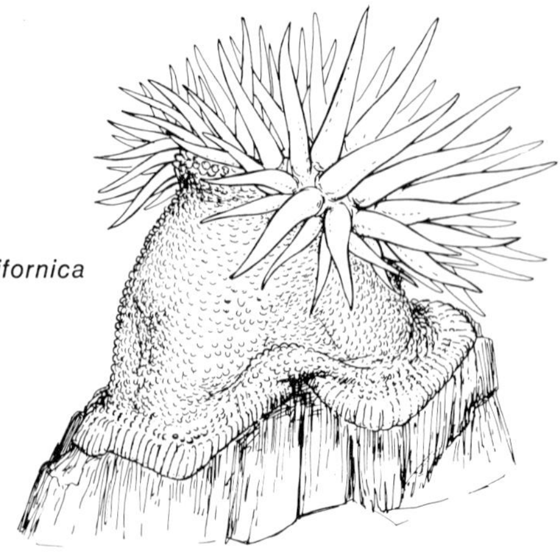

Fig. 2.6
Bunodosoma californica

Bunodosoma californica Carlgren

The column warts of this species, unlike the two species described above, are not adhesive, hence this anemone will not be found covered with sand or bits of shell. The warts are arranged in rows only at the base of the column. In fact, only the first 3 to 4 vesicles are in distinct rows, the rest being dense and randomly placed over the column. The column is usually red or orange, but occasionally brown. The tentacles are red, brown, or green, and may number nearly 150. There is often a series of yellow hashmarks around the mouth. The well-developed pedal disk is narrower than the oral disk. This species is possibly the most abundant large anemone of the Gulf.

Family Phyllactidae

Phyllactis cocinnata (Drayton, in Dana, 1846)

Phyllactis cocinnata and the following species are quickly distinguished from all the other anemones treated here by the presence of a ruff or collar, resembling a circle of short fused tentacles, below the region of free tentacles. The true tentacles are short and set inside the ruff, very near the mouth, and usually number 40 or 48. This species is generally found attached to rocks in the mid- and low intertidal by its well-developed pedal disk, often burrowed into a deep crack or hole in the rock. It has unmarked tentacles, and a pink or white column. The ruff bears numerous short filiform or dendritic appendages. The upper part of the column, below the ruff, is provided with 40 to 48 longitudinal rows of verrucae to which sand may adhere.

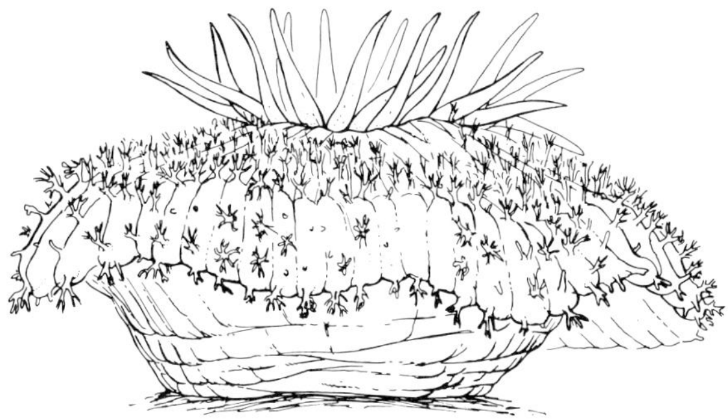

Fig. 2.7 *Phyllactis cocinnata*

Phyllactis bradleyi (Verrill)

This anemone greatly resembles the one above, differing principally in color, having an orange column and long brown stripes on the tentacles. The verrucae of this species occupy only a small space below the ruff. The tentacles usually number 48. It is generally smaller than *Phyllactis cocinnata,* and is not found attached to rocks, but rather buried in the sand or sand-mud of tidal flats. Pickens (1970) stated that these may be two ecotypes of the same species.

Subtribe Acontiaria

Family Diadumenidae

Diadumene leucolena (Verrill)

A very small anemone, often occurring in large numbers in low intertidal pools. It has a smooth, yellowish brown translucent column and many grayish brown tentacles. This anemone has acontia (as do all members of the subtribe Acontiaria). Members of the genus are known to reproduce by pedal laceration. This is a form of budding in which the basal disk breaks off one to many small pieces, each piece remaining attached to the substrate to grow into a new anemone and itself divide to form several anemones. MacGinitie and MacGinitie (1968) give a good description and photograph of this process in *Diadumene* (pp. 139–40).

Family Sagartiidae

Anthothoe carcinophila (Verrill)

This medium to small-sized anemone is always found (as far as I have seen) attached to the shells of living gastropods, usually mud-dwelling snails. It has many fine white tentacles, and longitudinal gray and white stripes on the column. Pickens (1970) reports this anemone from Bahía Cholla on the shells of the moon snails *Polinices* and *Natica,* and the auger and tower shells *Terebra* and *Turritella.* I have collected it also in shrimp trawls from 6 fathoms to 30 fathoms off Puerto Peñasco and in El Golfo de Santa Clara, on the snails *Polinices, Hexaplex, Fusinus,* and *Terebra.* C. Cutress (personal correspondence) also reports *Anthothoe panamensis* (Verrill) from the Gulf of California.

Family Aiptassidae

Aiptasia californica Carlgren

A small anemone, usually found in large numbers (when present) circling the edges of tidepools in the low intertidal region. It is often translucent, and may have iridescent patches on the column. The column and tentacles range from white to a dark or pale greenish brown. The column is fairly thin and the tentacles long.

Order Zoanthidea (colonial and solitary "anemones")

Palythoa ignota Carlgren

This species and the following two belong to the small order Zoanthidea, whose members are mostly colonial forms, the individuals of the colony resembling the "true" anemones (order Actiniaria), but differing in the arrangement of the septa. A pedal disk is absent, and the aboral ends are united by basal stolons or a basal coenenchyma. All members of

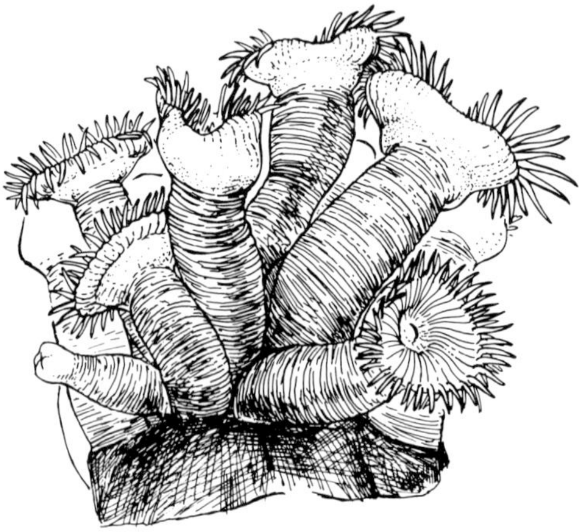

Fig. 2.8 *Palythoa ignota*

this order have only one siphonoglyph. The oral disk of this species bears a marginal circlet of very short tentacles, consisting of an exocoelic and an endocoelic cycle. The column epidermis is covered with a thick cuticle. The species is light to dark brown in color, although the oral disk occasionally has a faint green tinge due to symbiotic single-celled algae. The polyps are generally over 20 mm long, and set very close together. Colonies of these "anemones" occur in great masses in the low intertidal and subtidal, and are occasionally found in permanent tidepools and drainage channels of the midintertidal. *Palythoa ignota* is one of the commonest inhabitants of the rocky low intertidal of the northern Gulf. C. Cutress (personal correspondence) and Carlgren (1951) report six species of *Palythoa* from the Gulf of California: *P. complanata* Carlgren; *P. ignota* Carlgren; *P. insignis* Carlgren; *P. pazi* Carlgren; *P. praelonga* Carlgren; and *P. rickettsi* Carlgren. Species recognition in this genus is a difficult and time-consuming task.

Epizoanthus gabrieli Carlgren

This is a small solitary zoanthid anemone, with many fine, banded tentacles. The columns are generally encrusted with sand grains. The species is found occasionally under rocks and ledges in the low intertidal. A second species of this genus, *E. californicus* Carlgren, is known from the La Paz area.

Zoanthus danae (LeConte)

This "anemone" is similar to *Palythoa* but smaller, with thinly set polyps, a bright green or blue green oral disk and tentacles, and a generally yellowed mouth region. The polyps are rarely over 20 mm long. Students familiar with the Puget Sound fauna may at first mistake this species for the similar-appearing *Epizoanthus asper* of that region. Most members of *Zoanthus* habitually grow on other animals, such as sponges, hydroids, corals, gorgonians, bryozoans, worm tubes, and shells inhabited by hermit crabs. I have found this species growing on the common sponge *Geodia,* and on rocks in the low intertidal throughout the Gulf.

Order Ceriantharia (anemonelike anthozoans)

Pachycerianthus sp.

This genus of anemonelike cerianthids is characterized by its smooth, muscular, elongate body, that lies buried in the sand with only the oral disk exposed. The aboral end is rounded, not flattened. The oral disk is provided with simple, slender tentacles arranged in two groups, the outer *marginal set* and the inner *oral set.* Each set may be arranged in one to four circles. Two distinguishing histological features of the cerianthids are the highly developed longitudinal epidermal muscle layer and subepidermal nerve plexus throughout the column, and the presence of an anal pore.

Cerianthids live in vertical, cylindrical cavities in the substrate, inside a case formed of a hardened slimy secretion in which are embedded sand grains, shed nematocysts, and bits of shell bound to the mucus-lined tube. When disturbed, the animals quickly draw themselves into the tube by contraction of the body musculature. The "anemone" extends itself by water intake through the siphonoglyph (as in the true anemones). These creatures exhibit a number of behavioral features adapting them to their particular mode of existence. They are gravity negative, that is, they continually attempt to assume an upright position; they are strongly thigmotactic, with a strong "desire" for the body to remain in contact with some object (like the wall of their tube); and they are photonegative, retracting in strong light, and directing the aboral end toward the region of weakest light intensity. There are at least two species of *Pachycerianthus* in the Gulf intertidal, *P. aestuari* (Torrey and Kleeberger) and *P. insignis* Carlgren. The former has a short stubby column and short tentacles, both splotched with brown. The latter has a long column and many long, thin tentacles, not blotched with brown (Pickens, 1970). The column of the latter species is dark gray, while the tentacles generally have longitudinal brown stripes.

Isarachnanthus panamensis

Another anemonelike cerianthid. This species is a thin, wormlike form that has great powers of contraction and extension. It is white, with medium length tentacles. This species, like *Pachycerianthus,* is commonly found in the low intertidal zone of mud and sand-mud bays.

Order Madreporaria (the true or stony corals)

Family Seriatoporidae

Porcillopora capitata Verrill

This coral forms foliose encrustations, or, more commonly, is erect and branching. The coralla are composed of clusters of large, irregular, stout branches, often an inch or more in diameter, arising from a massive or encrusting base. The surface is covered with small, rough spinules, more prominent around the individual corallites. The corallites (or cells) are small, circular, and deep. The space between each corallite (the inter-cell space) is about the same width as the corallite diameters. A columella is either absent or minute, usually not visible. The zooids are more widely spaced than in *Porites,* and have 12 very narrow septa. This species is known to form massive colonies. Steinbeck and Ricketts mentioned a reef between Cabo San Lucas and Isla Cerralvo, probably part of Bajio Plumo, that is a mile-long stand of this coral. The species ranges from La Paz to Panamá, and is largely an insular species occurring on Isla Socorro and Pearl Island, but also recorded from Acapulco and Panamá.

Porcillopora robusta Verrill

Another species of coral barely falling into the range of this handbook, reported from La Paz to Oaxaca, Mexico. The coralla forms large, hemispherical clumps of stout, angular, flattened dichotomous branches. Each branch is totally covered with numerous ascending verrucae that bear 12 or more polyp corallites or zooids. The corallites are small, and the intercell space equals the corallite diameter. Each corallite has six or 12 narrow septa. The coralla reaches heights of 18 inches, the length of the undivided branches being 2 to 4 inches each. The corallite diameter is .02 to .03 inches. The columella of this species is well-developed.

Family Poritidae

Porites californica Verrill

A beautiful, brilliant, emerald green coral that is found encrusting on the rocks of the low intertidal and subtidal throughout the Gulf and as far south as Panamá. The corallites are large, separated by porous walls. The columella is rudimentary, spongy, and quite often entirely absent. The encrustation reaches thicknesses of 3 to 5 inches, and the corallite diameter is about .04 inches. The calices are irregularly polygonal, usually about 0.5 mm in depth. The septa are in two sizes (cycles). This coarse hemispherical coral forms colonies several inches to nearly a foot across. It is not only the commonest intertidal coral in the Gulf, but usually the only one encountered without the aid of diving gear, the others being largely subtidal.

Porites nodulosa Verrill

This coral is probably restricted to the extreme lower Gulf, and possibly to the La Paz-Cabo San Lucas insular fauna. Its corallum is subdivided into numerous small, crowded branches. The corallites are moderately large, shallow, and separated by thin, porous walls. There are usually 12 spinulose septa. The columella is poorly developed and the corallite diameter is .09 to .13 cm. This species has shallower calices, lower walls, and shorter pali than *P. californica,* and is a coarsely branching form.

Family Astrangiidae

Astrangia pedersenii Verrill

At first, this coral will appear to be a mass of isolated solitary corallites, somewhat resembling the California *Balanophyllia elegans* Verrill, but close examination will reveal the individuals are usually united at their bases to form a loose, encrusting colony. The individual zooids are minute, 3 to 4 mm across at their bases. There are 30 to 36 thin, unequal, narrow septa descending to the bottom of the corallite. The columella is narrow and papillose. Corallites are .10 to .25 inches high and .10 to .15 inches across. This species is probably restricted to the middle and lower Gulf, and has been reported from La Paz and Guaymas.

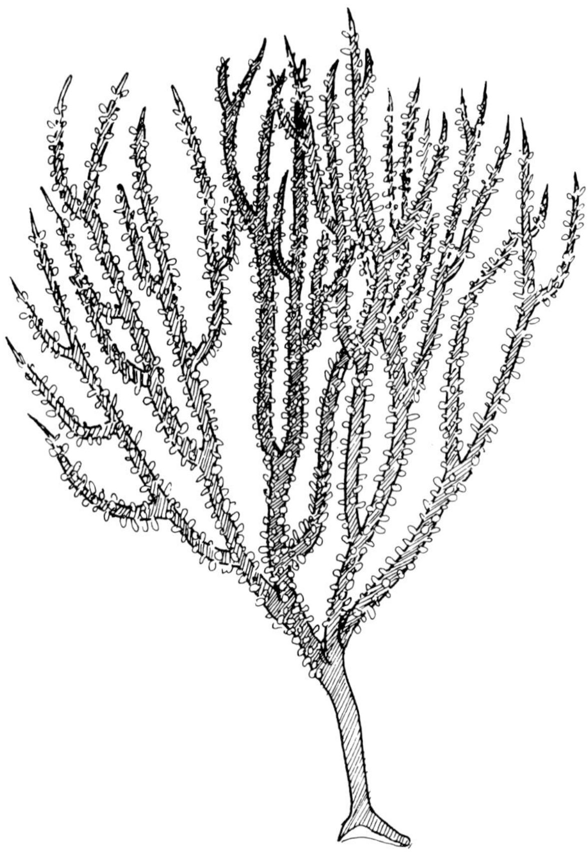

Fig. 2.9 *Muricea californica*

SUBCLASS ALCYONARIA

Order Gorgonacea (the sea fans or gorgonians)

Suborder Holaxonia

Family Gorgoniidae

Muricea californica Aurivillius

A highly branched, brick red to brown gorgonian, with white zooids. The numerous elevated polyp cones make the surface of the branches of this species very rough to the touch. The main stalks are thick, but the side branches are much thinner and taper to a thin, pointed end. This gorgonian attains heights of 20 to 30 cm, and it ranges as far north as southern California, where it represents one of the few members of this order in these cool northern waters. I have collected this gorgonian at Puerto Peñasco in the Gulf, as well as at Dana Point and Newport Bay in southern California.

Psammogorgia arbuscula Verrill

A bright red gorgonian, with the stalks all growing straight upward. The branches are robust, thick, and blunt-tipped. The species may attain heights of 30 to 40 cm.

Gorgonia spp.

Species of this genus generally have a growth pattern such that the branches all grow on one plane, interlocking to form a flat latticework fan, resembling a doily or piece of colored crochet work. "Sea fans" are generally 10 to 30 cm wide, and about the same height. Farmer (1968) reported *Gorgonia adamsi* (Verrill), a purple sea fan, offshore at Puertecitos and "points south," and I have collected this or a similar form from the shrimp trawlers at Puerto Peñasco (from 15 to 20 fathoms).

Eugorgia ampla Verrill

A smooth-surfaced, orange gorgonian. It is a highly branched species, attaining heights of 40 to 50 cm. It is commonly brought up by shrimp trawlers in the northern Gulf.

Eugorgia aurantica Verrill

A beautiful red and yellow gorgonian, with short bumpy branches, all about the same diameter. This species, like most members of the genus *Gorgonia,* branches in only one plane, leaving the colony very flat and wide.

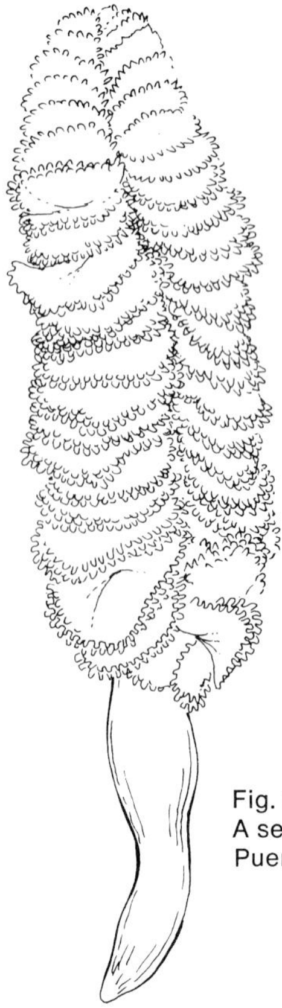

Fig. 2.10
A sea pen from
Puerto Peñasco

Order Pennatulacea (the sea pens)

These are stalked, colonial alcyonarians. The stalk is usually embedded in sand or mud, while the free portion (the rachis) stands erect to bear the zooids or polyps. The arrangement of the zooids varies from genus to genus. I have seen only one type of sea pen in the Gulf (from Puerto Peñasco and El Golfo de Santa Clara). My specimens are members of the family Pennatulidae, the colonies of which are featherlike, with the polyps borne each side of the rachis axis on leaflike structures. These beautiful creatures are brought up in shrimp trawls, or occasionally found in several feet of water at a very low tide. The Gulf specimens I have observed may be *Leioptilus* sp. or *Ptilosarcus* sp.

Platyhelminthes (Flatworms)

CLASS TURBELLARIA

Order Polycladida

Preservation: The flatworms are delicate animals, requiring special attention if one wishes to obtain good collections. They tear easily, and must be lifted gently from the rocks or substrate with a thin knife blade or similar tool. They should be kept in a separate container during the collecting period, to prevent other animals from destroying them. If they are put into a container with empty shells or other material, they will often crawl into small holes and crevices, becoming inaccessible. Preliminary examination of the animal should be made as it crawls about in a watch glass, or other small container of seawater. This will allow you to make notes on color and body form. When the flatworm clings to the bottom of a dry watch glass or petri dish, the entire dish can usually be turned upside down under a dissecting microscope to observe the ventral surface. Careful taxonomic studies of the turbellarians require that they be mounted on slides, usually both as whole mounts and sections. Ideally, three or more individuals of each species should be collected, two for slides, and the others to be preserved in 5 to 10 percent formalin. However, preparation of both slides and preserved specimens requires proper fixation and relaxing. Knudsen lists a number of fixing and staining techniques for slide mounts. Preparation of serial sagittal sections of the copulatory apparatus is a method occasionally required, but rarely attempted by anyone but the professional taxonomist. Relaxing for preservation may be done with chloretone, chloral hydrate, or menthol. Fixing for slide preparation may be done with F.A.A. (formal-acetic alcohol) or Gilson's fixative. Careful notes should always be made of live individuals, noting body form and color; presence or absence of ventral suckers, tentacles, and tubercles; and body size. Preserved animals will shrink and often curl up, making identification difficult or impossible, and dead individuals, if not preserved immediately, will begin deteriorating within hours.

Taxonomy: Although the phylum Platyhelminthes consists of three classes, only members of the class Turbellaria are generally free-living and collected commonly in the intertidal zone. Within this class, only the polyclads are abundant and large enough to be included in this handbook. The taxonomy of the group is based on a variety of anatomical features, mostly internal, requiring proper slide preparation for conclusive identification. Important external features include: presence or absence of ventral sucker, and anterior tentacles; number and arrangement of eyespots; and general body form and size. The systematics used are taken from Hyman (1953). It is important to note that all color, size, and morphological characters used in the key are for live material only. Most people will find the flatworms nearly impossible to identify (by external characteristics alone) once they are preserved, and I strongly suggest these worms be collected sparingly, and returned *alive* to the seashore after they have been examined.

Polyclads of the Gulf: Hyman (1953), in her excellent monograph of the polyclad flatworms of the Pacific coast of North America, listed about 20 species known from the Gulf of California. Steinbeck and Ricketts (1941) recorded seven to 12 species from the Gulf (unfortunately most of them were misidentified), noting *Alleena mexicana* (reported as *Latocestus* sp.), *Mexistylochus tuberculatus* (as *Stylochus* sp.), and *Pseudoceros mexicanus* (as *Pseudoceros* sp.) as the most frequently encountered flatworms of the intertidal zone. Hyman (1953) notes the following species as being common in the Gulf: *Alleena mexicana, Marcusia ernesti, Phaenocelis mexicana, Alloioplana sandiegensis,* and *Pseudoceros bajae.* I have found the following species to be common between Puerto Peñasco and Guaymas: *Mexistylochus tuberculatus, Prosthiostomum multicelis, Pseudoceros bajae,* and *Pseudoceros mexicanus.*

The various species of polyclads appear to be quite restricted in their distribution by factors of latitude, as most species have a very limited range. Hyman lists seven species limited to California and the Gulf of California, and nine species restricted to the Gulf itself. However, it is important to note that the marine flatworms, as a group, have been little studied in the field. Nearly all of the Gulf species of polyclads described by Hyman were collected by the MacGinities, Ricketts, or the Allan Hancock Expeditions. As more field work is accomplished, many of the present distributions for the various species will no doubt be greatly extended.

Flatworms are most readily found clinging to the undersides of rocks, or hidden in dense fronds of algae and in the crevices and holes of reefs and boulders.

KEY TO THE COMMON POLYCLAD FLATWORMS OF THE GULF OF CALIFORNIA

1 Dorsal surface highly tuberculate **2**
-Dorsal surface not tuberculate **3**

2(1) Ventral surface with a small median sucker; dorsal tubercles brown with white specks; anterior margin of body folded up to form 2 marginal tentacles .... ***Pseudoceros* sp.**
-Ventral sucker wanting; dorsal tubercles pale, often entirely white; without anterior, marginal tentacles ... ***Mexistylochus tuberculatus***

3(1) Ventral surface with a small median sucker **4**
-Midventral sucker wanting **7**

4(3) Anterior margin folded to form 2 short tentacles; body obovate **5**
-Anterior margin without tentacles; body long and slender **6**

5(4) Lateral margins of body strongly ruffled; body brown, black or reddish purple, with a red-orange margin; white dorsal spots do not extend to body margins (fig. 3.2) .... ***Pseudoceros mexicanus***
-Lateral margins of body without ruffles or with weak ruffles; body brown to black, without red margin; white dorsal spots extend to body margin .... ***Pseudoceros bajae***

6(4) Body white or cream-colored, with black or brown spots .... ***Enchiridium punctatum***
-Body not as above, usually tan (fig. 3.3) ... ***Prosthiostomum multicelis***

7(3) Possessing 2 short, anterior tentacles; body brown with brown or green patches, speckles or reticulations **8**
-Without anterior tentacles; body not as above **10**

8(7) Eyes in 2 lateral, anterior streaks; posterior end of body forms a blunt point; body tan with brown or green patches or speckles .... ***Alloioplana sandiegensis***
-Eyes in a circular patch around the base of each tentacle, and a median patch between the tentacles; posterior end of body rounded; body tan with brown speckles or reticulations (or occasionally pure white) **9**

9(8) Lateral margins of body highly folded or ruffled; never pure white, body flaccid; length usually less than 25 mm .... ***Paraplanocera oligoglena***
-Lateral margins of body not ruffled; occasionally pure white; body firm; length usually more than 25 mm .... ***Pseudostylochus burchami***

10(7) Body unicolored tan or gray **11**
-Body rarely unicolored, usually olive, brown, black, or gray with white spots **13**

11(10) Body with eyes restricted to small clusters near the anterior end (fig. 3.1) .... ***Stylochoplana***
-Body with or without anterior clusters of eyes, but always with marginal rows of eyes in anterior third of body **12**

12(11) Posterior end of body rounded, not tapering ...... ***Alleena mexicana***
-Posterior end of body tapers to a blunt point .... ***Phaenocelis mexicana***

13(10) Margin highly ruffled; posterior end of body rounded; length to 40 mm .... ***Marcusia ernesti***
-Margin weakly ruffled; posterior end of body tapers to a blunt point; length to 80 mm .... ***Mexistylochus levis***

SYSTEMATICS AND SPECIES DESCRIPTIONS

Polycladida

CLASS TURBELLARIA

Order Polycladida

Suborder Acotylea (Polyclads without a sucker behind the female gonopore; tentacles, if present, of the nuchal type).

Section Craspedommata

Family Latocestidae

Alleena mexicana Hyman

A long, slender polyclad, to approximately 50 mm in length. The body is smooth and firm, and tends to curl up when preserved. It is gray, and has two groups of eyespots and a median band, both restricted to the anterior end. Steinbeck and Ricketts found some slate-colored specimens at Puerto Escondido, and darker ones at Bahía de los Angeles and Guaymas (reported as *Latocestus* sp.). Hyman (1953) reports it from Puerto Peñasco and throughout the Gulf. This long, slender, uniform dark gray flatworm is one of the most common polyclads of the Gulf.

Family Stylochidae

Mexistylochus tuberculatus Hyman

This is one of two polyclads of the Gulf bearing dorsal tubercles. It is oval and thick, about 21 mm long (preserved), and a pale brown color, the tubercles usually being nearly white. A band of marginal eyes almost completely encircle the body. This species is quite common in, and restricted to, the Gulf. It is reported as *Stylochus* sp. in Steinbeck and Ricketts.

Mexistylochus levis Hyman

A very large, thick polyclad, to 67 mm in length, tapering to a blunt point posteriorly. It is brown, spotted with white, and the eyes form a large fan at the anterior end, and continue as a complete margin around the body. Hyman (1953) reported it from Puerto Peñasco.

Family Cryptocelidae

Marcusia ernesti Hyman

An elongate-oval form with rounded ends, and a body length to 35 mm. This species has ruffled margins, and is black or gray, usually dotted with white spots. A marginal band of eyes completely encircles the body. Steinbeck and Ricketts reported it from Punta Marcial Reef and Plumo Reef (as *Stylochus* spp.), and Hyman (1953) reported it from Puerto Peñasco, Guaymas, and Clarion Island.

Phaenocelis mexicana Hyman

An elongate form, with a smooth margin and a body length to 26 mm, preserved. It is gray, with a few scattered eyespots anteriorly. Hyman (1953) reported it from Guaymas, Sinaloa, and southern California. Steinbeck and Ricketts reported it from Guaymas (as species reminiscent of the genus *Leptoplana*).

Section Schematommata

Family Leptoplanidae

Stylochoplana panamensis (Plehn)

A thin, elongate to obovate form, tapering gradually posteriorly to a blunt point. This species is tan or gray and approximately 22 mm long (preserved). There are two patches of eyespots near the anterior end. Hyman (1953) reported the distribution of this species as southern California to the Gulf of Panamá, including records from Guaymas.

Stylochoplana longipenis Hyman

A slender, elongate species with rounded ends, tapering slightly posteriorly. Very small, about 16 mm long (preserved). It is gray, and the eyes are similar to *S. panamensis*. Hyman (1953) recorded the distribution as southern California to the Gulf of California.

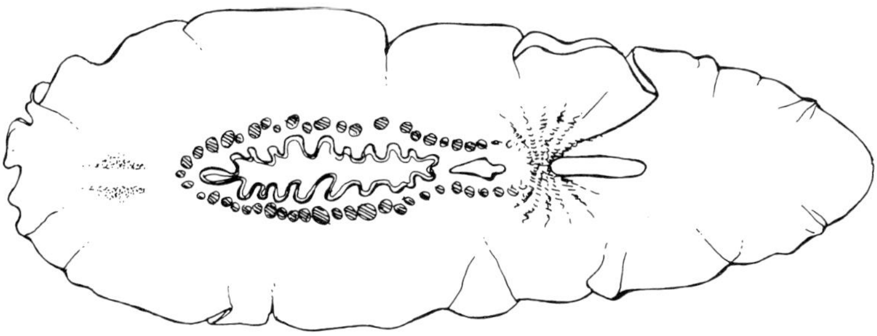

Fig. 3.1 *Stylochoplana longipenis* (modified from Hyman, 1953)

Family Planoceridae

Alloioplana sandiegensis (Boone)

This species is oval, rounded anteriorly, slightly pointed posteriorly, is up to 20 mm long, and possesses a pair of nuchal tentacles. Boone's (1929) original description says it is a light tan with dark brown and olive green patches, or brown patches only; however, MacGinitie and MacGinitie (1968) describe it as brownish, speckled. Hyman (1953) stated that the figure of *Planocera burchami* in Johnston and Snook (Pl. 4, fig. 3) is actually *Alloioplana sandiegensis,* even though this watercolor is nearly pure white. The eyes are in a large cluster at the anterior end. Hyman reported the distribution as Corona del Mar, California, to the Gulf of California.

Paraplanocera oligoglena (Schmarda)

This species is broadly oval, thin, with a ruffled margin. It is about 20 to 25 mm long and 15 to 20 mm wide. The nuchal tentacles are stout and conspicuous, and the body color is tan, speckled with dark brown. The eyes consist of a small girdle surrounding each tentacle. Hyman (1953) stated this species is ". . . probably cosmopolitan in tropical and subtropical waters." Steinbeck and Ricketts reported it as *Stylochus* (?) sp., from Bajio de Punta Marcial.

Family Callioplanidae

Pseudostylochus burchami (Heath & McGregor)

A firm, oval species, to 40 mm or more in length and 25 to 30 mm in width. It is one of the few species of polyclads from the Gulf with really conspicuous nuchal tentacles. The body is light brown, somewhat speckled or reticulated, and the eyes are restricted to small clusters near the tentacles. One of the most widely distributed species of polyclads on the west coast, known from Puget Sound to the Gulf of California.

Suborder Cotylea (polyclads with a ventral sucker; tentacles, if present, formed by folds of the body margin)

Family Pseudoceridae

Pseudoceros mexicanus Hyman

Described by Hyman (1953) from individuals sent to her by the MacGinities, collected at Puerto Peñasco. This is the "brown flatworm, over two inches in length" discussed in MacGinitie and MacGinitie

(1968, p. 151). *Pseudoceros mexicanus* is a large, obovate form, tapering slightly posteriorly. It is thin, with strongly ruffled margins, and reaches nearly 100 mm in length and 40 mm in width. It is one of the two species of Gulf flatworms possessing anterior tentacles formed by folds of the body margin. The body is brown to dark purple or reddish purple, with a reddish orange margin. The dorsum has scattered white spots, but these spots do not extend all the way to the body margins. I have found this species in great abundance under rocks in the low intertidal throughout the northern Gulf. Its colorful body and wide swimming undulations make it one of the most beautiful animals of the Gulf intertidal, and a prized aquarium pet.

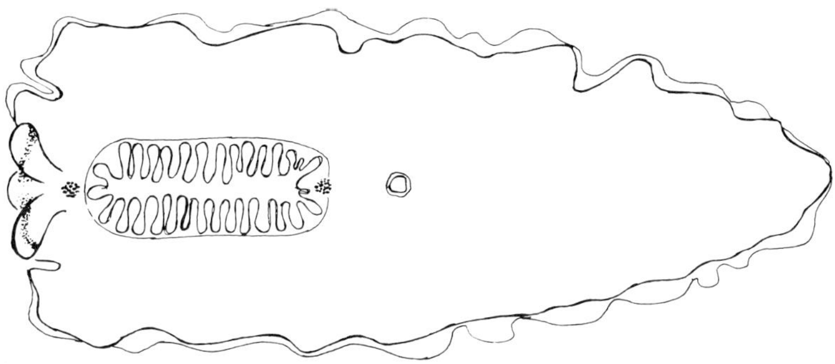

Fig. 3.2 *Pseudoceros mexicanus* (modified from Hyman, 1953)

Pseudoceros bajae Hyman

Another species described by Hyman (1953) from material sent to her by the MacGinities and Ricketts from Puerto Peñasco, Guaymas, and La Paz. This distinctive creature has marginal tentacles and a ventral sucker. It is tan to black, with white spots on the dorsum. It is obovate, tapering gradually posteriorly, and reaches lengths of 50 to 55 mm. It is distinguished from *Pseudoceros mexicanus* by its black color, lack of a red margin, smaller size, non-ruffled margins, and the fact that the white dorsal spots extend all the way to the body margin. When placed in water *P. bajae* will crawl along the bottom to find the shelter of a rock's underside, while *P. mexicanus,* on the other hand, will swim rapidly through the water. This species is common throughout the Gulf, in the mid- and low intertidal, where it may be found clinging very tightly to the undersides of rocks. Its ultra-thin body is easily torn if one attempts to remove it from rock.

Pseudoceros (?) sp.

This unusual flatworm is apparently an undescribed species of Cotylea from the Gulf. It appears to be the only intertidal member of this suborder (in the Gulf) possessing dorsal tubercles. Individuals collected by the author are small (20 to 30 mm long), obovate, and have a rounded posterior end. The tubercles are dark brown to purple on a pale background that is mottled white and purplish brown. The body margins are reticulate, with white and lavender streaks. The dark tubercles often bear numerous, small, white specks, while the median tubercles are more yellowish. The anterior margin is thrown up into two large tentacles. There are two clusters of eyespots, one between the tentacles, and another immediately behind the tentacles. The pharynx is long and narrow, occupying the anterior third of the body. This species has been collected only at Puerto Peñasco, where is was found under rocks in the mid-intertidal.

Family Prosthiostomidae

Prosthiostomum multicelis Hyman

A long, slender form, to 22 mm (preserved), tapering to a blunt point posteriorly. The eyes are in two small medial rows anteriorly, and also along the anterior margin. Hyman (1953) reported the distribution of this species as southern California and the Gulf of California. The body is usually pale, with a tan overcast and slightly yellowed margins.

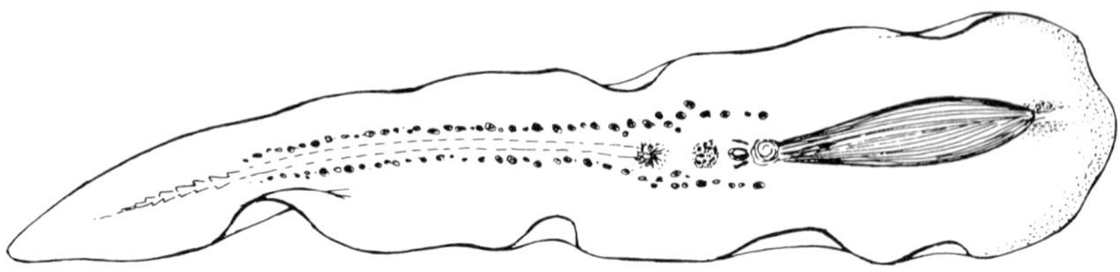

Fig. 3.3 *Prosthiostomum multicelis* (modified from Hyman, 1953)

Enchiridium punctatum Hyman

A long, slender species, 40 or more mm in length and 8 to 10 mm in width. It is white or cream, covered with brown to black spots. It has two small clusters of anterior eyes, and marginal eyes that completely encircle the body. The distribution is southern California to the Gulf of California.

Nemertea (Ribbon worms)

Preservation: Nemerteans are difficult to collect and preserve for several reasons: they secrete a slimy mucus substance covering the animal that becomes opaque when placed in preservative; many species tend to fragment when picked up or even touched; most species will contract into a shriveled-up knot when placed in formalin or alcohol; and they are relatively scarce in the first place. Specimens dropped randomly into a collecting bucket with other animals are rarely seen again, as they are chewed to pieces by other animals, or simply fragment their own bodies in response to the physical irritation. Proper narcotizing should be carried out for these worms, but, if materials are not available, an alternate but less effective method is that described for the holothurian echinoderms (see p. 337). Good narcotizing agents include chloretone, menthol, and Epsom salts. Five to 10 percent formalin is a good preservative for general collections, but if critical taxonomic examinations are to be made, the hoplonemerteans should be preserved in alcohol to prevent deterioration of the calcareous stylets (hoplonemerteans are rarely encountered in the Gulf intertidal).

Taxonomy: Field identification of the nemerteans is somewhat restricted, because these taxa are separated on the basis of internal structures (position of brain, proboscis armature, body wall musculature, etc.). Fortunately, all of the common Gulf nemerteans belong to the order Heteronemertea, which are those ribbon worms without proboscis armature and with three layers of muscle in the body wall, the innermost being longitudinal muscle. Three species of Hoplonemerteans are known to occur in the Gulf intertidal: *Paranemertes peregrina* Coe; *Oerstedia dorsalis* (Abildge); and *Zygonemerte virescens* (Verrill). They have not been included in the key because of their scarcity. Characters used in the key include the *caudal cirrus* (a little fleshy tail, visible under a dissecting microscope), a *proboscis sheath* (which requires dissection to examine), *ocelli* (visible with the naked eye, or under a low-power microscope, as rows of black dots along the lateral margins and on the head), body color, and body form. This treatment follows the systematic arrangement of Coe (1940 and 1944).

Nemerteans of the Gulf: Coe (1944) listed 14 species of ribbon worms from the Gulf intertidal. Steinbeck and Ricketts reported finding only one species in the Gulf, *Baseodiscus mexicanus,* which they concluded to be the most abundant and common Panamic nemertean of this region. The affinities of the Gulf nemertean fauna appear to be equally divided between the temperate species (six of the above 14), and Panamic or subtropical species (seven of the above 14).

Nemerteans are found under rocks and other stable objects, or buried in the sand and mud of bays and estuaries. They normally occur in the mid- and low intertidal zones.

I have found *Baseodiscus mexicanus, B. dilineatus, B. punnetti,* and *Lineus geniculatus* to be the most common nemerteans of the Gulf intertidal.

KEY TO THE INTERTIDAL NEMERTEANS OF THE GULF OF CALIFORNIA

1 Head end with shallow, longitudinal slits—the cephalic grooves (Lineidae) 2
-Head end without cephalic grooves (Baseodiscidae) 8

2(1) Lateral edges of body folded back over dorsum to form a median, longitudinal groove .... ***Diplopleura vivesi***
-Lateral edges of body not folded over dorsum as above 3

3(2) Caudal cirrus present; ocelli minute or absent; proboscis sheath nearly as long as body 4
-Caudal cirrus lacking; ocelli usually present and large; proboscis sheath much shorter than body 6

4(3) Body soft; lateral margins of body not thinned, and adapted for swimming .... ***Micrura wilsoni***
-Body firm; lateral margins of body thinned, and adapted for swimming ... 5

5(4) Body with conspicuous longitudinal markings .... ***Cerebratulus lineolatus***
-Body without conspicuous longitudinal markings .... ***Cerebratulus californiensis***

6(3) Tip of head same color as remainder of body; body drab, without distinct markings .... ***Lineus flavescens***
-Tip of head a different color than body; body with distinct markings 7

7(6) Body green or brown, with numerous narrow white or yellow rings at irregular intervals; middle of head with a white ring (fig. 4.2) .... ***Lineus geniculatus***
-Body dark brown or black, without pale rings as above; middle of head without a white ring .... ***Lineus pictifrons***

8(1) Body with numerous transverse white rings (fig. 4.1) .... ***Baseodiscus mexicanus***
-Body without transverse white rings 9

9(8) Body gray with black or dark brown anastomosing longitudinal striations .... ***Baseodiscus delineatus***
-Body pale brown, pink, red, or red-orange, without longitudinal striations as above .... ***Baseodiscus punnetti***

SYSTEMATICS AND SPECIES DESCRIPTIONS

Nemertea

CLASS ANOPLA

Order Heteronemertea

Family Baseodiscidae

***Baseodiscus delineatus* (Delle Chiaje)**

This nemertean is easily recognized by its numerous, anastomosing, black or dark brown longitudinal lines. The head is broad and rounded, and capable of being withdrawn far into the body. This is one of the most widely distributed species of nemerteans. It has been recorded from the Mediterranean, the South Pacific, the west coast of South America, and in subtropical North America on both sides of the continent. Coe (1940) reports a subspecies, *Baseodiscus delineatus curtus,* differing from the typical form in having small red or brown dots or larger mottlings, instead of the longitudinal lines. Both are reported from the Gulf of California. This may well be the most common ribbon worm of the Gulf intertidal.

***Baseodiscus mexicanus* (Burger)**

A long (20 to 80 cm) and slender Nemertean, moderately flattened dorso-ventrally. The head is set off by annular constrictions. Small ocelli occur on the corners of the head. The body is firm, brownish green, brownish violet, or red, with numerous white rings encircling the body at irregular intervals. The head is usually bordered by a narrow white band. One of the most common nemerteans of the tropical Pacific coast, ranging from

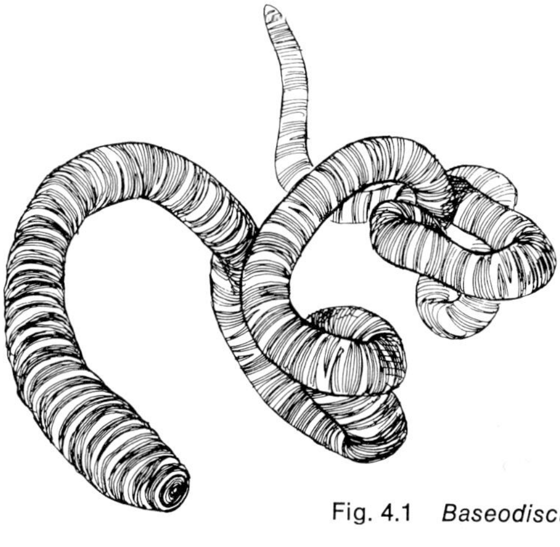

Fig. 4.1 *Baseodiscus mexicanus*

the Gulf of California to the Galapagos Islands. I have collected it at Guaymas, Isla Tiburón, and near La Paz.

Baseodiscus punnetti (Coe)

Dark brownish red to bright red-orange in color, with a truncate snout, tipped with white. The body is often covered with a whitish bloom. This species, and others of the genus, are without longitudinal cephalic grooves or a caudal cirrus. *Baseodiscus punnetti* has numerous small ocelli along the sides of the head. It is common among red algae, corallines, corals, and other growths of the intertidal zone. It ranges from central California to the Gulf of California.

Family Lineidae

Lineus pictifrons Coe

A dark brown nemertean, usually with pale yellow or white markings. The dorsal surface has numerous, fine, longitudinal streaks, and occasional transverse streaks. The head has a narrow white border. Individuals have longitudinal cephalic slits, and are about 5 to 15 mm long. The body is soft, often with a velvety sheen, with the anterior and posterior ends more narrow than the middle. It lacks a caudal cirrus. This species is found in rocky habitats, often among algae and kelp holdfasts, or in mud bottoms of bays. It is known to occur from Puget Sound to the Gulf of California.

Lineus geniculatus (Delle Chiaje)

A green or brown nemertean, with numerous narrow white or yellow rings at irregular intervals. The caudal cirrus is absent. The middle of the head is marked by a white ring. This species occurs in the rocky intertidal from the Gulf of California to Panamá (also reported from the Mediterranean and Black seas and the west coast of Africa). *Lineus geniculatus* is reminiscent of *L. vegetus* of the California coast.

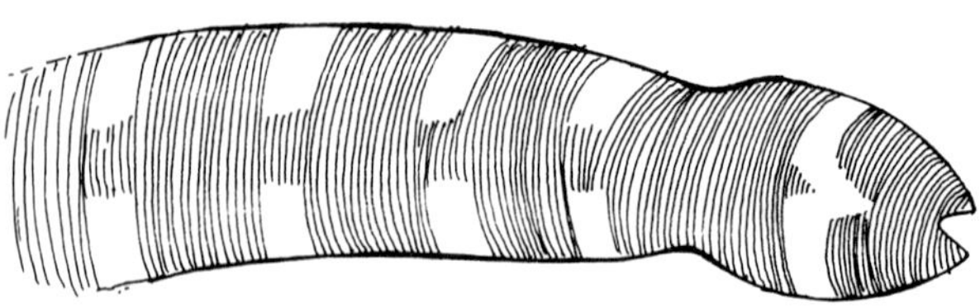

Fig. 4.2 *Lineus geniculatus* (head end only)

Lineus flavescens Coe

A rather bland, unmarked nemertean found in algae and other growths in the low intertidal of rocky shores or wharf pilings. The body is usually a pale yellow to red-orange, with a white longitudinal line down the back. The caudal cirrus is wanting. This species ranges from southern California to the Gulf of California and the lower west coast of Mexico. It, as well as many other nemerteans, seems to prefer living in entanglements of various species of red algae.

Micrura wilsoni (Coe)

This nemertean is long and ribbonlike, with a very soft body. The caudal cirrus is minute and white. The tip of the head is white, while the body is brown or blackish, lighter ventrally. This species is found beneath rocks and on kelp holdfasts in the intertidal zones, and it has also been reported from sandy mud of bays. It ranges from central California to the Gulf of California.

Cerebratulus californiensis Coe

A highly variable species, ranging from dark brown or greenish brown in adults to gray, pink, red, yellow, or white with a brown head in juveniles. Age, sexual maturity, and substrate all affect the color. The nerve cords, and occasionally the muscle layers, are often a conspicuous diffuse red color. This species possesses longitudinal cephalic grooves and a caudal cirrus. The anterior region of the body is generally firm and stout, while the posterior portion is flattened and soft. These worms live in mud, sandy mud, or pure sand, mainly in the intertidal zones of bays and harbours. They range from Puget Sound to the Gulf of California.

Cerebratulus lineolatus Coe

A slender, pale gray nemertean with numerous fine, irregular, dark olive-brown longitudinal lines along the body. The head is without distinct markings. A caudal cirrus is present. This species lives in sandy mud in the low intertidal from southern California to the Gulf of California.

Diplopleura vivesi (Joubin)

This should be an easily recognized species because the lateral edges of the body are folded back over the dorsal surface to form a longitudinal groove in the dorsomedian line. It is reported to occur in fine sand near the low water line, but is known from only a single specimen collected at La Paz.

Annelida (Polychaetes)

by Jerry D. Kudenov*

CLASS POLYCHAETA

Preservation: The primary concern in the preservation of polychaetous annelids is to produce an undistorted or unaltered specimen that can be regarded as morphologically representative of a particular species. If a specimen were to be placed directly into a fixative such as formalin, it would immediately become contracted and impossible to examine. The worms must therefore be narcotized prior to preservation. Because there are so many undescribed polychaetes in the Gulf of California, improper fixing techniques would be wasteful not only to the collector, but especially to the taxonomist.

The following methods are recommended. First obtain a *fast-acting* narcotizing agent that will prevent degradation and deterioration of the specimens. For instance, a 0.15% solution of propylene phenoxytol (P.P.O.; Goldschmidt Chemical Corp., N.Y.) or a 0.50% solution of Finquel (MS-222, tricaine methonsulphonate; Ayerst Laboratories Inc., N.Y.) are rapid narcotizing agents that take about 15 minutes to narcotize a specimen. The narcotizing fluid is added directly to the container with the specimens. Please note, however, that all tubicolous worms must be removed from their tubes to allow for a fully narcotized specimen. Two less desirable techniques employ $MgSO_4 \cdot 6H_2O$ (7.3 grams per 100 ml of water) and isopropyl alcohol. In the latter technique, alcohol is added drop by drop into the container of specimens until they are insensitive to external stimuli. The latter two methods require from 1 to 6 hours, generally result in poorly preserved specimens, and are not recommended.

Once the worm is motionless and insensitive to external stimuli, it may be placed in a fixative. One may simply decant the relaxing solution from the container and slowly add the fixative. Note that propylene phenoxytol can be saved and reused several times, while Finquel must be prepared fresh each time. Two recommended fixing solutions are 10 percent formalin and Bouin's solution. With the latter, care must be taken with the picric acid crystals, as they are highly explosive upon impact. In general, the worms should remain in these fixatives for at least 24 hours,

*Department of Biological Sciences, University of Arizona.

but a longer period of time will not harm them. Afterwards, the fixing solution is decanted and saved (or the specimens removed with forceps), and 70 percent isopropyl alcohol is added to the container. After another 24 to 48 hours, the alcohol will have become diluted by the residual fixative in the worm and should be discarded. Fresh alcohol is then added to the container. Alcohol should not be used to fix worms, because it tends to leave specimens soft and their appendages poorly defined.

A way to limit the bulkiness of many jars is to purchase some vials (or prescription tubes) into which specimens can be placed. About 12 cotton-plugged vials can be placed into a pint-sized jar. Alcohol is added to the vials and jar in order to prevent the specimens from drying out.

A final point in the collecting-preservation technique is the collecting notes. These are as important as the specimens and should never be overlooked. The following information applies to all phyla. These notes should include the date of collection, time of day, location, height in the intertidal, local habitat, color of specimen prior to preservation, animal associations (if any), method of capture, and collector's name. The notes should be written in indelible ink on good quality paper (high rag content), and deposited directly into the jar of specimens. It is also a good idea to maintain a field notebook in which all of the above information is duplicated.

Taxonomy: The major classes of the phylum Annelida are: Polychaeta (bristleworms), Oligochaeta (earthworms), and Hirudinea (leeches). The polychaetes are principally marine animals and represent the largest class of annelids, with nearly 10,000 described species. The other two classes are smaller, largely restricted to terrestrial and freshwater habitats, and altogether approximate 3,500 described species. The annelids have a repetitious series of body parts called segments, and this condition is referred to as *metamerism*.

The polychaetes are arbitrarily divided into two orders — the free-living (order Errantia) and the tube-dwelling (order Sedentaria) forms. Actually this classification represents a gradation of adaptations, because there are tube-dwelling errant and burrowing sedentary polychaetes. Errantiate worms are considered to be the more primitive of the two orders and are partially characterized by having undifferentiated body parts and macrophagous feeding habits. Sedentariate worms, on the other hand, are typified by differentiated body regions and microphagous feeding habits. These two orders will be more fully defined following a discussion of their external morphology.

Although the present chapter is not intended to be a full treatment of polychaete systematics, its purpose is to discuss basic taxonomic criteria that are reflected in the diversity of form in polychaetes in the area studied. For greater treatment of the polychaetous annelids, the reader is referred to Berkeley and Berkeley (1948, 1952), Day (1967), Fauvel (1923, 1927, 1932), Hartman (1939, 1940, 1941b, 1944a, 1944c, 1947a, 1947b, 1950, 1957, 1968, 1969), and Uschakov (1955).

External Morphology of Polychaetes: The general shape of a polychaete's body varies from a cylindrical to a flattened elongate tube. For convenience, the body may be divided into three anatomical regions, consisting of the head, trunk, and tail. These are discussed below.

Fig. 5.1 Anterior segments of selected polychaetes. (a) Amphinomidae. (b) Lysaretidae. (c) Nereidae. (d) Phyllodocidae. (e) Maldanidae. (f) Sabellidae. (g) Spionidae. (h) Terebellidae. (a) to (c) and (g) modified from Day, 1967; (f) and (h) modified from McIntosh, 1923. C, Caruncle; Pl, Palp; Pr. Prostomium; PT, Prostomial Tentacle; Ps, Peristomium.

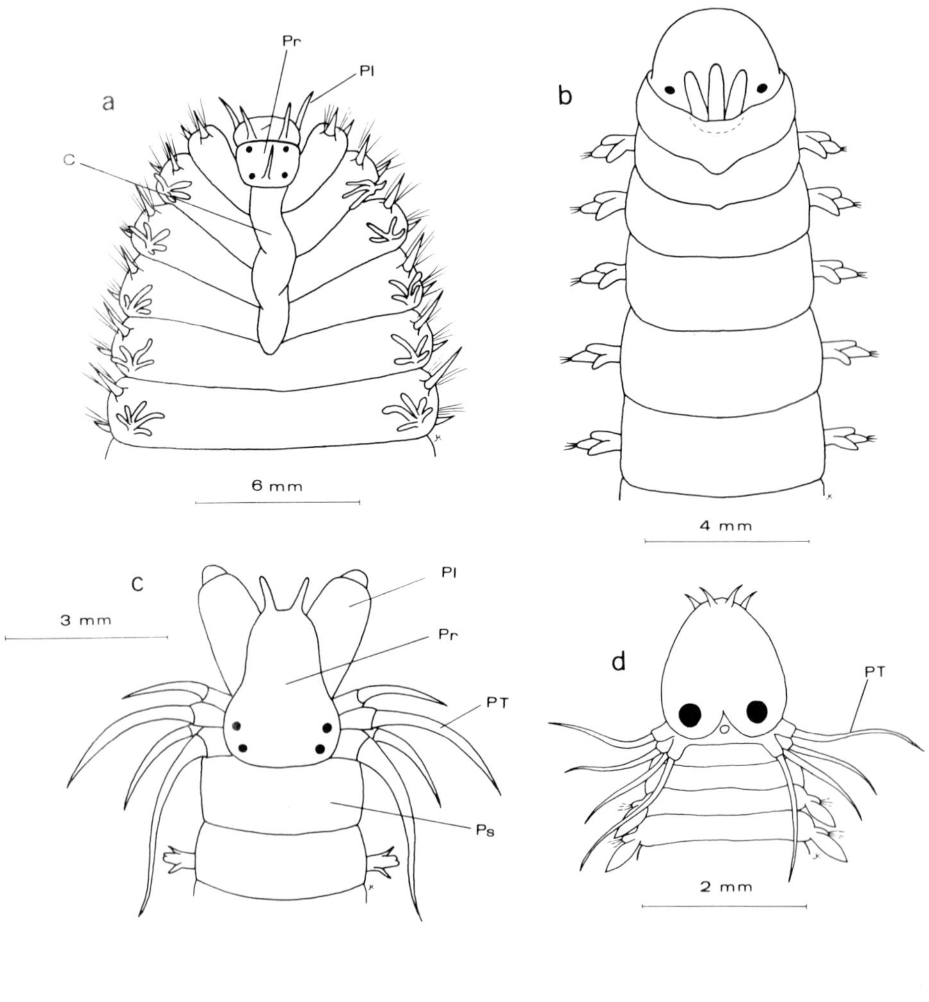

Head: The head consists of a preoral area (the prostomium) and one or more fused segments (the peristomium) around the mouth. The prostomium often bears sensory appendages and eyes, while the peristomium is equipped with tentacles. Together the prostomium and the peristomium display various degrees of cephalization, ranging from a simple cone with or without tentacles to a plumose array of filaments.

Figure 5.1 portrays diverse head types of four errant families (Amphinomidae, Lysaretidae, Nereidae, Phyllodocidae) and four sedentary families (Maldanidae, Sabellidae, Spionidae, Terebellidae). For example in nereids (fig. 5.1c), the prostomium bears two biarticulated (jointed) sensory structures, called palps. These palps originate from the ventro-lateral surfaces adjacent to the mouth. In spionids (fig. 5.1g), the palps arise from the peristomium, are elongate with a ciliated groove, and function as feeding organs. Note that the palps in each of these two families are not homologous structures.

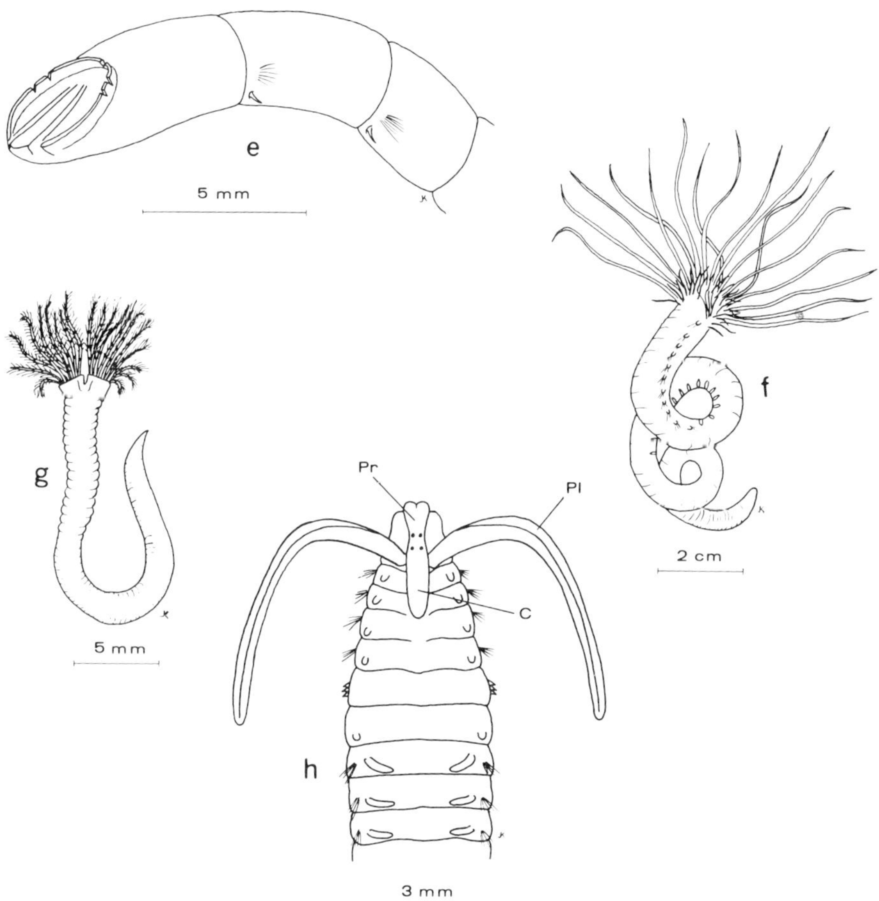

Peristomial tentacles are typical of many free-living errant families, such as the nereids and phyllodocids (fig. 5.1c, d). These tentacles may be short or long, smooth or annulated. Furthermore, a cylindrical annulated base called a ceratophore may be present at the proximal regions of peristomial tentacles in some errant families.

The heads of some sedentary polychaetes may be obscured by buccal tentacles, such as those found in terebellids (fig. 5.1h). These tentacles function as ciliated grooves that transport particulate matter towards the mouth. Sabellids and serpulids (feather-duster worms) have a plumose array of numerous bipinnate structures called radioles (fig. 5.1f). Radioles arise from the peristome, and, with the aid of cilia and mucus, collectively function to filter food particles suspended in water. Serpulids differ from sabellids, in that one radiole has been modified into a plug (operculum) that is used to seal their calcareous tube.

In other families, such as errant lumbrinerids and arabellids, tentacles may be totally absent. Amphinomids have small prostomial antennae, and a median keel (caruncle) that extends posteriorly from the prostomium to the second or third setigerous segment (fig. 5.1a). The caruncle serves as a sensory organ in the amphinomids. The sedentary maldanids (fig. 5.1e) have a reduced and truncated head, in which there is a median lobe bordered on each side by a parallel groove which collectively comprise a nuchal sensory organ.

Other features of the head include eyes. When present in errant polychaetes the eyes are usually paired. Most of the tube-dwelling worms have reduced eyes, called eye spots, located on ventral peristomial surfaces. Sabellids, however, may have well-developed eyes positioned on the radioles (fig. 5.1f).

Proboscis: The structure of the proboscis and its accessory organs are of high systematic importance in the identification of many polychaetes. The proboscis is a feeding organ that is everted through the mouth and is defined as the eversible part of the gut. The proboscis may be armed with chitinous jaws, or it may be unarmed. In the latter case, the proboscis can be a thick muscular tube or a thin flexible sac. Some sedentary families lack a proboscis. Dales (1962) has examined the feeding organs of many polychaete families, and the reader is referred to his article for greater detail on the structure and function of the proboscis.

Some errant polychaetes, such as nereids, have a single pair of serrated jaws accompanied by many small conical teeth called paragnaths. Paragnaths are arranged in specific areas of the pharynx, and their distributions are important in nereid systematics. For example, figure 5.2 represents the dorsal and ventral aspects of an everted nereid stomodaeum (the proboscis). The pair of jaws is exposed at the anteriormost end of the proboscis. A maxillary ring is adjacent to the jaws and anterior to the oral ring, which is just ahead of the body. Each ring is divided into six areas designated by Roman numerals. Paired areas of paragnaths are arbitrarily assigned even numerals (II, IV, VI, VIII); unpaired areas of paragnaths are given uneven numerals (I, III, V, VII).

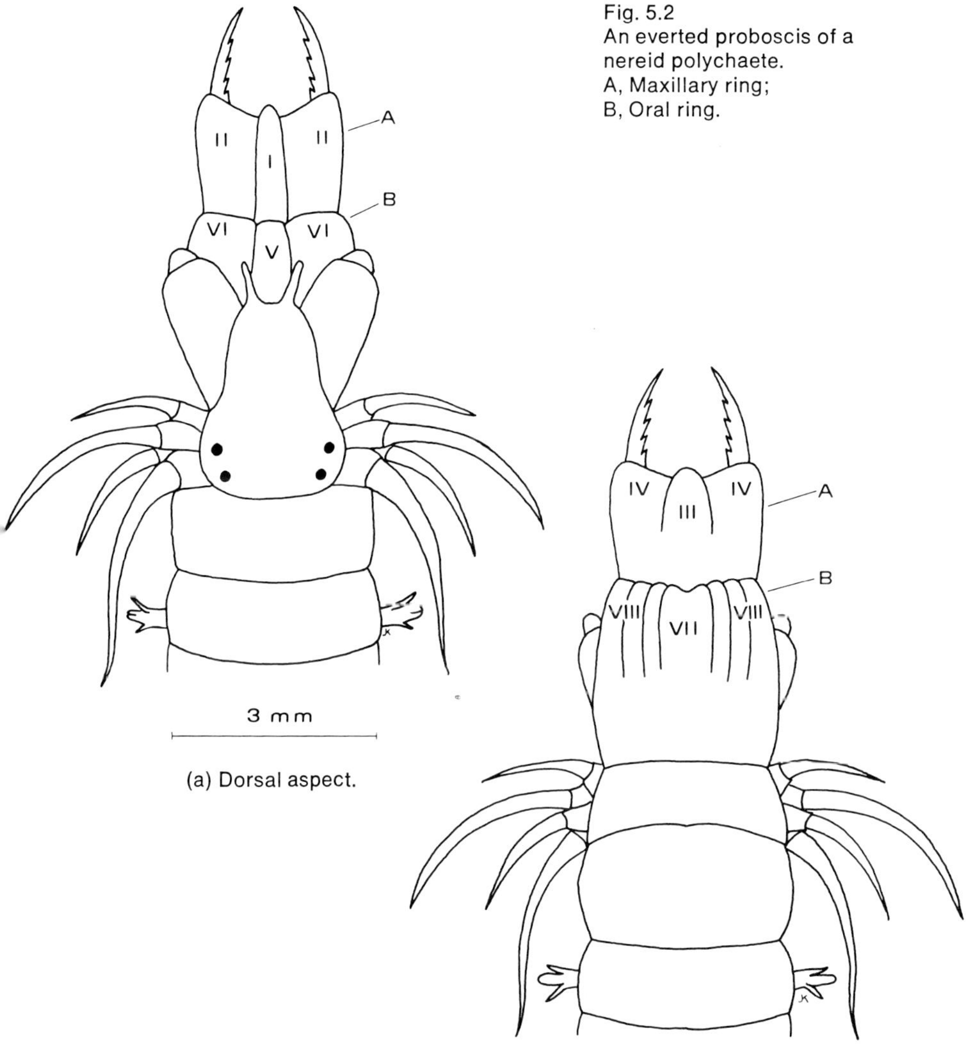

Fig. 5.2
An everted proboscis of a nereid polychaete.
A, Maxillary ring;
B, Oral ring.

(a) Dorsal aspect.

(b) Ventral aspect.

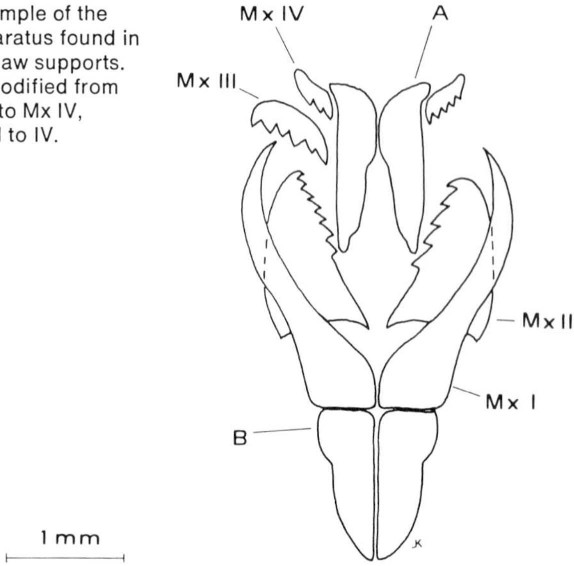

Fig. 5.3 An example of the mandibular apparatus found in the Eunicea. A, Jaw supports. B, Mandibles. (modified from Day, 1967). Mx I to Mx IV, Maxillary plates I to IV.

The superfamily Eunicea contains the errant families, such as Eunicidae, Onuphidae, Lysaretidae, Lumbrineridae, and Arabellidae, all of which have an elaborate mandibular apparatus (fig. 5.3). This apparatus consists of numerous chitinous or calcareous upper jaws (maxillae) and a lower jaw (mandible). There may be more than five pairs of maxillae, but usually only one pair of mandibles. Each piece of the upper jaw is referred to as a maxillary plate. Included in this apparatus are pieces of chitin that support the plates, called maxillary carriers, or supports.

An arbitrary dental formula using the symbol Mx has been developed to identify the various maxillary plates. In the apparatus shown in figure 5.3, Mx I denotes the largest pair of plates, called the forceps; Mx II is the next largest pair, with up to five or six teeth on each plate; Mx III is represented only by a smaller left plate with six teeth; Mx IV, V, VI, etc., are represented by progressively smaller plates. The precise number of maxillary plates depends on the species. A dental formula can therefore be described for the apparatus shown in figure 5.3 as follows: Mx I = 1 + 1, Mx II = (5–6) + (4–5), Mx III = (6) + (0), Mx IV = (3) + (4–5). Arabic numerals denote the number of teeth in each plate. The features of the left plate are always given first. If the number of teeth on a plate is variable, the extremes are entered in brackets. If a plate is absent, it is so indicated by a zero.

A simpler pharyngeal apparatus exists in the family Glyceridae, where four jaws are present at the distal end of the everted proboscis (fig. 5.4a). Each jaw has a jaw support called an aileron (fig. 5.4b). The size and shape of ailerons provide useful characters in glycerid systematics. In addition, the proboscis is equipped with many small papillar processes whose shape, size, and surface features are important characters.

The proboscis of errant worms like phyllodocids, most hesionids, and pilargiids is unarmed (fig. 5.5). Usually there are soft papillae on the feeding organs of these worms that may be arranged terminally, or in straight or alternating rows on the proboscis.

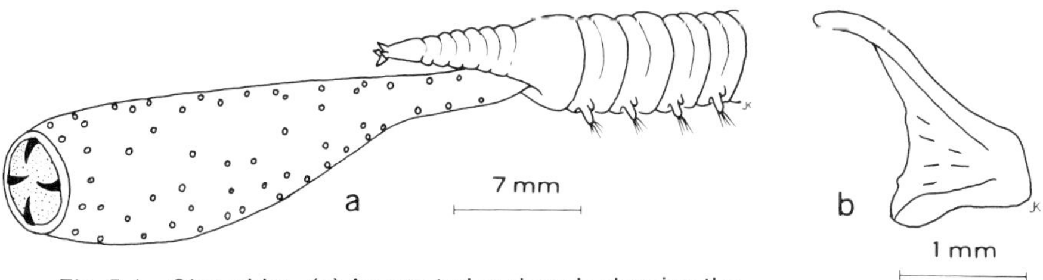

Fig. 5.4 Glyceridae. (a) An everted proboscis showing the four distal jaws. (b) A single jaw support (aileron). (modified from Berkeley & Berkeley, 1948).

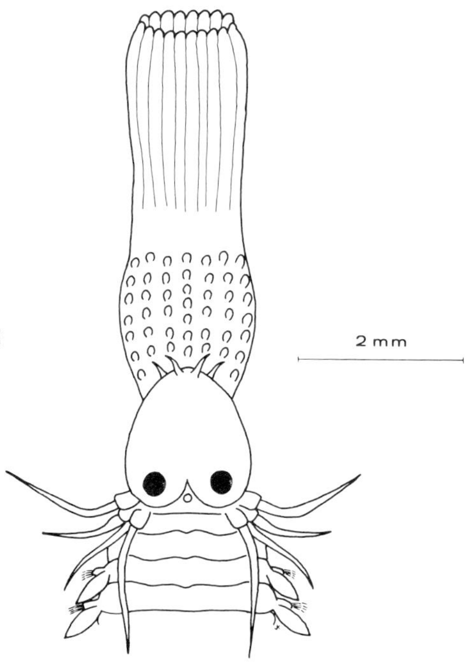

Fig. 5.5
An example of an everted proboscis found in the Phyllodocidae.

The proboscis of sedentary polychaetes is usually thin, flexible, and unarmed. It is also usually ciliated and may have buccal papillae. Upon proboscideal eversion, the proboscis of opheliids may be saclike, or it may appear as a ciliated funnel like those found in arenicolids and maldanids (fig. 5.6).

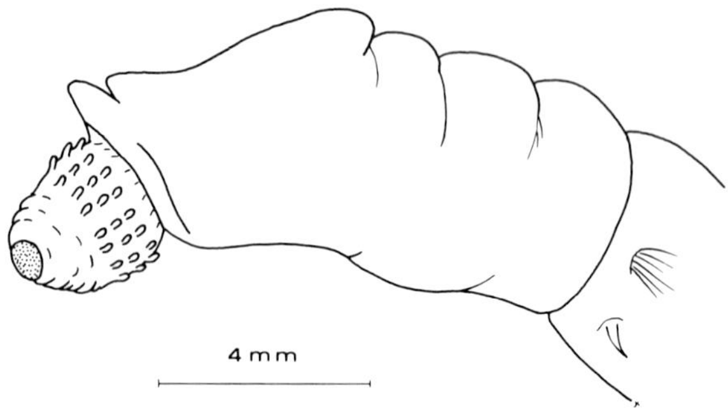

Fig. 5.6 A papillate proboscis of a maldanid polychaete.

Some tubicolous worms like pectinariids, sabellids, and serpulids lack a proboscis. Pectinariids use thick digging combs to loosen the sediment, and short palps to ingest it. The mouth of sabellids and serpulids is surrounded by a series of radioles which are used to filter food particles out of the water. Trapped food is then conveyed to the mouth by a complex sorting system.

Trunk: The trunk, or metastomium, generally represents a linear series of morphologically similar segments. Generally, errant polychaetes have similar trunk segments, while sedentary worms may have at least two distinct anatomical regions in which the segments are different. These two regions are called the thorax and abdomen. Various external features such as the number of segments, branchiae (gills), types of bristles (setae), length, and sometimes color are used to distinguish between species. Basic external characters such as parapodia, setal types, and body coverings are discussed below.

Parapodia: Each segment has a pair of parapodia which function in locomotion and increase the surface area of the body for respiratory purposes. Each consists of an upper lobe (notopodium) and a lower lobe (neuropodium) (fig. 5.7a). Each lobe has a cirral process that takes on a variety of forms and may be important in the identification of a species. For example, phyllodocids have foliaceous (leaflike) cirri (fig. 5.7b) while nepthyids have interramal cirri (fig. 5.7a).

Fig. 5.7 Parapodia
ac, aciculum;
dc, dorsal cirrus;
irc, interramal cirrus;
nrp, neuropodium;
ntp, notopodium;
sf, setal fascicles;
vc, ventral cirrus.

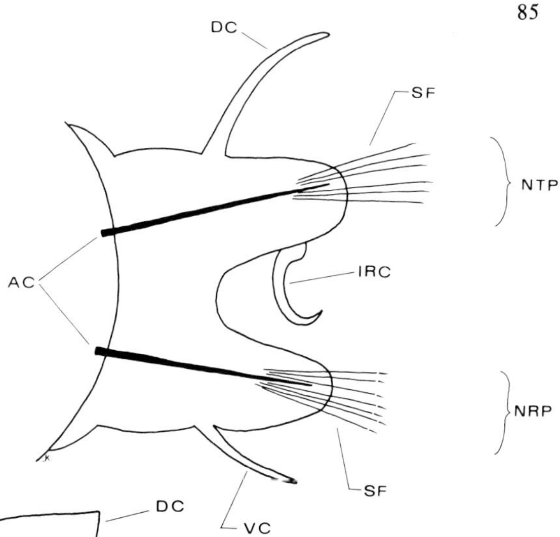

(a) Biramous parapodium.

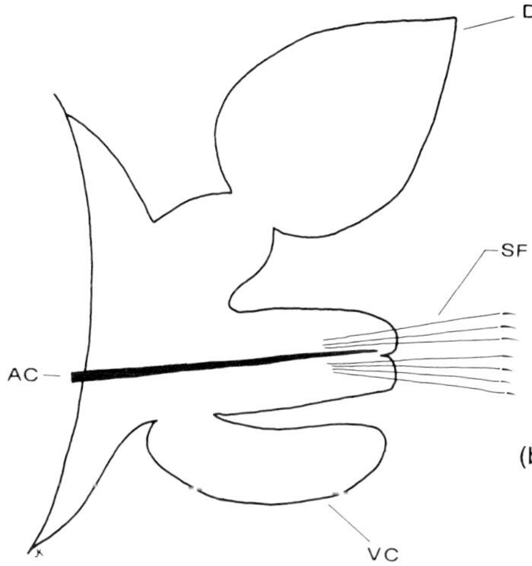

(b) Uniramous parapodium.

Within a parapodium there may be one or two thick chitinous supporting rods called acicula. An aciculum supports the setal fascicle and confers some degree of rigidity to it. Associated with the acicula are bundles of setae (bristles) called fascicles. If a parapodium has a pair of acicula and a pair of fascicles, it is termed biramous; if a parapodium has a single aciculum and one fascicle, it is uniramous.

Setae: Setae represent important taxonomic criteria, and their size, shape, and number per fascicle are used extensively in polychaete systematics. Most setae are solid, composed of a chitin-protein complex secreted by specialized cells in the parapodia. The amphinomids, in contrast, have hollow calcareous setae which are glistening white, filled with a toxin, and very brittle to the touch.

There are two basic kinds of setae — simple and composite. Generally, errant worms have composite (jointed) setae, while sedentary worms have simple (unjointed) setae. Although there are numerous types of setae, familiarity with the common types will facilitate the handling of keys in a species identification.

Basic kinds of simple setae include: capillary setae, that are long and hairlike (fig. 5.8a); bilimbate setae, which are distally broad and oarlike (fig. 5.8b); limbate setae, having only a single blade (fig. 5.8c); falcate setae, which are sickle-shaped (fig. 5.8d); and paleal setae, which are broad and flat (fig. 5.8e). More complex setae are the hooks (also known as crochets) that may be distally entire, bifid, or trifid (fig. 5.8 f–g). Rostrate hooks have a single large primary fang and many secondary teeth (fig. 5.8h). If a hook is embedded in a clear matrix, it is classified as a hooded hook (fig. 5.8i). Another general type of seta that is very characteristic of the sedentary worms is the uncinus. Such setae are rectangular in shape, with a single serrated edge, and are set very close together in one to many rows in a parapodium (fig. 5.8j–k).

Composite setae have a multiarticulate appendage as in some flabelligerids, to the usual biarticulate setae found in most errant worms. Biarticulate setae consist of a shaft and a distal blade. If the blade tapers to a point, it is termed a composite spiniger (fig. 5.8l); if it terminates in a hook, it is referred to as a composite falciger (fig. 5.8m). In nereid polychaetes, however, compound setae are termed either homogomph or heterogomph. If both sides of the shaft joint are even (symmetrical), the seta is called homogomph; if the joint is uneven (asymmetrical), it is called heterogomph. Both heterogomph and homogomph falcigers and spinigers occur in nereids (fig. 5.8n–q).

Branchiae: Each segment of the worm may have a pair of gills or branchiae. The branchiae are usually placed on the dorso-lateral sides of the animal adjacent to the parapodia, and assume a variety of shapes, ranging from a simple lobe to an elaborately dendritic or spiraled gill.

Nephridiopores: Opening at the base of a parapodium is a conical excretory pore called a nephridiopore. They may be present on all setigerous segments of errant worms, but are confined to a few of the anterior segments in sedentary worms.

Fig. 5.8 Setal types. (a) Capillary. (b) Bilimbate. (c) Limbate. (d) Falcate. (e) Paleal Seta. (f) Distally entire hook. (g) Bifid hook. (h) Rostrate hook. (i) Hooded hook. (j) Avicular uncinus. (k) Ampharetidae uncinus. (l) Composite spiniger. (m) Composite falciger. (n) Heterogomph falciger. (o) Homogomph falciger. (p) Heterogomph spiniger. (q) Homogomph spiniger. Approx. size 0.1 to 0.5 mm. (modified from Fauvel, 1923).

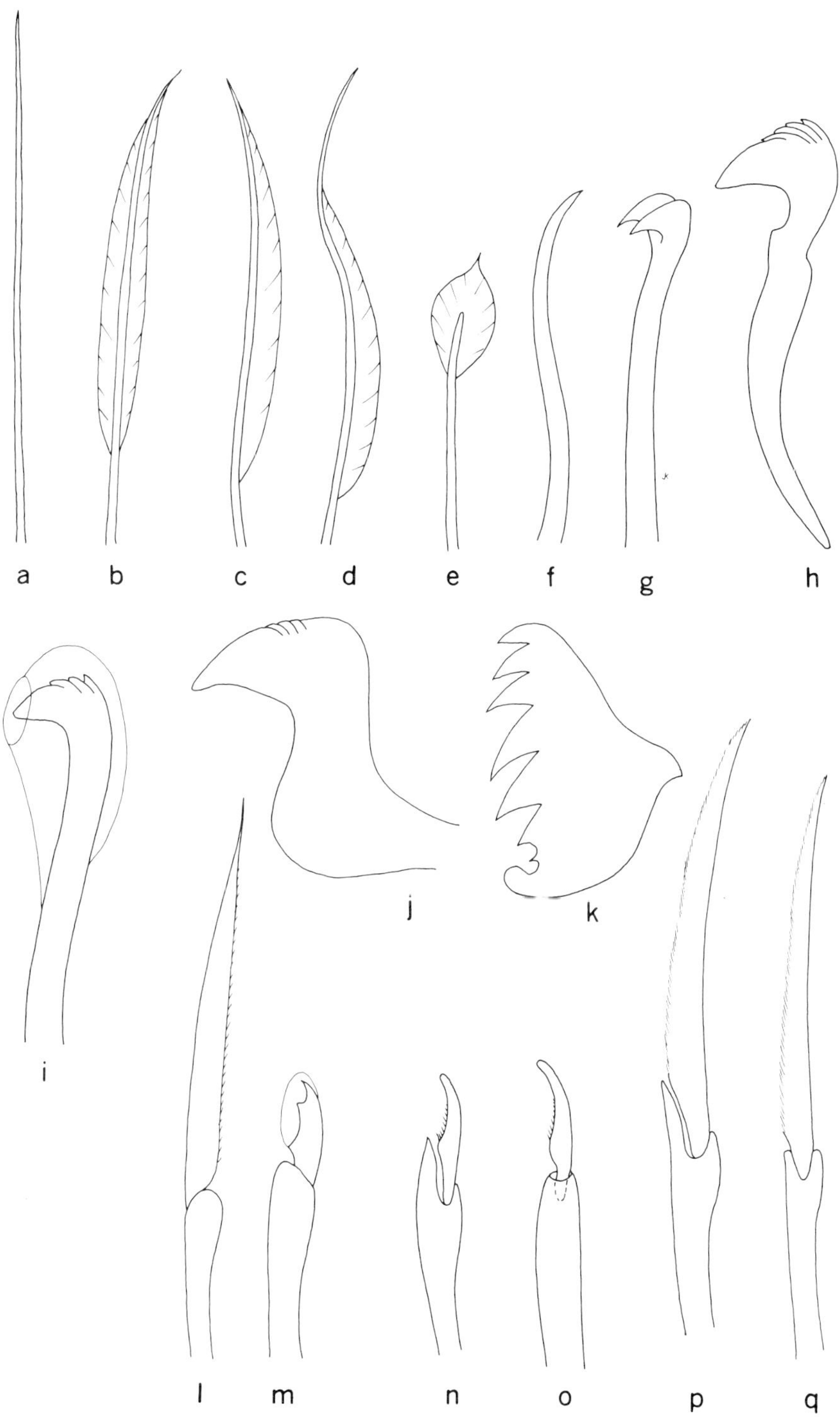
a
b
c
d
e
f
g
h
i
j
k
l
m
n
o
p
q

Elytra and Filaments: Occasionally the body is covered with scales (elytra). This is characteristic of errant worms belonging to such families as the Aphroditidae, Polynoidae, and Sigalionidae. The placement, size, and shape of these elytra are important. In the aphroditids, however, the elytra are obscured by long notopodial setae that intertwine with one another to produce a feltlike appearance very much like the hairs of a mouse. As a result, these worms are commonly called sea mice (fig. 5.10a).

In the case of cirratulids, an extensive series of filaments usually projects from numerous trunk segments. These filaments are respiratory structures and are characteristic of this family (fig. 5.25).

Tail: The final section of the body is the tail (pygidium), which contains either a dorsal or terminal anus. The pygidium may be truncated or tapered in shape, with or without anal cirri (fig. 5.9).

Fig. 5.9 Pygidia of Amphinomid and Maldanid polychaetes.

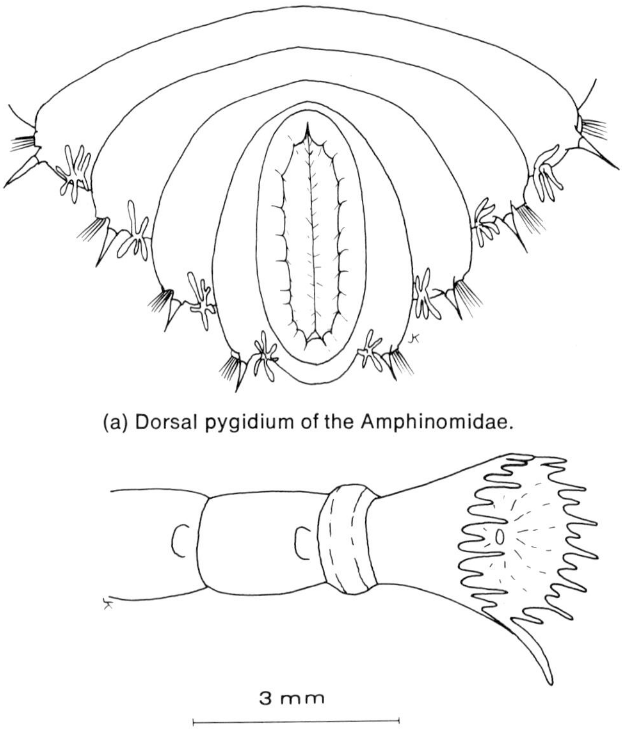

(a) Dorsal pygidium of the Amphinomidae.

(b) Terminal pygidium of the Maldanidae.

Techniques for Examining the Proboscis and Setae: In order to identify a species of polychaete, it is often necessary to dissect open the proboscis and setae. The proboscis can be exposed with a small, sharp scalpel and mounted between a microscope slide and a cover glass for examination. Mounting media found useful for this technique are Diaphane and CMC-9AF (Turtox, Chicago, Illinois). Setae can be obtained by pinching off the parapodia with a pair of fine-pointed forceps (such as jewelers' forceps). The setae are then placed on a slide with a drop of absolute alcohol and properly oriented under a dissecting microscope. Once this process has been completed, a drop of Diaphane followed by a cover glass will produce a permanent slide. The reader is referred to Day (1967) for additional techniques.

Feeding in the Polychaeta: Polychaetes chiefly inhabit marine and estuarine environments in which they have evolved pelagic, crawling, burrowing, and tube-dwelling modes of existence. Concomitant with habitat diversification has been the development of different types of feeding. The two principal types of feeding in the Polychaeta are macrophagy and microphagy. Macrophagy is typical of errant worms and involves the consumption of large, coarse pieces of plant and animal tissue. Microphagy, on the other hand, is widespread in sedentary polychaetes and entails the ingestion of particulate matter.

For example, raptorial feeding habits are typical of crawling errant worms like nereids, and burrowers like glycerids, lumbrinerids, and arabellids. These worms prey upon small invertebrates (including other polychaetes), and are aided in their task by powerful jaws and teeth. There are also herbivorous nereids and nepthyids that tear off pieces of plant tissue with their jaws, while others are omnivorous and consume whatever is available.

Sedentary polychaetes generally exhibit microphagous methods of deposit, filter, and suspension feeding. Deposit feeding worms are subdivided into direct deposit and indirect deposit feeders. Arenicolids, maldanids, and opheliids are examples of direct deposit feeding worms that have a proboscis and actively ingest sediment for their food. Indirect deposit feeders like terebellids usually lack a proboscis and depend upon ciliated tentacles to glean particulate matter from the surface sediment and convey it to the mouth. An exquisite example of filter feeding is to be found in chaetopterid polychaetes. These tubicolous worms force a current of water through a mucus net formed by the worm within the tube. Particulate matter captured in this net is periodically ingested while another net is being produced. Suspension feeding is found in the feather-duster worms (sabellids, serpulids), which strain food particles out of suspension by forcing water through the branchial plumes.

Gulf of California Polychaetes: In general, little is known about the extent and diversity of polychaetous annelids in the Sea of Cortez. Based upon previously collected specimens, the Gulf polychaete fauna appears to be a unique combination of Panamic, Caribbean, Indo-Pacific, temperate, cosmopolitan, and many endemic species. Similar biogeographic affinities have been found for other invertebrates in the Gulf of California. However, the extent of each provincial group of polychaetes in the Gulf is currently unknown, because only a small portion of the estimated 1,000 species of worms has been described.

Definition of Orders: Although various morphological characters enumerated below will overlap between the two orders of polychaetes, a specimen can usually be keyed out to the correct order if it has a majority of the following characters:

1. Prostomium with sensory appendages such as palps and peristomial tentacles. The proboscis is usually quite muscular and may be armed with jaws. Parapodia and parapodial cirri are well-developed; parapodia generally bear composite setae. The metastomium is not divided into different anatomical regions. Nephridiopores are usually located on all segments . . . . **Errantia**
2. Prostomium is usually reduced and lacks sensory appendages. The proboscis, when present, is unarmed and quite flexible. Parapodia and parapodial cirri are reduced; parapodia bear simple setae. The metastomium may be divided into two distinct morphological regions called a thorax and abdomen. Nephridiopores are usually restricted to a few anterior segments . . **Sedentaria**

Keys to Selected Gulf Polychaetes: The following keys (pp. 93 and 114–15) are intended for use in the northern Gulf, and particularly for the Puerto Peñasco region, where the families selected are all abundant and easily found. Many common species that can be found throughout the Gulf have also been listed. Please note, however, that there are approximately 75 known families of polychaetes in the world, of which only 24 are included in the key. The chances of finding families not listed here are therefore good, and the reader is referred to Berkeley and Berkeley (1939, 1948, 1952), Day (1967), Fauchald (1968, 1970), Fauvel (1923, 1927, 1932), Hartman (1939, 1940, 1941b, 1944a, 1944c, 1947a, 1947b, 1950, 1957, 1968, 1969), Reish (1968), Rioja (1941b, 1942a, 1942b, 1947a, 1947b, 1962)and Uschakov (1955), for further information regarding polychaetes. Reish (1968) provides a summary of all the reported Gulf polychaetes known to that year. In the descriptions that follow the keys, recent references are cited, in which a more detailed account of a particular family or species may be obtained.

Glossary of Selected Terms: The immense amount of terminology present in polychaete systematics necessitates a glossary, separate from that in the back of the handbook, of selected terms to aid in the understanding and handling of keys and species descriptions that follow. Additional terms can be found in Day (1967).

Abdomen. The posterior portion of the metastomium, behind the thorax.
Aciculum(a). A thick chitinous rod conferring internal support to a setal fascicle.
Anal cirrus(i). An elongate protrusion arising from the pygidium and usually adjacent to the anus.
Annulate. Furnished with rings.
Antenna(e). Dorsal sensory projections of the prostomium.
Apodous. A segment without parapodia.
Asetigerous. A segment without setae.
Auricular. Ear-shaped.
Avicular seta. A seta with a beak.
Biarticulate. Two-jointed.
Bidentate seta. A seta with two teeth.
Bifid. Diverging distally into two parts.
Bilimbate seta. A simple pointed seta with two lateral edges.
Bipinnate. Featherlike, with a main axis and a row of branches on each side.
Biramous. A parapodium with two setigerous lobes and two acicula.
Branchiae. Respiratory organs found on the dorsal surface of the trunk.
Campanulate. Bell-shaped.
Caruncle. A median sensory lobe of the prostomium.
Capillary seta. A simple, slender, hairlike seta.
Ceratophore. The basal portion of a tentacle, which may be ringed.
Cordate. Heart-shaped.
Cirrus(i). A threadlike appendage.
Composite seta. A compound or jointed seta composed of a basal shaft and a distal blade or hook.
Crochet. A long-shafted seta with a hooked end.
Cupules. Small cups found on the median segments of chaetopterids.
Dentate. Toothed.
Distal. The end part of a structure; that part farthest from the point of attachment.
Elytron(a). Platelike scales found on the dorsal surface of scale worms.
Entire. Applied to the marginal edge of a seta which is smooth and not interrupted with protrusions.
Falcate seta. Sickle-shaped.
Falciger. A composite seta with a hooked blade.
Fascicle. A bundle of setae.
Filiform. Filamentous or threadlike.
Foliaceous. Leaflike.
Gills. See Branchiae.

Heterogomph. Pertaining to the uneven shaft-joint of nereid setae.
Homogomph. Pertaining to the even shaft-joint of nereid setae.
Hooded hook. The apex of a hooded seta embedded in a clear chitinous matrix.
Hooks. A general term applied to simple setae with short shafts and recurved ends.
Inferior. The more ventral of two structures.
Jaws. A chitinous pair of curved, serrated, or smooth pinchers.
Lamella(e). A flattened structure.
Limbate seta. A seta with a single blade.
Mandible. The lower jaw in the mandibular apparatus of the Eunicea.
Maxillae. The upper jaws in the mandibular apparatus of the Eunicea.
Metastomium. The trunk region of polychaetes.
Neuropodium. The ventral or lower half of a biramous parapodium.
Notopodium. The dorsal or upper half of a biramous parapodium.
Occipital. Pertaining to the posterior aspect of the prostomium.
Operculum. A stalked radiole with a distal plug to seal the opening of its calcareous tube.
Palps. Sensory appendages associated with the prostomium and mouth in errant worms; feeding organs in sedentary worms.
Papillate. Bearing many small conical processes.
Paragnaths. Small conical, chitinous denticles in the pharynx of nereids.
Parapodia. Paired segmental locomotory appendages which bear setae.
Pectinate. Arranged like the teeth on a comb.
Peristomium. The region just behind the prostomium and around the mouth which often has tentacles.
Proboscis. That part of the gut which is everted through the mouth.
Prostomium. The anteriormost portion of an annelid that may bear sensory appendages, like eyes, palps, and antennae.
Proximal. The part of a structure nearest its base; that portion nearest the point of attachment.
Pygidium. The posteriormost portion of a polychaete, bearing the anus.
Radioles. Bipinnate processes arising from the peristome of sabellids and serpulids which function in feeding and respiration.
Scale. See Elytron.
Scute. Platelike.
Sessile. Without a stalk.
Setae. Chitinous bristles produced by specialized cells in the parapodium.
Setiger. A segment bearing setae.
Spiniger. A composite seta with a distally tapered blade.
Superior. The dorsalmost of two structures.
Tentacle. A general term applied to a head appendage.
Trifid. Diverging distally into three parts.
Tuberculate. Equipped with many small fingerlike processes.
Tubicolous. Tube-dwelling.
Uncinus(i). Small rectangular seta with a serrated edge that are widespread in sedentary polychaetes.
Uniramous. A parapodium with a single setigerous lobe and aciculum.

KEY TO SELECTED POLYCHAETE FAMILIES FROM THE NORTHERN GULF OF CALIFORNIA: ERRANTIA

1 Dorsal surface covered with elytra or long feltlike notopodial spines **2**
-Dorsal surface not covered with elytra or feltlike spines **3**

2(1) Dorsal surface covered by a feltlike layer of overlapping setae that obscures the elytra; elytra present on 15 segments; body short and elliptical .... **Aphroditidae**
-Dorsal surface not covered by a feltlike layer of setae; elytra present on segments 2,4,5,7,9 . . . to 23, and on every third segment thereafter; body elongate and flattened .... **Polynoidae**

3(2) Prostomium with conspicuous cephalization **4**
-Prostomium without conspicuous cephalization **10**

4(3) Prostomium with palps **5**
-Prostomium without palps **9**

5(4) Palps biarticulate **6**
-Palps not articulate **7**

6(5) Proboscis usually a muscular tube, but may be equipped with a styliform denticle; notopodia reduced to jointed dorsal cirri with internal acicula .... **Hesionidae**
-Proboscis armed with a pair of serrated jaws and many smaller teeth (paragnaths) that are arranged into specific areas; notopodia not reduced, and dorsal cirri smooth .... **Nereidae**

7(5) Palps cirriform; prostomium may be obscured; prostomium divided into an anterior and posterior lobe; caruncle present .... **Amphinomidae**
-Palps not cirriform; prostomium obvious; prostomium not divided into anterior and posterior lobes; caruncle absent **8**

8(7) Palps globular; prostomium with 2 small frontal antennae and 5 occipital tentacles, each of which has a thick ringed base (ceratophore); first segment apodous and asetigerous; simple setae present .... **Onuphidae**
-Palps cushionlike; prostomium with 1, 3, or 5 antennae without ceratophores; first 2 segments apodous and asetigerous; simple and composite setae present .... **Eunicidae**

9(4) Prostomium with 4 or 5 tentacles; peristomium with 2 to 4 pairs of tentacles; parapodia uniramous; dorsal and ventral cirri foliaceous .... **Phyllodocidae**
-Prostomium with 2 or 3 tentacles; peristomium with 6 to 8 pairs of tentacles; parapodia biramous; dorsal and ventral cirri cirriform .... **Hesionidae**

10(3) Prostomium strongly conical with secondary annulations; prostomium with 4 distal antennae; proboscis with 4 distal jaws and minute surface papillae .... **Glyceridae**
-Prostomium flattened and rounded without secondary annulations; prostomium without 4 distal antennae; proboscis without 4 distal jaws and minute surface papillae **11**

11(10) Prostomium flattened and pentagonal in shape; prostomium with 4 to 6 marginal tentacles; proboscis with a pair of jaws, with or without papillae; parapodia biramous, with interramal cirri..... **Nephtyidae**
-Prostomium somewhat rounded; prostomium without appendages; proboscis with an elaborate mandibular apparatus but without papillae; parapodia uniramous, without interramal cirri **12**

12(11) Peristomium with 3 tentacles; first 2 segments with parapodia; parapodia with simple setae .... **Lysaretidae**
-Peristomium without tentacles; first 2 segments lack parapodia; parapodia with composite or simple setae **13**

13(12) Prostomium with eyes; parapodia with simple setae only .... **Arabellidae**
-Prostomium without eyes; parapodia with simple and composite hooks .... **Lumbrineridae**

Fig. 5.10. Aphroditidae.

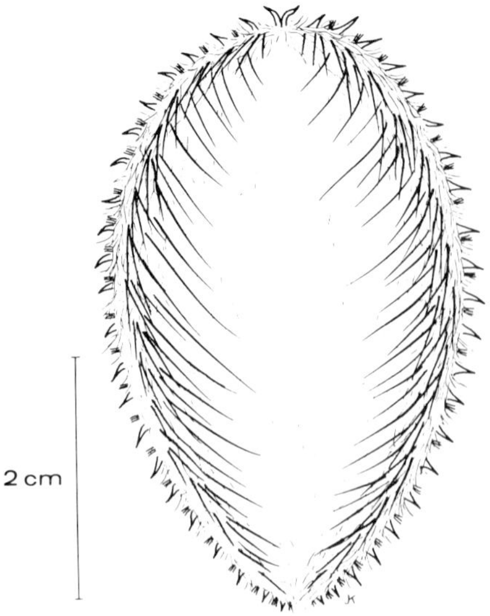

(a) Dorsal aspect of *Aphrodita*.

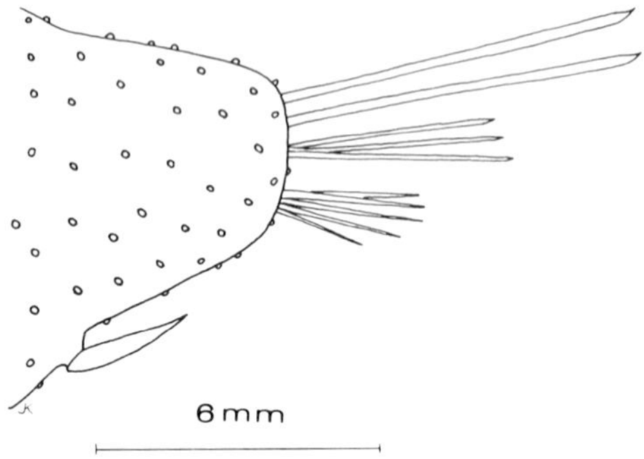

(b) Neuropodium of *Aphrodita* sp.

Family Aphroditidae

The body is oval to elliptical and tapered posteriorly; the dorsum is convex, and the ventrum is flat. These worms have approximately 30 to 40 segments, with 15 pairs of elytra on the anterior segments. The dorsum is covered with a dense feltlike mat of overlapping notosetae that cover the elytra (fig. 5.10a). Prostomium with one or two pairs of eyes, two long palps, and a median tentacle placed between the eyes. Two pairs of peristomial tentacles are present. Parapodia are biramous. Notosetae are of two kinds; stout, smooth gold-colored setae and fine capillary setae. Neurosetae are present in three tiers and are thick and acicular with lateral spines (ref: Hartman, 1939).

The aphroditids are subtidal polychaetes that inhabit muddy and silty bottoms. The thick dorsal felt gives them a mouselike appearance; hence the name "sea-mice." Specimens of this family are often obtained from shrimp trawlers.

Aphrodita sp.

A body of 42 segments is approximately 100 mm long by 40 mm wide. The acicular, stout notosetae are copper-brown in color, and the fine capillary setae are greenish yellow. Neuropodia have three tiers of setae: the superior level with two dark, large, and stout acicula; the middle tier with three intermediate-sized copper-colored acicula; and the inferior level with five smaller, fine, light gold-colored setae (fig. 5.10b). The dorsum of these animals is often clean and lacks a dense layer of slit and mud particles (ref: Hartman, 1939).

Family Polynoidae

The body is dorso-ventrally flattened, and oval to elongate in outline (fig. 5.11a). The prostomium bears two frontal horns and two pairs of eyes in a trapezoidal arrangement. Two lateral tentacles and one unpaired median tentacle are also present on the prostomium. Two pairs of peristomial tentacles with ceratophores are present. Elytra number from 12 to 18 pairs, occurring on segments 2, 4, 5, 7, 9, 11, 13, 15, 17, 19, 21, and 23, thereafter alternating on every third segment. Parapodia are biramous and all setae are simple. Notosetae are slightly bent with sharp or blunt tips (fig. 5.11b); neurosetae all have a distal blade (fig. 5.11c). Further family characters can be obtained from Hartman (1939, 1968) and Day (1967).

Polynoids are abundant at Puerto Peñasco, where they can be found either under rocks, or associated with other invertebrates. Quite often these worms are distinctively colored.

Fig. 5.11 Polynoidae.

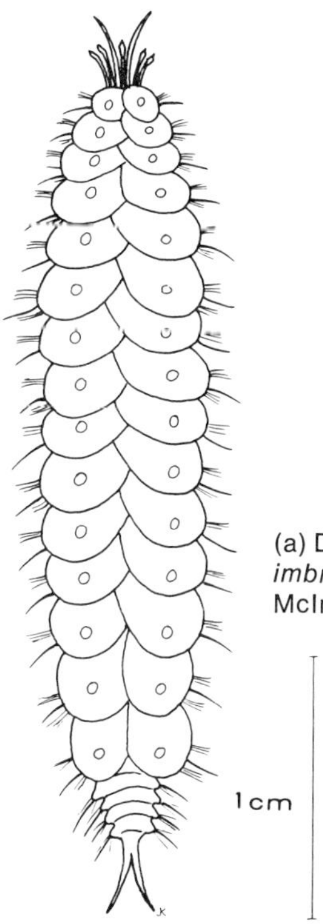

(a) Dorsal aspect of *Harmothoe imbricata* (modified from McIntosh, 1900).

(b) Notoseta.

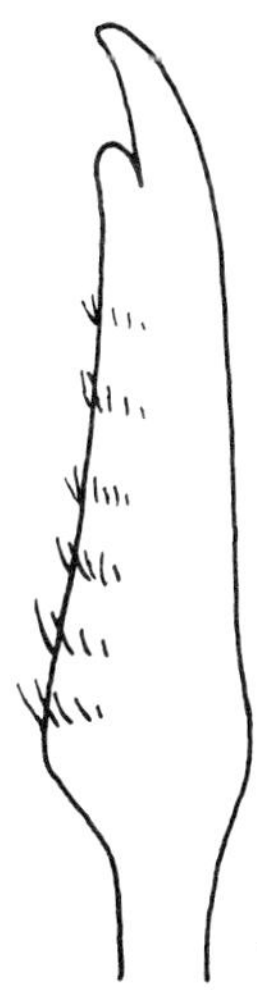

(c) Neuroseta.

Halosydna johnsoni (Darboux)

Thirty-seven segments are 48 mm long by 7.25 mm wide. The dorsum usually has transverse bars of dark pigment that alternate with light bands. The prostomium has two pairs of eyes. The elytra are ear-shaped. This worm is free-living under rocks, or found as a commensal with terebellid polychaetes or certain gastropods. It is distributed from southern California into the Gulf of California. It has been recorded in the Gulf from San Felipe and Puerto Peñasco. See Hartman (1939, 1968) and Berkeley and Berkeley (1939) for further information on this species.

Family Amphinomidae

The body is usually short and stout, but can be elongate, and is rectangular in cross section (fig. 5.12a). The prostomium is often obscured by the anterior segments and setae, but consists of an anterior and a posterior lobe. The anterior lobe is rounded with a pair of eyes, two lateral cirriform palps, and two dorsal antennae. The posterior lobe is equipped with another pair of eyes, a median tentacle, and a caruncle that extends posteriorly to the second or third setigerous segment. The caruncle has a median ridge that is bordered on each side by scalloped edges. The proboscis is muscular and unarmed. Parapodia are biramous, with simple

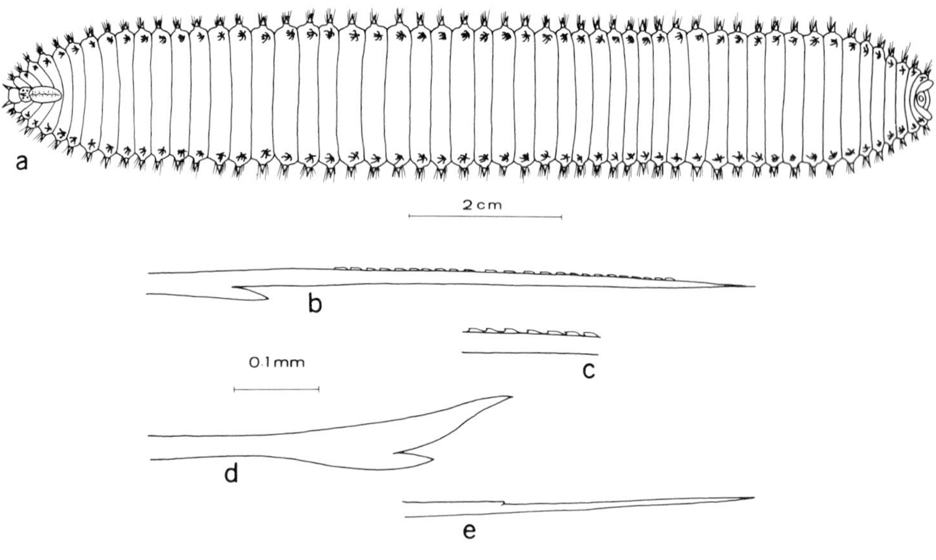

Fig. 5.12 Amphinomidae. (a) Dorsal view of *Eurythoe complanata.* (b) Posterior notoseta of *Chloeia entypa.* (c) Detail of (b). (d) Notoseta of *Chloeia viridis.* (e) Notoseta of *Eurythoe complanata.* (b) to (e) modified from Hartman, 1940.

tubular calcareous setae. Parapodial lobes have cirriform cirri, and branchiae are well-developed and present on the dorsal surface.

Amphinomids are thought to feed on soft-bodied animals, such as sponges and ascidians. These worms are often referred to as "fireworms," because the setae are filled with poison and are brittle, being easily detached from the worm if it is handled. This is particularly true for *Eurythoë complanata;* it is recommended that this species be collected with forceps. In any case, these worms are abundant in rocky midintertidal regions of Puerto Peñasco. The reader is referred to Hartman (1939), Rioja (1947a), and Day (1967) for additional information on this family.

KEY TO SELECTED GENERA OF AMPHINOMIDS

1 Branchiae are bipinnate structures .... ***Chloeia***
-Branchiae are present as bushy tufts **2**
2(1) A single dorsal cirrus per notopodium; body elongate and rectangular in cross section .... ***Eurythoë***
-Two dorsal cirri per notopodium; body stout and hemispherical in cross section .... ***Notopygos***

Chloeia entypa Chamberlin

Dorsum with one wide longitudinal reddish brown stripe that is present throughout the length of the worm. Posterior bifid notopodial setae (fig. 5.12b, c) are serrated on the outer side of the primary tooth and are directed downward. This is a subtidal polychaete that usually can be obtained from the shrimp trawlers at Puerto Peñasco. These worms inhabit sandy or rocky benthic surfaces and are distributed from southern California south to Ecuador. In the Gulf, they have been reported from La Paz, Topolobampo, and the northern Gulf. (ref: Treadwell, 1937).

Chloeia viridis Schmarda

Dorsum with three longitudinal stripes that are usually restricted to the anterior end of the worm. Posterior bifid notopodial setae are smooth and without serrations. The secondary fang is inconspicuous and small (fig. 5.12d). This is another subtidal worm that can be obtained from the shrimp trawlers. These worms also appear to inhabit sandy or rocky benthic surfaces, and are common in the Gulf. The species has also been reported from the Caribbean (Jamaica). (ref: Hartman, 1939).

Eurythoe complanata (Pallas)

The body is elongate, approximately 300 mm long; the width approaches 20 mm. Color in life is pinkish gray to blue, with deep red branchiae. Setae are smooth, white, and bifid distally (fig. 5.12e). This species is extremely common throughout the Gulf, usually being found under rocks in the midintertidal zone. *Eurythoë complanata* is a highly gregarious, slow-moving polychaete. (ref: Hartman, 1939; Rioja, 1947a).

Notopygos ornata Grube

Body is stout and about 50 to 70 mm long and 10 mm wide. The dorsum has a pair of broken stripes, and the caruncle is conspicuous with a high median lobe and folded lateral lobes to either side. *Notopygos ornata* is a Panamic species, common to the Gulf, and usually found under rocks in the mid- and low intertidal zones. It ranges from the northern Gulf to Colombia and the Galapagos Islands. (ref: Hartman, 1939).

Family Phyllodocidae

The body is elongate, narrow, and distinctively colored. The prostomium is rounded to cordate, with a single pair of eyes and four to five prostomial tentacles. The proboscis is cylindrical and tubular, with or without soft papillae. Up to four pairs of cirriform peristomial tentacles may be present on the first two to three fused anterior segments. These segments usually have reduced parapodia. In normal body segments, the parapodia are uniramous with foliaceous or leaflike dorsal and ventral cirri. Setae are composite, and acicula are generally present. These worms take on a brilliant color pattern, which in some cases is important to the identification of a species. Additional information on this family can be found in Treadwell (1937), Rioja (1941b), Day (1967), and Hartman (1968).

Phyllodocids are common in the Puerto Peñasco region. Members of this family are found in rocky crevices, under rocks, or in sandy beaches.

Anaitides species 1

Body color is dark bluish gray, with brilliant orange dorsal and ventral cirri. Body of 195 segments is elongate and about 70 mm long by 2.5 mm wide. Prostomium is cordate, with four frontal antennae (fig. 5.13a). A small nuchal papilla is present between the eyes in the mid-dorsal incision of the prostomium. The proboscis has a circlet of 17 terminal papillae and six pairs of proximal longitudinal rows of orange papillae. Papillae number about 19 to 20 per row. Parapodia are uniramous, with foliaceous dorsal and ventral cirri. Setae are composite spinigers that are marginally toothed and distally spinose. This unnamed species resembles *A. madeirensis* Langerhans, except that it differs in

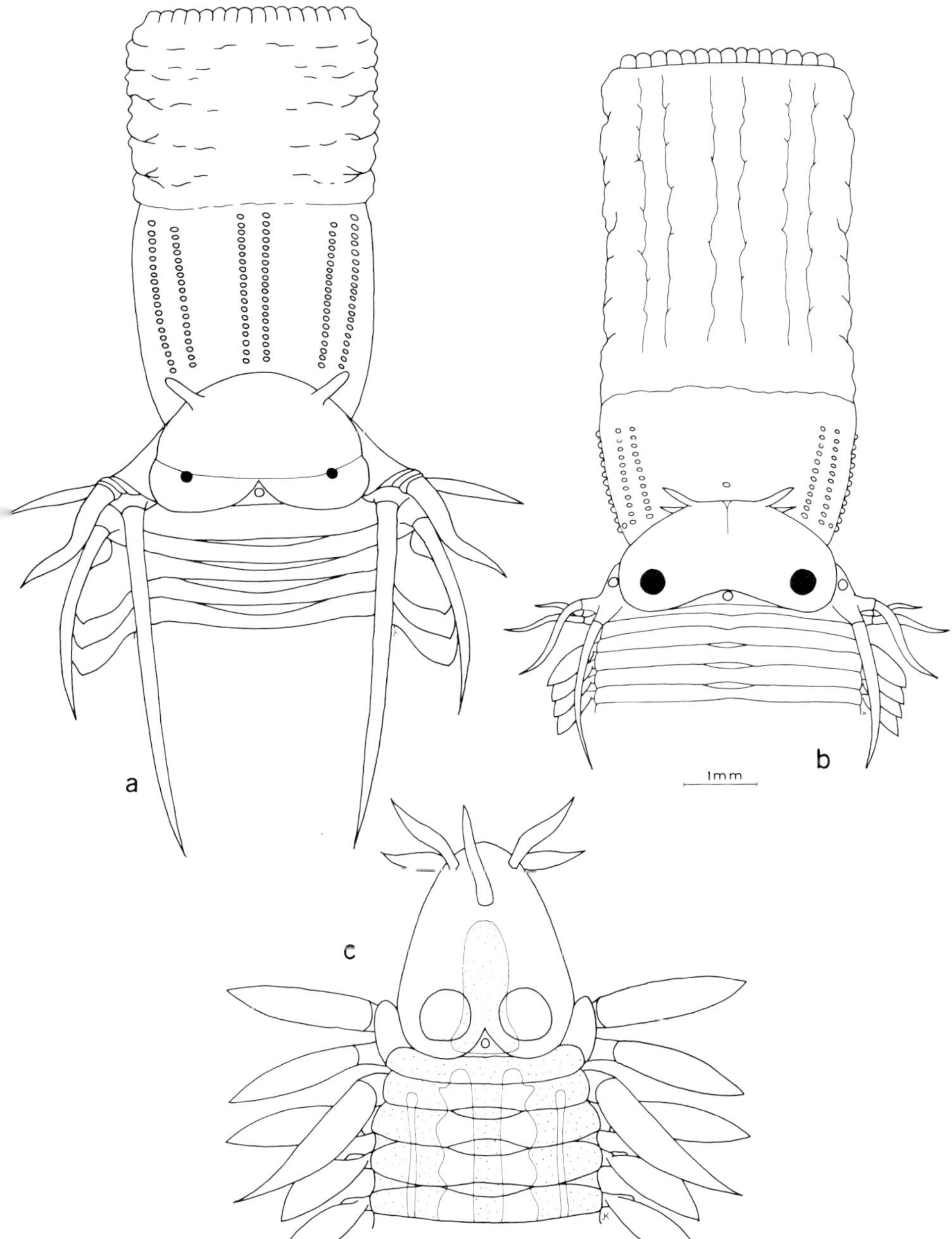

Fig. 5.13 Phyllodocidae, anterior segments. (a) *Anaitides sp. 1.* (b) *Anaitides sp.* 2. (c) *Eulalia myriacyclum.*

size, number of terminal proboscideal papillae, and the number of proximal proboscideal papillae. This species has been taken from Bahía Kino. *A. madeirensis,* however, occurs throughout the Gulf. (ref: Treadwell, 1937; Hartman, 1968).

Anaitides species 2

Color in life of this unnamed species is purplish brown. Its length is 150 mm and width 5 mm, including the parapodia. There are approximately 200 to 250 segments. The prostomium is cordate, with one pair of eyes, two pairs of frontal tentacles, and a nuchal papilla located in the mid-dorsal incision of the prostomium (fig. 5.13b). Proboscis terminates in a circlet of 17 papillae; the proximal surface has six pairs of longitudinal rows of papillae. Each row consists of five to six soft brown papillae. A dorsal, unpaired proximal papilla is present just anterior to the prostomium. This species is superficially similar to *A. mediopapillata* Moore except for the number of terminal proboscideal papillae and the presence of an unpaired proximal papilla. This species has been found in Bahía San Agustín, Sonora, near the south side of the bay. (ref: Hartman, 1968).

Eulalia myriacyclum (Schmarda)

Color in life is brilliant green, with five dark green to black longitudinal stripes on dorsum (one median, two submedian), and three similarly colored stripes on ventrum. Length of body is up to 450 mm, by 3 mm wide, including parapodia. Prostomium is round with four frontal antennae, an unpaired median antenna, and a pair of eyes (fig. 5.13c). Peristome with four pairs of cylindrically tapered tentacles. Proboscis is densely papillate over its entire surface. The first three segments lack parapodia. This worm secretes enormous quantities of mucus when disturbed or irritated. It can be found under rocks in midintertidal zones of Puerto Peñasco. Circular green egg masses are deposited in January and February. (ref: Rioja, 1962).

Phyllodoce sp.

Small worms that are approximately 50 to 60 mm long and 2 mm wide, although several larger individuals have been found. These worms have a characteristic green body and clear to translucent parapodia and cirri. The prostomium is cordate in shape and bears a nuchal papilla. The proximal region of the proboscis is diffusely papillate; the distal region is smooth. The proboscis also terminates in a circlet of 19 terminal papillae. *Phyllodoce* sp. can be found on intertidal sandflats near the entrance to Estero Morua.

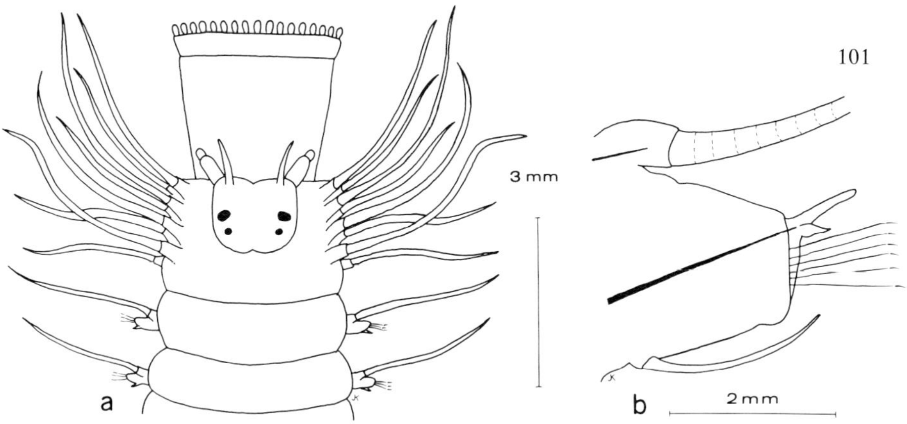

Fig. 5.14 Hesionidae. (a) Anterior segments of *Kefersteinia* (adapted from Day, 1967). (b) Left tenth parapodium of *Hesione intertexta* (modified from Hartman, 1940).

Family Hesionidae

The body is short and stout, with about 16 setigerous segments. The prostomium is equipped with two pairs of eyes and generally a pair of biarticulate palps (fig. 5.14a). The first one to four segments are apodous, reduced, and fused to the prostomium such that from six to eight pairs of ventral and dorsal cirri are present. Notopodia of other body segments are reduced to long, jointed dorsal cirri, each with an internal aciculum. Parapodia are either biramous or uniramous. Notosetae are simple, and neurosetae composite.

Hesionids are relatively active worms found under rocks and in crevices. Commensals of this family are also found in the starfish *Heliaster kubinijii* at Puerto Peñasco. (ref: Hartman, 1940; Day, 1967).

KEY TO SELECTED GENERA OF HESIONIDS

1 Eight pairs of tentacular cirri present; prostomium with a pair of frontal antennae, but no median antenna .... ***Hesione***

-Six pairs of tentacular cirri present; prostomium with a pair of frontal antennae and a median antenna .... ***Ophiodromus***

Hesione intertexta Grube

Up to 50 mm long and 7 mm wide, Parapodial segments number 16. Body is truncated at both ends. Four eyes are present on the prostomium, arranged in a rectangular fashion. The proboscis is unarmed. Notopodia reduced to dorsal cirri with black acicula (fig. 5.14b). Neuropodia with composite setae. *Hesione intertexta* has been reported from the Philippine Islands and western Mexico. This species is common in the Gulf and has been found at La Paz, Roca Consag, and at Puerto Peñasco under basaltic boulders. Gravid specimens have been collected in June and July. (ref: Hartman, 1940).

Ophiodromus sp.

Color in life is dark reddish brown on the anterior dorsum and brick red on the posterior dorsum and ventral surface. A pair of biarticulate palps and two pairs of eyes are present on the prostomium. This species occurs in the ambulacral grooves of the sun star *Heliaster kubinijii* at Puerto Peñasco. (ref: Hartman, 1940; 1968).

Family Nereidae

The body is elongate and cylindrical. The prostomium bears two pairs of eyes, a pair of biarticulate palps, and usually a pair of frontal antennae. The peristomium has four pairs of elongate tentacles. The proboscis is equipped with a pair of serrated jaws and numerous paragnaths. Parapodia from the third one are biramous and setae are composite.

Many nereids inhabit rocky crevices or dense algal mats in the intertidal zone. Some nereids undergo a morphological change to a "heteronereis" upon attaining sexual maturity. Heteronereids are adapted for swimming. At night many of these forms will swarm at the water's surface to spawn gametes. Once this reproductive process is completed the heteronereid usually dies. (ref: Hartman, 1940, 1968; Day, 1967).

KEY TO SELECTED GENERA OF NEREIDS

1 Paragnaths conical and present only on the maxillary ring; posterior notopodial setae without homogomph falcigers .... ***Ceratonereis***
-Paragnaths conical and pectinate and present on both rings; posterior notopodial setae with composite falcigers .... ***Platynereis***

Ceratonereis mirabilis Kinberg

Length of body is 12 to 20 mm, with 46 to 50 segments. The prostomium has a median notch, and four eyes of which the anterior pair are farther apart than the posterior pair (fig. 5.15a). The chitinous jaws of the proboscis have five to six teeth. Pharyngeal areas with teeth are I (a single cone), II (7 to 8 cones), III (a patch of small cones), and IV (small numerous cones) (fig. 5.15b, c). Peristomium with eight elongate tentacular tentacles, with the first pair being the longest. Notopodia lack homogomph falcigers; neuropodia with hetergomph falcigers.

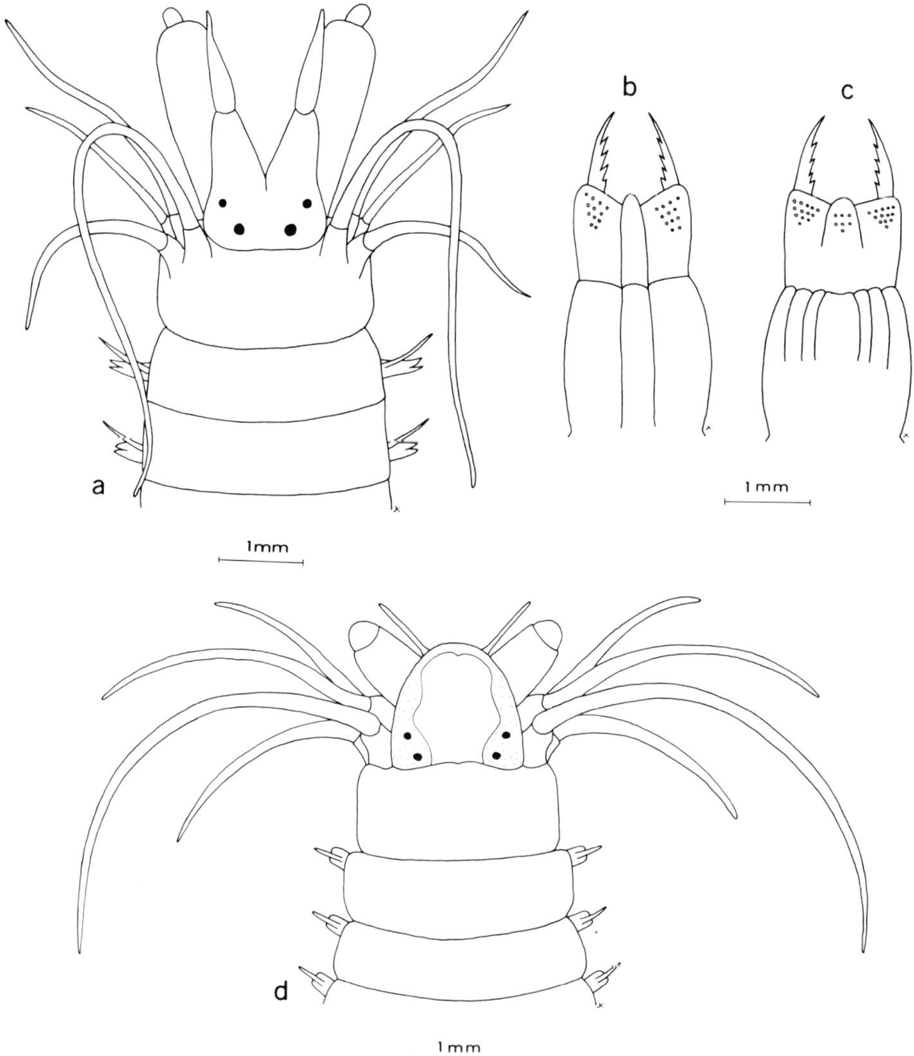

Fig. 5.15 Nereidae. (a) Anterior segments of *Ceratonereis mirabilis.* (b) Dorsal aspect of a *C. mirabilis* proboscis. (c) Ventral aspect of a *C. mirabilis* proboscis. (d) *Platynereis bicanaliculata.* (a) to (c) modified from Day, 1967; (d) modified from Hartman, 1968.

This species is common in the Gulf and is found on both sides of tropical America. It is associated with algae, particularly the dense algal mats present in the rocky intertidal environments of Puerto Peñasco. (ref: Hartman, 1968).

Platynereis bicanaliculata (Baird)

Length of 95 to 100 segments is 60 to 70 mm, width is 3 to 5 mm. The prostomium has four eyes in a trapezoidal arrangement, and has dark pigment distributed about its edges (fig. 5.15d). The proboscis lacks paragnaths on areas I, II, and V; area III bears an elliptical patch of cones; area IV has a crescent of seven to ten rows; area VI bears two to three rows of pectinate paragnaths; and areas VII and VIII have five rows of pectinate and two lateral rows of conical paragnaths. Each jaw has seven teeth and a prominent distal fang. The peristomium has four pairs of elongate tentacles. Parapodia with triangularly shaped lobes. Notopodia with spinigers and simple falcigers; neuropodia with spinigers and short-shafted falcigers. *Platynereis bicanaliculata* is distributed from western Canada south to Mexico. It has been found in the Gulf at Isla Tiburón and Puerto Peñasco, in the rocky intertidal. (ref: Hartman, 1968).

Family Nepthyidae

Worms with an elongate and rectangular body that is usually white in color (fig. 5.16a). The prostomium is pentagonal in shape and has four small antennae and sometimes a pair of eyes. The proboscis (fig. 5.16b) is muscular and may be smooth or papillate, with a pair of jaws. Parapodia are biramous and the setae are simple. A characteristic interramal cirrus is present between the notopodium and neuropodium.

These worms are typical inhabitants of sand- and mudflats like those found in Bahía Adair (Bahía Cholla), Estero Morua, and El Golfo de Santa Clara. They are active burrowers and swimmers with characteristic sinusoidal movements. (ref: Hartman, 1940, 1950, 1968; Day, 1967).

KEY TO SELECTED GENERA OF NEPTHYIDS

1 Interramal cirri are recurved inwards, towards body .... ***Aglaophamus***
-Interramal cirri are recurved away from body .... ***Nepthys***

Aglaophamus dicirris Hartman

Length of body is 44 mm. The rectangular prostomium has a pair of eyespots at its posterior region, and four frontal antennae. The proboscis terminates in 22 distal bifid papillae and 22 subterminal rows of papillae. Each row has about nine papillae each, except the median rows which have about 6 to 7. Interramal cirri are recurved inwards and begin from the fifth setiger (fig. 5.16c). This species occurs commonly in the intertidal sandflats of Bahía Cholla and ranges from southern California to Panamá. (ref: Hartman, 1950; 1968).

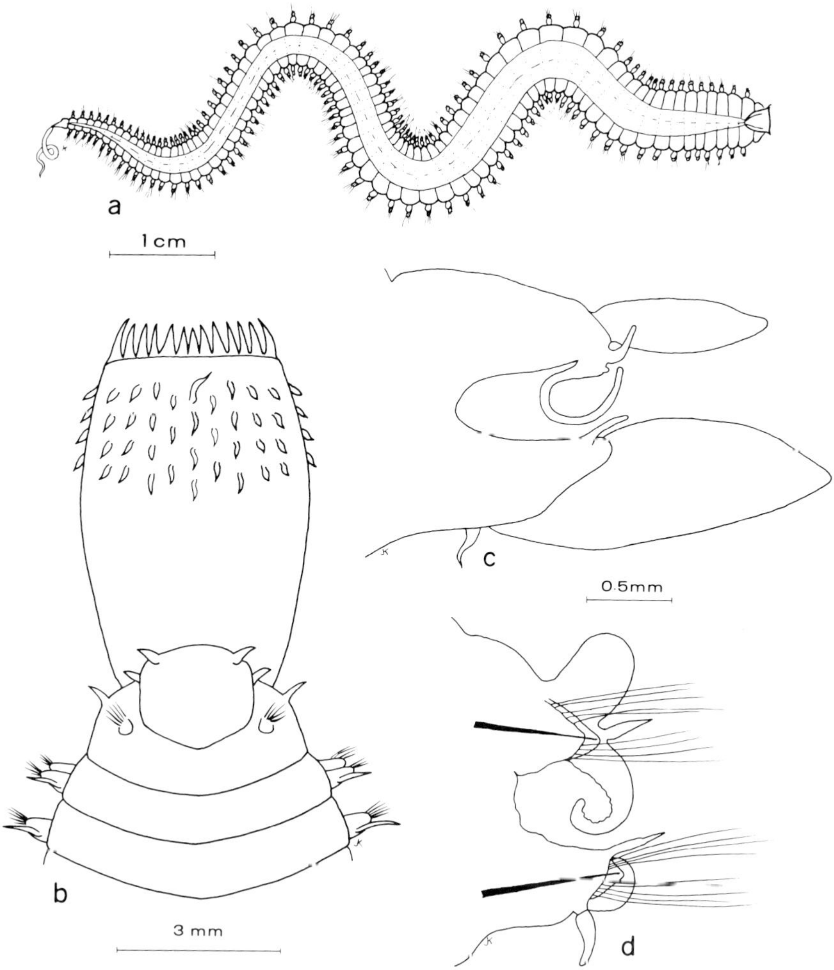

Fig. 5.16 Nepthyidae. (a) *Nepthys caeca* (modified from McIntosh, 1908). (b) Anterior segments of *Nepthys.* (c) Parapodium of *Aglaophamus dicirris.* (d) Parapodium of *Nepthys squamosa,* with setae omitted. (b) and (c) modified from Hartman, 1950.

Nepthys squamosa Ehlers

Color in life is gray to brownish. Length of 117 segments is 50 mm and width is 3 mm. Prostomium is almost trapezoidal in shape, with four small antennae arising from the frontal and lateral sides. The proboscis has 22 rows of distal papillae with five to six papillae per row, and also 22 terminal papillae. There are two conspicuous distal proboscideal papillae, each of which is mid-dorsal and mid-ventral. Interramal cirri begin from the fourth setigerous segment and are best developed around setiger 25 (fig. 5.16d). Characteristic dorsal lamellae above the notopodium are inflated and resemble plates or scales that appear to imbricate with one another. This species ranges from southern California to Ecuador and is common in the Gulf, where it is known mostly from subtidal records. It is also found in southern Florida and West Africa. (ref: Hartman, 1940, 1968).

Nepthys sp.

This species is white to reddish in color. The body length is approximately 90 mm and the width is 5 to 6 mm. The prostomium is rectangular to ovate, with a rounded anterior edge. The proboscis has 22 distal rows of papillae, with five to six papillae per row. There are no mid-dorsal or mid-ventral papillae. The proboscis terminates in 22 rows of terminal papillae with two papillae per row. The outer papilla is about three times the length of the inner one. The proximal surface of the proboscis is smooth. The first parapodium is biramous, with the notopodium larger than the neuropodium. Interramal cirri are first present from setiger 4 and continue to the end of body. Preacicular setae are transversely barred, and postacicular setae are transversely serrated.

This species is common in Bahía Cholla and at El Golfo de Santa Clara in intertidal sandflats.

Family Glyceridae

The body is elongate, cylindrical, and distally tapered at both ends. The prostomium is conical, with four small distal antennae, and is usually multiannulate. The proboscis bears four jaws; each jaw has a V-shaped jaw support called an aileron. The proboscis also has many small tuberculate papillae strewn over its surface. Parapodia are typically biramous but may be uniramous. There may be up to two presetal and two postsetal lobes accompanying the parapodia.

There are only two genera in this family. The genus *Hemipodus* is partially characterized as having parapodia with a single aciculum and only composite setae. The genus *Glycera,* on the other hand, has two acicula per parapodium, and both simple notosetae and composite neurosetae. Only members of the latter genus are described here. Hartman (1940, 1950) provides useful information on the systematics of this family.

Glycerids are found in sandflats and in sandy crevices between and and under rocks. The proboscis may be up to one-third the length of the worm. These polychaetes are thought to be raptorial feeders that prey upon tubicolous worms and other small invertebrates.

Glycera americana Leidy

The body is up to 350 mm long and 13 mm wide with about 300 segments. Setigers are biannulate. The prostomium has eight to ten annulations. The proboscis is diffusely papillate, with many tongue-shaped papillae and a few large globular papillae. Parapodia have two presetal and two postsetal lobes that are about equal in size (fig. 5.17a). Branchiae are digitate, with few to many lobes, and eversible from setiger 16 to end of body. Superior setae are simple capillaries, while inferior setae are composite spinigers. *Glycera americana* is distributed on both coasts of America. It has also been recorded from Peru, New Zealand, and Australia. It is common in the Gulf at such locales as Isla Tiburón and Roca Consag. (ref: Hartman, 1968).

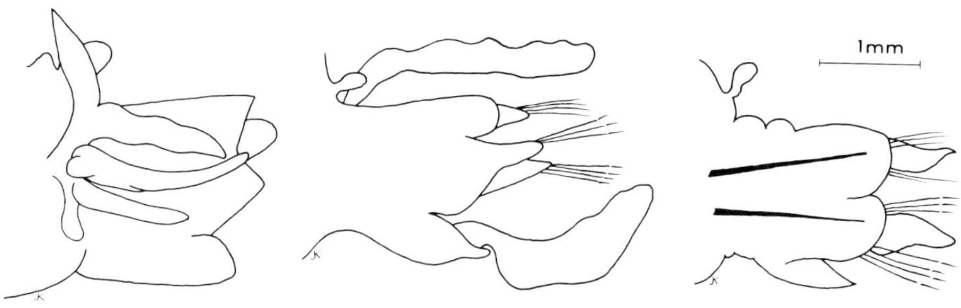

Fig. 5.17 Glyceridae, parapodia of *Glycera*. (a) *G. americana*, posterior aspect (modified after Berkeley & Berkeley, 1948). (b) *G. dibranchiata*, anterior aspect. (c) *G. tesselata*, posterior aspect. (b) to (c) modified from Hartman, 1968.

Glycera dibranchiata Ehlers

The length of the body ranges from 100 to 240 mm, with a width of 4.5 to 8.5 mm. About 310 segments are present. The prostomium is long and conical with 12 to 13 annulations. The proboscis is diffusely papillate, with small tuberculate papillae that have six transverse ridges. The parapodia have two presetal and two postsetal lobes, the upper lobe being the shortest (fig. 5.17b). Branchial lobes are present on the dorsal and ventral surfaces of the parapodia and extend beyond the parapodia. Superior setae are simple capillaries, and inferior setae are composite falcigers.

This species is found on both coasts of America. In the Gulf it is common at El Golfo de Santa Clara and at Mazatlán, in midintertidal sandflats. (ref: Hartman, 1940; 1950; 1968).

Glycera tesselata Grube

Body is up to 25 mm long by 1.5 mm wide. Segments number about 100. The prostomium has 12 to 14 annulations and bears a pair of dark eyes. Parapodia have two long and equal presetal lobes and two much shorter postsetal lobes (fig. 5.17c). Branchiae are absent. The proboscis bears long, erect, and slender papillae with longitudinal ridges. The species has been recorded from sublittoral depths throughout the Gulf. It is a circumtropical species that occurs as far north as the Pacific Coast of Canada. (ref: Hartman, 1940, 1968).

Family Onuphidae

The body is elongate and cylindrical. The prostomium bears a pair of globular palps, two short frontal antennae, and five long occipital antennae with ringed ceratophores (fig. 5.18a). Only the buccal segment is apodous. Anterior parapodia have conical ventral cirri, while posterior parapodia have padlike ventral cirri. Branchiae range from a cirriform to a spiraled shape. Parapodia are uniramous with simple and composite setae. All members of this family are tubicolous.

Onuphids are thought to be scavengerous feeders that consume dead animal tissue. These worms inhabit tough durable tubes to which shell

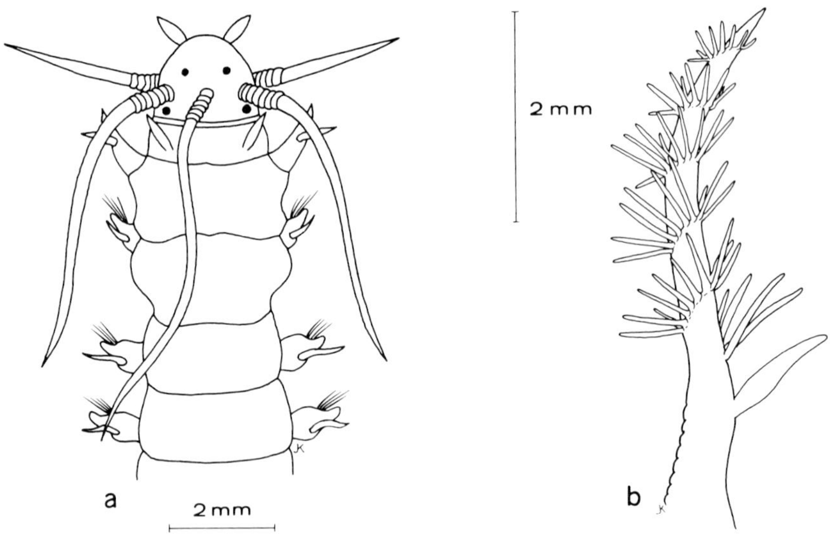

Fig. 5.18 Onuphidae. (a) Anterior segments of *Onuphis* (modified after Berkeley & Berkeley, 1948). (b) Branchia of *Diopatra* (modified after Hartman, 1944a).

fragments, sediment, and algal debris have become attached. Onuphids are abundant in the Puerto Peñasco region, where their presence is betrayed by their characteristic tube. (ref: Hartman, 1944a; Fauchald, 1968).

KEY TO SELECTED GENERA OF ONUPHIDS

1 Branchial filaments spirally arranged .... ***Diopatra***
-Branchial filaments not spirally arranged .... ***Onuphis***

Diopatra splendidissima Kinberg

Branchiae first present from setiger 4 or 5. Anterior branchiae are strongly spiraled (fig. 5.18b). Ventral cirri of the first four setigers are cirriform. Thereafter ventral cirri are padlike. Hooded hooks in anterior segments are bidentate. The tube built by this worm is covered with shell fragments, sand grains, and algal debris. *Diopatra splendidissima* is a Panamic species distributed from Ecuador to southern California. It has been recorded from La Paz and Topolobampo, and its range is here extended to Guaymas and Puerto Peñasco (ref: Hartman, 1944a; Fauchald, 1968).

Onuphis microcephala Hartman

The anterior dorsal surface has five to six reddish brown transverse stripes across the middle of the peristomium and the first few setigers. Length of approximately 250 segments is 120 mm. The body is wide and flattened. Branchiae are first present from the sixth setiger as a single filament; segments 7 to 9 with two filaments per branchia; segment 10 has three filaments. The number of filaments per branchia increase to eight or nine in the vicinity of setiger 50. Filaments are pectinately arranged. All four anterior parapodia have tridentate hooks. The tube is about 40 cm long and 15.5 mm wide. It is soft, thin, and flexible with a leathery base covered by fine white sand grains and a few small shell fragments.

This species is distributed from southern California into the Gulf of California, where it has been found at Bahía Cholla, Puerto Peñasco. *Onuphis microcephala* has also been recorded from the Gulf of Mexico. (ref: Hartman, 1944a).

Family Euncidae

The body is elongate and cylindrical. The prostomium has two cushionlike palps and one, three, or five antennae without ringed cerato-phores (fig. 5.19a). There are two anterior apodous segments with ten-

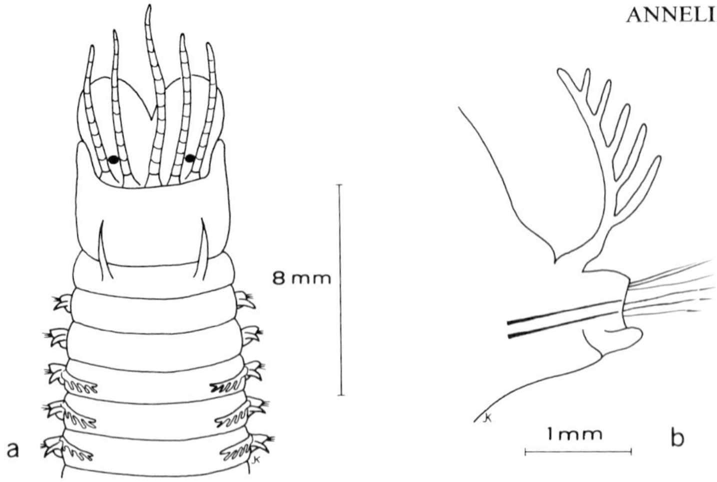

Fig. 5.19 Eunicidae. (a) Anterior segments of *Eunice.* (b) Parapodium and branchia of *Eunice* (modified after Berkeley & Berkeley, 1948).

tacular cirri present on the second. Branchiae are either cirriform or pectinate (fig. 5.19b).

Eunicids have diversified into a variety of rocky and sandy intertidal habitats. All types of macrophagy occur in this family. Also found are the so-called palolo worms that swarm at the water's surface for reproductive purposes at precise times of the year. (ref: Hartman, 1944a; Day, 1967; Fauchald, 1970).

KEY TO SELECTED GENERA OF EUNCIDS

1 Acicular hooks and pectinate setae absent; branchiae restricted to posterior segments .... ***Palola***
-Acicular hooks and pectinate setae present; branchiae present on anterior segments .... ***Eunice***

Palola paloloides (Moore)

Prostomium is bilobed anteriorly with five short smooth antennae. Branchial filaments are absent from the first 100 to 150 segments. Branchiae are cirriform, or branched into two or three lobes. Right maxillae II with two large teeth and one distal tooth. This species ranges from southern California to Acapulco. It has been found in rocky intertidal habitats at Mazatlán, La Paz, Topolobampo, and Puerto Peñasco. (ref: Fauchald, 1970).

Palola siciliensis (Grube)

This species is similar to *Palola paloloides* except that the right maxillae II is equipped with only two teeth. It has a circumtropical distribution. It has been recorded from Puerto Peñasco and Tastiota, Sonora, living among granitic rocks in intertidal zones. (ref: Fauchald, 1970).

Eunice antennata (Savigny)

Moniliform branchiae are present from anterior segments where they are best developed. Branchiae are reduced in median segments. The prostomium is bifid with two eyes and moniliform antennae. Acicula and tridentate subacicular hooks are pale in color. *Eunice antennata* occurs throughout the Gulf of California. (ref: Hartman, 1940).

Eunice sonorae Fauchald

Length of 586 setigers is 345 mm, width 12.1 mm. Occipital tentacles are white with reddish brown ceratophores. The fourth setiger has a broad white dorsal transverse bar. Branchiae begin from setiger 41 and terminate at about setiger 480. The first pair of branchiae is cirriform; the second pair has three lateral filaments; latter pairs are best developed with five to six lateral filaments in a pectinate array. Notopodial setae with 10 to 20 straight capillaries, and a varying number of pectinate setae. Neuropodia with composite bidentate hooded hooks. Dark subacicular hooks are present from setiger 54 to the end of body. The type locality is Puerto Peñasco, from the rocky intertidal. It has also been found in Guaymas. (ref: Fauchald, 1970).

Family Lysaretidae

Worms with elongate cylindrical bodies. The prostomium lacks palps but has three small tentacles that may be hidden by the folded peristomium. Peristomial tentacles are absent. All segments have parapodia. Parapodia are uniramous with straplike dorsal cirri; ventral cirri are absent. All setae are simple.

This family is circumtropical in distribution. It is represented in the Gulf by at least one species at Puerto Peñasco, where it is found in rocky intertidal habitats. (ref: Day, 1967; Fauchald, 1970).

Oenone fulgida (Savigny)

Length of body to 350 mm. Color in life is uniformly reddish orange. The prostomium is rounded, with two pairs of eyes and three short tentacles that are often concealed by a biannulate peristomial fold (fig. 5.20). The jaw apparatus is either symmetrical or asymmetrical. Bidentate subacicular hooks are present from setigers 14 to 24. Anterior parapodia with truncated presetal lobes and foliaceous postsetal lobes; median parapodia rounded, with presetal and fingerlike postsetal lobes; posterior parapodia similar to those of median segments.

Fig. 5.20 Lysaretidae. Anterior segments of *Oenone fulgida* (modified after Day, 1967).

This species has a circumtropical distribution and is also found in Japan. It has been found at Mazatlán and within the Gulf in rocky habitats at Puerto Peñasco. (ref: Day, 1967; Fauchald, 1970).

Family Lumbrineridae

Polychaetes with an elongate and cylindrical body. The prostomium generally lacks appendages, eyes, and palps, but is rounded to conical in shape (fig. 5.21a). The two anterior segments are apodous. Parapodia are small and uniramous. Dorsal cirri are rudimentary, while ventral cirri are absent. Simple capillary setae and simple or composite hooks are present.

Lumbrinerids are commonly found in sediment, algal holdfasts, and under rocks. These worms are common in the Puerto Peñasco area. (ref: Hartman, 1944a; Day, 1967; Fauchald, 1970).

Lumbrineris erecta (Moore)

The prostomium is short and rounded. The worm is large. Anterior setigers 35 to 49 are equipped with pointed limbate setae. Hooded hooks begin on setigers 35 to 46 and continue to end of body (fig. 5.21b). Postsetal lobes are longer than presetal lobes. Acicula and setae are yellow. *Lumbrineris erecta* ranges from southern California to Acapulco, Mexico, and has been found at Mazatlán, and in the Gulf at Bahía Kino and Puerto Peñasco. (ref: Fauchald, 1970).

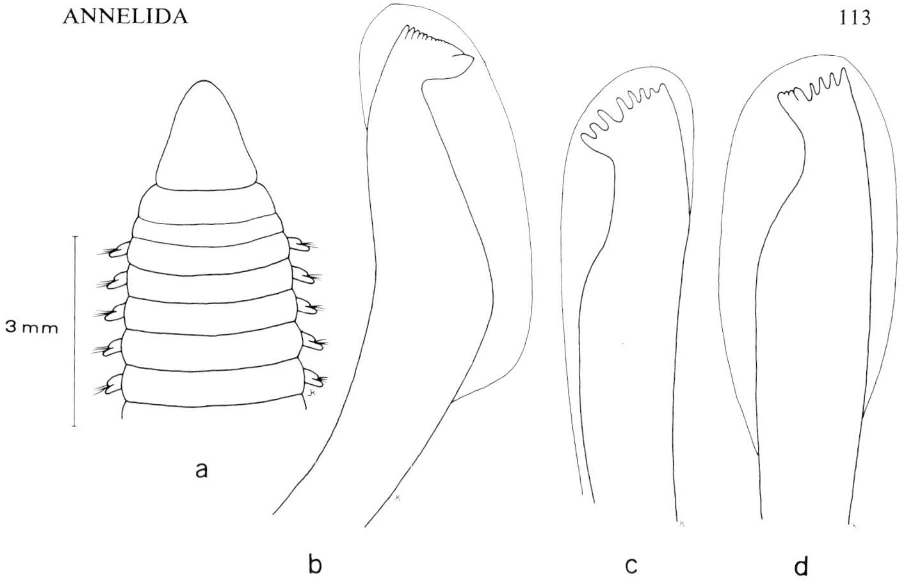

Fig. 5.21 Lumbrineridae. (a) Anterior segments of *Lumbrineris.* (b) Anterior hook of *L. erecta* (0.16 mm). (c) Anterior hook of *L. tetraura* (0.08 mm). (d) Posterior hook of *L. tetraura* (0.13 mm). (b) to (d) modified after Fauchald, 1970.

Lumbrineris tetraura (Schmarda)

The prostomium is pale, flattened, and blunt to conical in shape. Parapodia are conspicuous, with long auricular postsetal lobes and fan-shaped setal fascicles. Simple slender hooded hooks begin on the first setiger (fig. 5.21c, d); parapodium of setiger 5 with 7 to 10 hooks, superior and inferior bilimbate setae and two to three acicula. All acicula and setae are yellow. This species is distributed from southern California south to Chile. It has been found at various points within the Gulf, and has been found to be relatively abundant at Puerto Peñasco. (ref: Fauchald, 1970).

Family Arabellidae

Worms with elongate and cylindrical bodies. The prostomium lacks appendages, but eyes are often present (fig. 5.22a). The two anterior segments are apodous. Parapodia are uniramous and small. Dorsal cirri are rudimentary, while ventral cirri are absent. All setae are simple capillaries. Hooded hooks are totally absent. The members of this family superficially resemble those of the family Lumbrineridae in their feeding habits and ecological habitats, and can be found in sandy as well as rocky environments. (ref: Hartman, 1944a; Day, 1967; Fauchald, 1970).

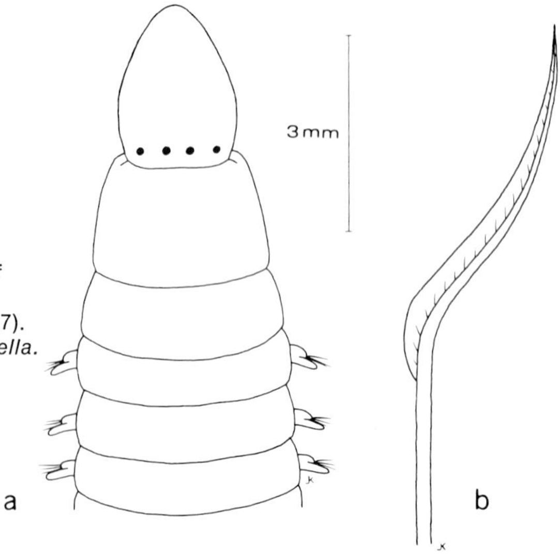

Fig. 5.22 Arabellidae. (a) Anterior segments of *Arabella iricolor* (modified from Day, 1967). (b) Falcate seta of *Arabella.*

Arabella iricolor (Montagu)

The body is 100 mm long. The prostomium is conical and has four eyes in a transverse row along its posterior border (fig. 5.22a). The parapodia have a low presetal lobe and a conical postsetal lobe. All setae are simple (fig. 5.22b). Anterior setigers have only smooth bilimbate setae; median and posterior setigers also have setae with teeth. This species is widespread in tropical seas and is common in western Mexico. *Arabella iricolor* has been found at Mazatlán, Topolobampo, Bahía San Carlos, and Bahía Cholla in intertidal sandflats. (ref: Fauchald, 1970).

KEY TO SELECTED POLYCHAETE FAMILIES FROM THE NORTHERN GULF OF CALIFORNIA: SEDENTARIATA

1 Head or anterior segments with tentacles, palps, filaments, or radioles ... **2**
-Head and anterior segments without tentacles, palps, filaments, or radioles ... **8**

2(1) Head with 2 grooved, elongate palps ... **3**
-Head without palps ... **4**

3(2) Body divided into 3 distinct regions; peristomium with a rounded lip and a pair of palps arising from sides of mouth; setiger IV with modified cutting setae ... ***Chaetopteridae***
-Body not divided into regions; peristomium without a rounded lip but with a pair of palps originating from the dorsum; setiger IV without cutting setae, although setiger V may have modified setae ... ***Spionidae***

4(2) Head or anterior segments covered with many soft flexible tentacles ... **5**
-Head and anterior segments not covered with soft flexible tentacles ... **6**

5(4) Body divided into a thorax and abdomen; branchiae confined to thoracic segments 2 to 4; thoracic parapodia biramous; abdominal parapodia lack notopodia; capillary setae and uncini present ... ***Terebellidae***
-Body not divided into a thorax and abdomen; branchiae usually confined to numerous abdominal segments; all parapodia biramous, without developed lobes; simple capillaries and acicular hooks present ... ***Cirratulidae***

6(4) Head covered with a flowerlike plume of radioles 7
-Head covered with a short frilly or dendritic feeding membrane .... ***Oweniidae***
7(6) Operculum present; thoracic membrane generally well-developed and capelike; tubes calcareous .... ***Serpulidae***
-Operculum absent; thoracic membrane poorly developed or absent; tubes leathery, gelatinous, or sandy .... ***Sabellidae***
8(1) Body divided into at least 2 regions **9**
-Body not divided into regions **10**
9(8) Body weakly divided into a thorax and abdomen, resembling an earthworm; numerous body segments, occasionally multiannulate; parapodia biramous, without lobes in thorax and with only ridges in abdomen; capillary setae and hooded hooks present .... ***Capitellidae***
-Body divided into an anterior setigerous region without branchiae, a middle region with bushy gills, and a posterior asetigerous region without gills; few body segments, all multiannulate; parapodia biramous, lacking lobes; notosetae are capillaries, neurosetae are rostrate hooks .... ***Arenicolidae***
10(8) Body stout, fusiform and ventrally grooved; segments wider than long; all setae are simple capillaries; branchiae usually present ...... ***Opheliidae***
-Body elongate and cylindrical, truncate on both ends (not ventrally grooved); segments longer than wide; simple setae include capillaries, rostrate hooks, and acicular spines; branchiae usually absent .... ***Maldanidae***

Family Spionidae

Worms with small and flattened bodies. The prostomium is slightly rounded to bifurcated, with two lateral horns (fig. 5.23a). A median caruncle is present and extends posteriorly to the second or third setigerous segment. One pair of long, grooved, ciliated palps present on peristomium. Parapodia are biramous; setae are winged capillaries in both notopodia and neuropodia or accompanied by hooded hooks.

Spionids use their palps to gather food from the surrounding sediment or from the water column, and can generally be found inhabiting tubes in a variety of substrata (i.e., intertidal mudflats, pilings, boat bottoms, shells). In some places around Puerto Peñasco, spionid tubes form dense carpet-like mats in intertidal sandflats. (ref: Day, 1967; Hartman, 1969).

KEY TO SELECTED GENERA OF SPIONIDS

1 Fifth setigerous segment bearing heavy spines that differ from all other setae; branchiae begin after setiger V .... ***Polydora***
-Fifth setigerous segment unmodified; branchiae begin from setiger I .... ***Scolelepis***

Polydora ligni Webster

Length to 25 mm; width to 1 mm. Prostomium is bifid with two pairs of eyes, a median antenna, and a caruncle that extends posteriorly through the first three setigers. Spines on setiger V pilose (brush-top) and distally divided and alternate with 10 to 11 distally bifid falcate

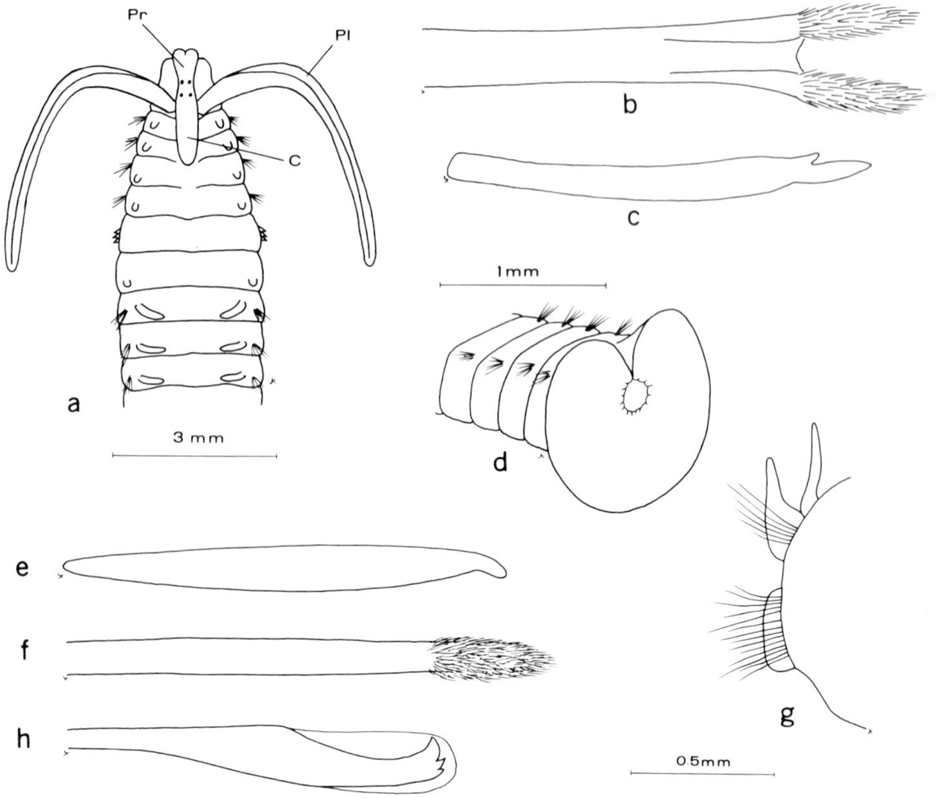

Fig. 5.23 Spionidae. (a) Anterior segments of *Polydora.* (b) Pilose seta from setiger V of *Polydora ligni.* (c) Falcate spine from setiger V of *P. ligni.* (d) Pygidium of *P. ligni.* (e) Spine from Setiger V of *P. nuchalis* (0.66 mm). (f) Pilose seta from setiger V. of *P. nuchalis* (0.1 mm). (g) Anterior parapodium of *Scolelepis indica.* (h) Hooded hook of *S. indica.* (a) to (c) modified from Hartman, 1969; (d) and (e) after Woodwick, 1953; (f) and (g) adapted from Day, 1967. C, Caruncle; Pl, Palp; Pr, Prostomium.

spines (fig. 5.23b, c). Branchiae begin on setiger VII and are present to end of body. Hooded hooks begin from setiger VII, each distally bifid. Pygidium circumscribed by a broad disk with a mid-dorsal notch (fig. 5.23d). Found on wood pilings, and on ship hulls as fouling animals. Also found in silty and muddy benthic surfaces. *Polydora ligni* ranges from western Canada to the Gulf of California and western Mexico. It has been taken from Guaymas, Topolobampo, and Bahía Cholla. This species also occurs in New Jersey. (ref: Hartman, 1969).

Polydora nuchalis Woodwick

Length to 20 mm, width 1 mm. Up to 110 segments present. Color in life ranges from translucent yellow to red. Prostomium is deeply notched mid-dorsally and has two pairs of eyes, a small nuchal papilla, and a caruncle that extends through setiger II. First setiger without notosetae, but having neurosetae. Spines of setiger V are distally falcate and alternate with distally pilose setae (fig. 5.23e, f). Branchiae and distally bifid hooded hooks commence from setiger VII. Pygidium is a broad flat disk with an open mid-dorsal notch. Some specimens collected differ from the holotype in being red in color, and in having branchiae continue to the end of the body.

Southern California, in lagoons. Found in the Gulf in the high intertidal of Bahía Cholla, in dense mats in sediment between boulders. (ref: Woodwick, 1953; Hartman, 1969).

Scolelepis indica Fauvel

Body is up to 60 mm long and 1.5 mm wide. Long cirriform branchiae begin on setiger I. Notopodia with capillary setae. Anterior neuropodia with five to six large capillaries and long bidentate hooded hooks (fig. 5.23g, h); posterior neuropodia with sabrelike setae. This spionid has been recorded from the Indian Ocean and western Mexico. It has also been taken from sandflats at Bahía San Carlos. (ref: Day, 1967).

Family Chaetopteridae

Chaetopterids are elongate worms, with bodies divided into three regions. Two long, grooved palps are present from the base of the mouth. Anterior segments are uniramous and bear simple setae. Segment IV, however, has characteristic modified cutting setae. The middle region has biramous parapodia, with simple notosetae and minute neuropodial uncini. Notopodia of this region are bilobed. In the posterior region the parapodia are again biramous, but the notopodial lobes are entire.

These worms are tubicolous and inhabit tough leathery or parchmentlike tubes. The tube may be U-shaped or vertical. These polychaetes are efficient filter feeders, pumping water into their tube and through a mucus net constructed by the worm. Periodically the net and all of its trapped food materials are ingested. (ref: Fauvel, 1927; Day, 1967).

KEY TO SELECTED GENERA OF CHAETOPTERIDS

1 Middle body region bears bilobed notopodia; tube is clear and annulated .... ***Spiochaetoperus***

-Middle body region without bilobed notopodia; tube is brownish, leathery, and coarse **2**

2(1) Body is large with short palps; worm inhabits a large U-shaped tube .... ***Chaetopterus***

-Body is generally small with long palps; worm inhabits a straight tube .... ***Mesochaetopterus***

Spiochaetopterus costarum (Claparède)

Body is elongate, narrow, and green. Length is approximately 50 to 70 mm. Palps are elongate and speckled with brown pigment. Anterior region with nine to ten setigers; middle with more than 30 setigers; posterior region with fingerlike notopodial lobes and reduced neuropodia.

This worm is common in the Puerto Peñasco region. It is easily recognized by its clear annulate tube. (ref: Day, 1967).

Chaetopterus variopedatus (Renier)

The length of this worm attains 250 mm, its width 25 mm. The anterior region in life is brown to cream-colored, the middle is green, and the posterior is yellow. There are nine to ten anterior segments, five middle segments, and over 20 posterior segments (fig. 5.24). Tubes are tough, leathery, and U-shaped. Commensal porcellanid crabs *(Polyonyx* sp.) are often found in the tube. This is a cosmopolitan species that is found at Puerto Peñasco, Loreto, and La Paz. (ref: Hartman, 1969).

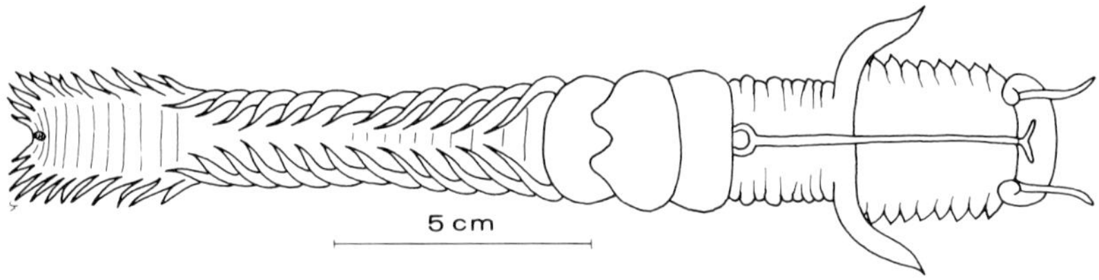

Fig. 5.24 Chaetopteridae. Dorsal view of *Chaetopterus variopedatus* (modified from Hartman, 1969).

Mesochaetopterus alipes Monro

Relatively large worms, up to 285 mm long and 10 mm wide. In life the anterior region is white to cream, and the posterior region is green. Anterior segments number nine to 11; middle region with three long segments, of which the second and third bear cupules; posterior segments number 68. This species was previously known from Taboguilla, Panamá. Its range is here extended into the Gulf, where it is known only from Bahía Cholla. (ref: Monro, 1928).

Mesochaetopterus sp.

Known only from one incomplete specimen collected in the low midintertidal at Bahía Cholla. Length is 100 mm, and width 10 mm. Anterior region with 13 setigers; middle with four segments, all of which bear cupules; posterior region with more than 15 segments. Tubes are parchmentlike and straight to somewhat J-shaped.

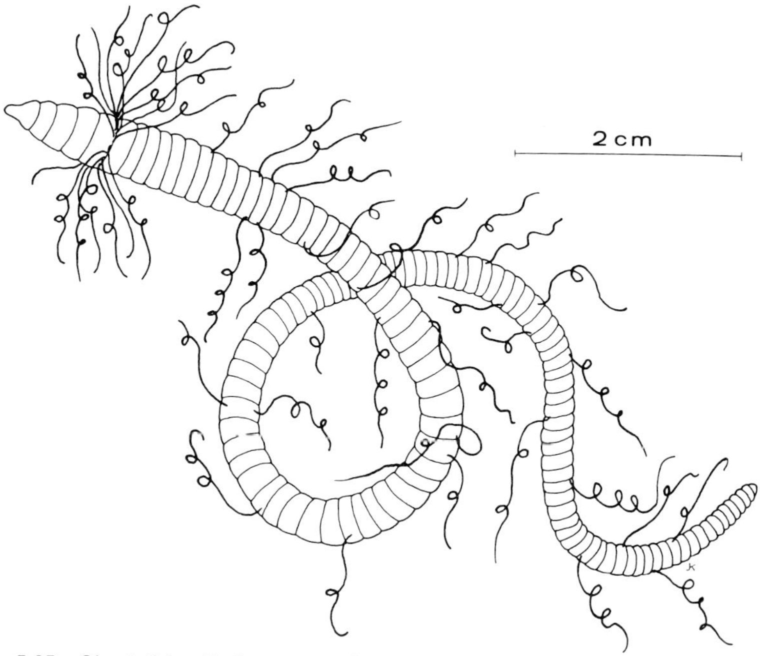

Fig. 5.25 Cirratulidae. Entire aspect of *Cirriformia* (modified from McIntosh, 1922).

Family Cirratulidae

Worms with a cylindrical body and many segments (fig. 5.25). Both ends of body are tapered. The prostomium is small and conical to bluntly rounded. A pair of grooved palpi or several transverse rows of filaments are present on the dorsum of a single anterior setigerous segment. Parapodia are biramous, but the lobes are rudimentary to absent. Simple capillary setae or also acicular hooks present in the parapodia. Branchial filaments may be present on trunk segments.

Cirratulids are deposit feeding polychaetes that inhabit sediment between or under rocks in intertidal zones. These worms also occur in intertidal sandflats and subtidal habitats. Usually cirratulids can be detected by their exposed red filamentous gills that wave about on top of the sediment. (ref: Day, 1967).

KEY TO SELECTED GENERA OF CIRRATULIDS

1 A pair of elongate anterior palps present on dorsal surface; all setae are simple capillaries .... ***Tharyx***

-Without elongate palps; capillary setae and acicular hooks both present .... ***Cirriformia***

Tharyx parvus Berkeley

The body is slender and 10 to 15 mm long and 0.4 to 0.6 mm wide, with approximately 300 segments. A mid-dorsal reddish brown strip is present on the anterior third of the body. The first setigerous segment bears the palps and a pair of branchiae. Trunk filaments are located just above the notopodia. The pygidium lacks appendages and bears a posterior pore. This species is distributed from western Canada south to the Gulf of California, where it has been taken from Bahía de los Ángeles. (ref: Hartman, 1969).

Cirriformia species 1

Color in life is reddish brown, with red branchiae and black spines in parapodia. Length of 250 to 300 setigers is 50 mm, and width is 3 mm. The body has an arched dorsum and a concave ventrum; segments are uniannulate (fig. 5.25). Branchiae are first present from setiger III; dorsal tentacles present on setiger IV with 12 to 15 pairs of tentacles that insert towards the mid-dorsum. Lateral filaments are not coiled. Simple notopodial setae present in the first 35 to 37 setigers, with acicular hooks and capillary setae thereafter to end of body. Neuropodia with hooks and capillaries from the first setiger that continue to end of body. This species has been found in the midtide zone of Puerto Peñasco, between and under granitic rocks.

Cirriformia species 2

Color in life is reddish, with brilliant green tentacles and red branchiae. This is a distinctively colored species found amongst conglomerate rock outcroppings at the base of Cerro Negro (Black Mountain) in Bahía Cholla.

Family Opheliidae

Worms with fusiform bodies and relatively few segments. The ventral body surface of these polychaetes is often grooved. Parapodia are biramous but poorly developed, and dorsal and ventral cirri are usually absent. Branchiae are cirriform when present. Only simple capillary setae are present in parapodia. Pygidium is elongate and usually has anal cirri.

These worms are deposit feeders that generally live in sandy habitats. They are particularly abundant in Bahía Cholla. (ref: Day, 1967).

Travisia sp.

Body is fusiform with 33 segments, is approximately 25 mm long, with a continuous ventral groove (fig. 5.26). Body is equipped with

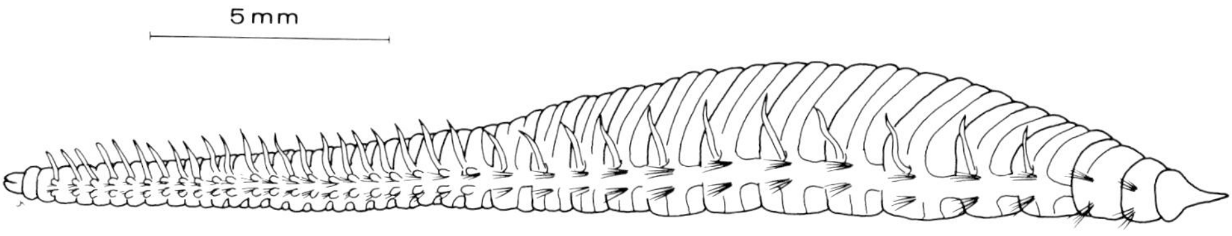

Fig. 5.26 Opheliidae. Lateral aspect of *Travisia* sp.

innumerable small papillae on dorsum and ventrum. Prostomium is conical and smooth. The first two segments are uniannulate and smooth; segments three to 11 are large and triannulate on dorsum and ventrum. Segments 12 to 33 are partially biannulate, such that there is a dorsal and ventral scute at the anterior aspect of each segment. All setae are simple capillaries. Cirriform branchiae begin from the second setiger and continue posteriorly to the end of body where they become quite long. The pygidium terminates in an elongate hemisphere with eight adjacent digitate anal cirri.

This species is found in the top 15 cm of the sediment in Bahía Cholla from low to high tide zones, although it seems to be most abundant in the low tide regions.

Family Capitellidae

Polychaetes that are earthwormlike in appearance, with a thin elongate body (fig. 5.27). The body is generally weakly divided into a thorax and abdomen. The abdomen may or may not have branchiae. Parapodia are biramous but each lobe is reduced to setal bundles and ridges. Capillary setae and hooded hooks are present.

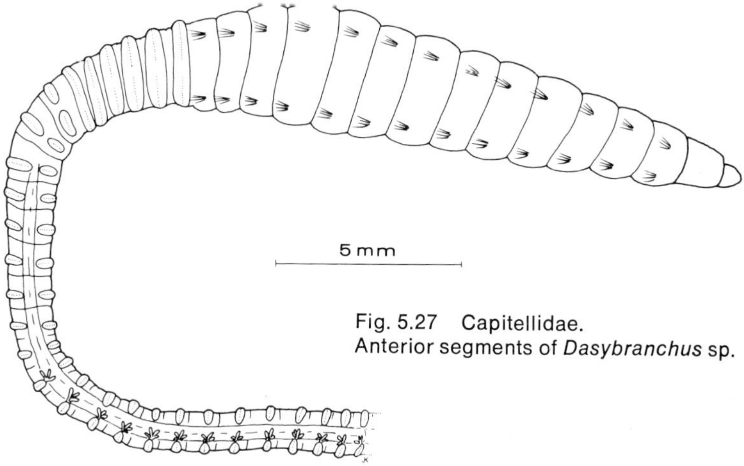

Fig. 5.27 Capitellidae. Anterior segments of *Dasybranchus* sp.

These worms are found in nearly every conceivable intertidal habitat, ranging from reduced (unoxidized) sediment to crevices under boulders. There are many capitellids in the Puerto Peñasco region, some of which are difficult to collect because they readily autotomize when handled. (ref: Hartman, 1947a).

KEY TO SELECTED GENERA OF CAPITELLIDS

1 Thorax with 11 setigerous segments, all of which bear capillary setae .... ***Notomastus***
-Thorax with 13 setigerous segments, all of which bear capillary setae .... ***Dasybranchus***

Notomastus sp.

This species is known from an anterior fragment consisting of 10 thoracic setigers and 10 abdominal segments. Length of thorax 20 mm; length of abdominal portion is 15 mm. Thorax 5 mm wide at widest point; abdomen 3 mm wide. Thoracic segments with capillary setae only and abdominal segments with long-shafted hooded hooks. Abdominal notopodia with 70 to 80 hooks and neuropodia with 80 to 90 hooks. Hooded hooks with a large primary fang, surmounted by three crescent-shaped rows of small denticles. Each crescent-shaped row has three to eight denticles. This species was taken from the high intertidal zone of Bahía Cholla, near conglomerate rock outcroppings at the base of Cerro Negro.

Notomastus tenuis Moore

Color in life is red; body is elongate and very slender. First setigerous segment has notopodia, but lacks neuropodia. The following ten segments have only notopodial and neuropodial capillary setae. Characteristic hooded hooks with a primary fang surmounted by a crescent-shaped row of four to five small denticles. Branchiae are absent.

These small worms are conspicuously present in Bahía Cholla when shovelfuls of sediment are dug up. Their red color and hairlike shape immediately give them away. The species is distributed from western Canada into the Gulf of California, where it has been taken subtidally at Bahía de los Angeles and intertidally in the sandflats of Bahía Cholla. (ref: Hartman, 1947a; 1969).

Dasybranchus platyceps Hartman

Length of worm with 180 segments is approximately 100 mm. Width to 8 mm in posterior thorax. Thoracic segments number 14, of which all but the first bear capillary setae. Nephridiopores are first present between thoracic setigers 9 to 10, and terminate at about the eleventh abdominal segment. Branchiae are retractile and branched, and are first present from the thirteenth abdominal segment almost to the end of body. Branchiae arise from a pore just above the neuropodium. Abdominal hooded hooks are present, with a primary and secondary fang surmounted by a transverse row of three small teeth. Type locality for this species is Gulf of California, Sonora, in intertidal situations. (ref: Hartman, 1947a; 1969).

Dasybranchus sp.

Color in life is blood red. Length approximately 125 mm by 3 to 5 mm wide, with approximately 120 segments (fig. 5.27). Thoracic segments number 14, the first of which is asetigerous while all others bear capillary setae. Nephridiopores begin from segment ten and continue to the end of body. Branchiae are retractile and branched, and are first present from abdominal segment 17. Abdominal parapodial ridges are conspicuous and elevated. Hooded hooks bear a primary and two secondary fangs that are surmounted by a longitudinal row of three small teeth. This species is found in silty-shelly sediment under basalt boulders at Puerto Peñasco in midintertidal zones. Specimens are gravid in late February and early March.

Family Arenicolidae

Polychaetes with a large body and superficially multiannulate segments (fig. 5.28a). Prostomium is small and lacks appendages. Proboscis is a large eversible sac. Body with few segments. Body divided into three regions: anterior setigerous section without gills; median setigerous region with bushy conspicuous gills; posterior portion without setae and gills. Parapodia are biramous, with capillary notosetae and rostrate hooks (fig. 5.28b) in neuropodia. These worms inhabit subtidal and intertidal sand and mudflats. (ref: Day, 1967).

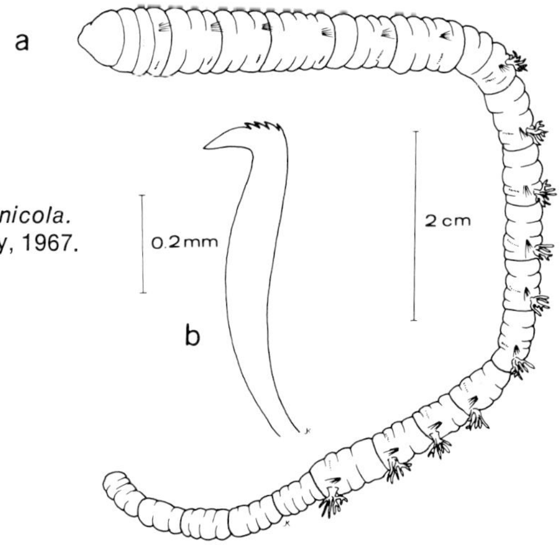

Fig. 5.28 Arenicolidae.
(a) Entire view of *Arenicola*.
(b) Neuropodial hook of *Arenicola*.
(a) and (b) adapted from Day, 1967.

Arenicola glasselli Berkeley and Berkeley

Worms to 110 mm with 16 setigers. Ten pairs of bushy conspicuous gills are present on segments 7 to 16. Seven pairs of nephridial openings found on setigers 5 to 11. This tropical species has been recorded from San Felipe and is a host for the pinotherid crab *Pinnixa pembertoni* Glassell. (ref: Berkeley and Berkeley, 1939).

Family Maldanidae

Polychaetes with a cylindrical body, usually having a few very long segments (fig. 5.29a). The prostomium is reduced and lacks appendages. A nuchal organ and eyespots are usually present. An anterior cephalic plate is present, with a raised margin about the nuchal organ. The proboscis is an eversible sac. Parapodia are biramous, having a tuft of capillary notosetae and dentate hooks in the neuropodia. Acicular hooks may be present, however, in the anterior neuropodia. Asetigerous preanal segments proceed the pygidium. The pygidium may be conical, funnel-shaped, or truncated, and may bear numerous anal cirri.

Maldanids look very much like bamboo canes. As a result their common name is "bamboo worm." Bamboo worms are common in the Puerto Peñasco area, where they are typically found under rocks in the intertidal region, and particularly on sandflats. (ref: Day, 1967).

KEY TO SELECTED GENERA OF MALDANIDS

1 Body segments number 90 to 100, with acicular neuropodial spines in the first 4 setigers .... ***Proclymene***

-Body with 21 segments, with or without acicular neuropodial spines in the first 3 setigers **2**

2(1) Body with 19 setigers and 2 preanal asetigerous segments; first 3 setigerous segments with acicular spines; a midventral anal cirrus absent .... ***Euclymene***

-Body with 18 setigers and 3 preanal asetigerous segments; first 3 setigerous segments with dentate hooks; a midventral anal cirrus present .... ***Axiothella***

Proclymene sp.

Acicular neurosetae on first four setigers. Nuchal organs are curved. Posterior segments are campanulate (bell-shaped). *Proclymene* sp. can be found in the southeast section of Bahía Cholla in the midtide zone, where it inhabits long, straight fragile tubes. It is extremely difficult to obtain an intact specimen, because the worms easily autotomize, and the tube is cemented to large shell fragments at its deepest point of penetration.

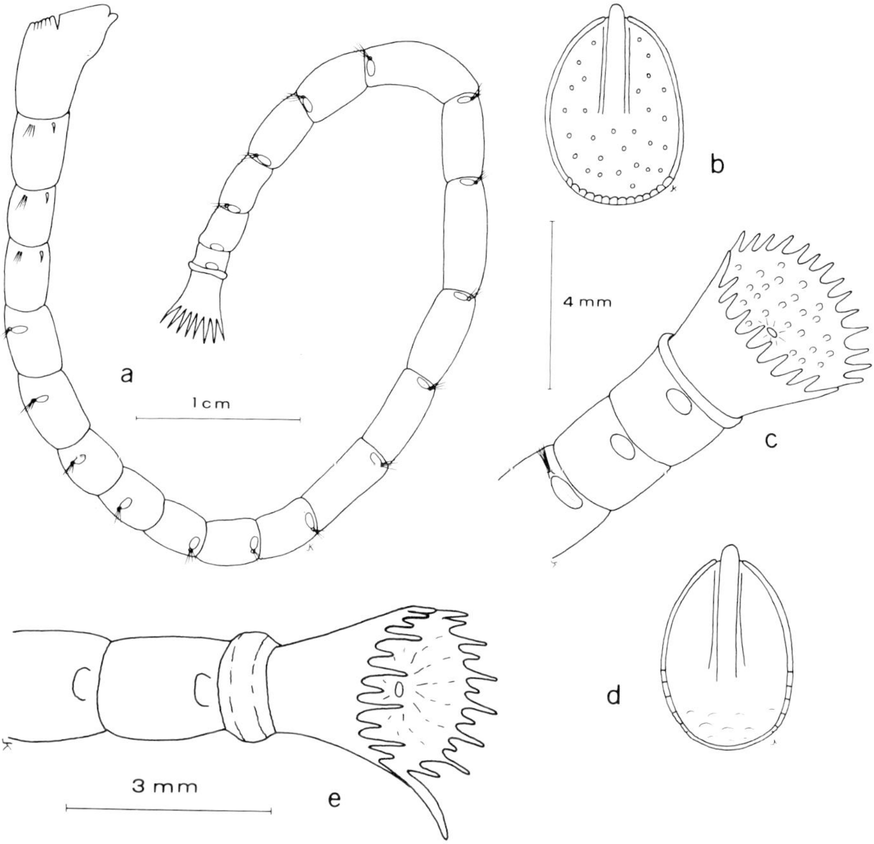

Fig. 5.29 Maldanidae. (a) Entire aspect of *Euclymene papillata.* (b) Cephalic plate of *E. papillata.* (c) Pygidium of *E. papillata.* (d) Cephalic plate of *Axiothella rubrocincta.* (e) Pygidium of *A. rubrocincta.*

Euclymene papillata Berkeley and Berkeley

Color in life is reddish anteriorly to pink posteriorly. Worms about 85 mm long and 2 to 3 mm wide. There are 21 segments that include 19 setigers and two preanal asetigerous segments (fig. 5.29a). The cephalic plate is oval and papillate (fig. 5.29b). Cephalic rim has about 15 dorsal notches. First three setigers with one or two acicular spines per segment. Anal funnel papillate and rimmed with 20 to 25 unequal cirri (fig. 5.29c).

This bamboo worm inhabits a U-shaped tube in rocky midintertidal situations at Puerto Peñasco, from where it was first described. The tubes can be found in crevices, or, in some instances, attached to the bottom of rocks, and are composed of mucus and cemented sand grains. (ref: Berkeley and Berkeley, 1939).

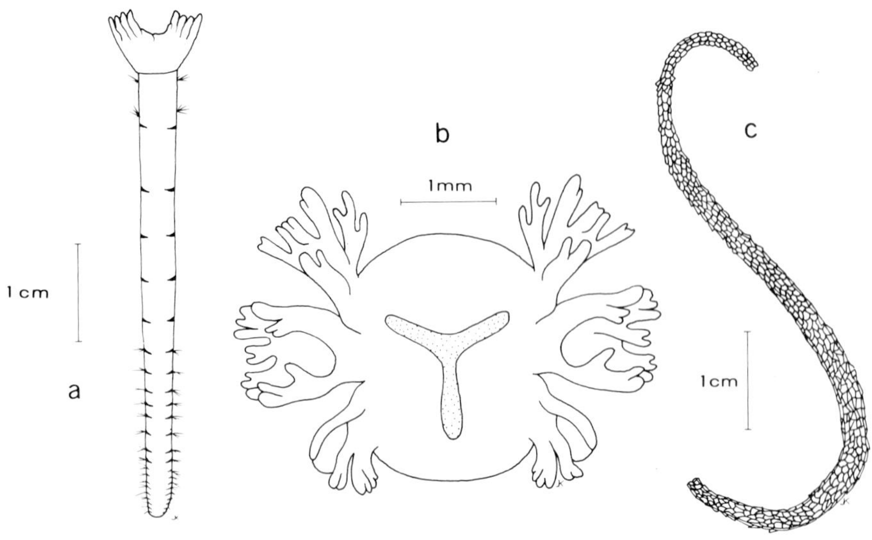

Fig. 5.30 Oweniidae. (a) Dorsal view of *Owenia.* (b) Feeding membrane of *Owenia.* (c) Tube of *Owenia.* (a) to (c) modified from Day, 1967.

Axiothella rubrocincta (Johnson)

Color in life is green to red, with deeply pigmented green or red parapodial ridges. Length up to 120 mm by 2 to 3 mm wide, but specimens from Bahía Cholla are usually smaller. The 21 segments include 18 setigers with two or three preanal asetigerous segments. Cephalic plate is oval and lightly scaled dorsally (fig. 5.29d). Cephalic rim is smooth except for two to three lateral notches. Notosetae are capillaries. Neurosetae are rostrate hooks, with a large primary fang and five to six secondary teeth. Anal plate has 18 to 30 alternating short and long cirri with a distinctly mid-ventral cirrus (fig. 5.29e).

Axiothella rubrocincta inhabits a U-shaped tube that usually has skewed upright arms. It is found in large numbers in low intertidal regions of Bahía Cholla. This species ranges from western Canada into the Gulf of California, and has also been collected at Bahía de los Ángeles. (ref: Hartman, 1969).

Family Oweniidae

Polychaetes with a small elongate body and few segments (fig. 5.30a). The prostomium and peristomium are fused and lack antennae. The head may be equipped with a branching membrane. The mouth is terminally located and surrounded by the aforementioned membrane. The parapodial lobes are poorly developed, but are biramous with capillary notosetae and minute notosetal hooks. Anal cirri are absent.

Oweniids are fairly common in the Gulf, and have been found regularly at Bahía Cholla and El Golfo de Santa Clara, in intertidal sandflats. (ref: Day, 1967).

Owenia collaris Hartman

Color in life is green. Length of entire body up to 100 mm. Feeding membrane has three bases, with each base divided into two branches such that a total of six branches surround the mouth (fig. 5.30b). The thorax consists of three short segments, each with capillary setae. The first five abdominal segments are long and gradually decrease in size thereafter. Abdominal notosetae are bladelike; neurosetae are long shafted hooks with two recurved apical teeth. This worm inhabits a tough, flexible tube, upon which shell fragments and sand grains have been stacked together like shingles on a roof (fig. 5.30c). It is a cosmopolitan species that has been collected from sand flats at Bahía Cholla and El Golfo de Santa Clara. (ref: Day, 1967).

Family Terebellidae

The body is generally divided into two parts: an anterior thorax and a posterior abdomen. Soft, non-retractile grooved tentacles usually cover the head (fig. 5.31). From one to three pairs of dorsally placed gills are present on segments 2 through 4. The thorax is equipped with biramous parapodia. Thoracic notosetae are capillaries, and neurosetae are uncini.

Fig. 5.31 Terebellidae. Anterior segments of *Streblosoma* sp.

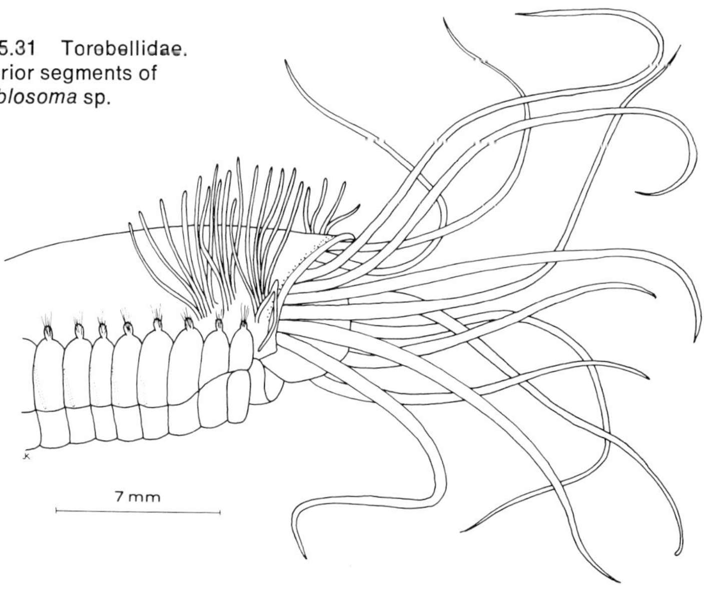

Abdomen without notopodia and notosetae; neurosetae are uncini. Pygidium lacks anal cirri.

Terebellids are indirect deposit feeding polychaetes, using grooved tentacles to ingest detritus. These worms usually inhabit surf-protected areas or live under rocks. Terebellids also can be found in the dense algal mats that cover rock surfaces. (ref: Day, 1967).

KEY TO SELECTED GENERA OF TEREBELLIDS

1 Notosetae and cirriform gills begin on segment 2; neurosetae begin on segment 5 .... ***Streblosoma***
-Notosetae begin on segment 3, cirriform gills begin on segment 2; neurosetae begin on segment 4 .... ***Thelepus***

Streblosoma sp.

Color in life is green, with cream colored tentacles. Worms are approximately 50 mm in length and 3 mm at widest point, with about 60 segments. Dorsal collar at base of tentacles with a band of eyespots (fig. 5.31). Notosetae present to posterior end of animal. Neurosetae begin on segment 5 as double rows of uncini until the fifteenth setiger, after which uncini are arranged as loops or circlets. Found in tubes under rocks and in dense algal mats on top of rocks at Puerto Peñasco. It is also found associated with the reef in Bahía Cholla.

Thelepus setosus (Quatrefages)

Color of body is reddish brown, with red branchiae and orange tentacles. Length of body is 200 mm, width is 10 mm. Notopodia are present on segment 3, and neuropodia commence on segment 4. Notosetae are bilimbate capillaries, and thoracic uncini are arranged in straight single rows. Nephridiopores are present on setigers 4 to 7. The tubes of this worm are covered with shell fragments and algal debris, and are attached to the bottom of rocks and hard surfaces. The species has been recorded from San Felipe and Bahía de los Ángeles. (ref: Hartman, 1969).

Family Sabellidae

These worms may be recognized by their two terminal branchial lobes that circumscribe and hide the mouth. Each lobe is composed of many bipinnate radioles, the bases of which may be fused by a web. An operculum is absent. The peristomium usually has a two- or four-lobed collar (fig. 5.32). The body is distinctly divided into a thorax and abdomen. Thoracic notopodia with bilimbate capillaries and thoracic neuropodia with avicular (beaked) uncini. Setae of abdomen similar to those of thorax, only inverted.

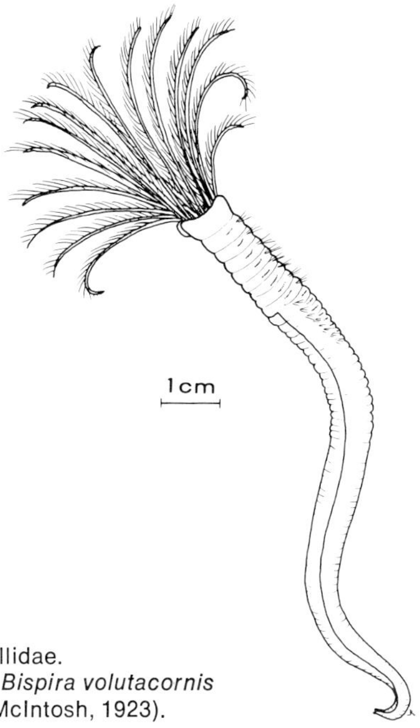

Fig. 5.32 Sabellidae.
Entire aspect of *Bispira volutacornis* (modified from McIntosh, 1923).

Sabellids are abundant in the northern Gulf, and are easily recognized by their brilliantly colored branchial plumes. These polychaetes inhabit a multitude of intertidal and subtidal habitats, such as sandflats, sponges, and sediment-filled crevices between rocks. The following species of sabellids are especially abundant in sandy crevices between rocks in the midintertidal of the Puerto Peñasco area. (ref: Day, 1967).

KEY TO SELECTED GENERA OF SABELLIDS

1 Radioles with terminal compound eyes; branchial plume purple to brownish orange in color; tubes with a rough surface texture .... ***Megalomma***

-Radioles with double rows of eyes; branchial plume banded orange and bluish white in color; tubes with a smooth surface texture .... ***Bispira***

Megalomma sp.

Length approximately 50 mm, width 8 mm, with about 120 body segments that include eight thoracic and about 112 abdominal segments. Collar is four-lobed. The first segment has only a tuft of bilimbate notosetae, while other thoracic notopodia have both palea and bilimbate notosetae. Thoracic neurosetae with an anterior row of pickax setae and a posterior row of short shafted uncini. Abdominal setae similar to thoracic setae, but inverted.

Bispira sp.

Length of body is approximately 50 mm, width is 10 mm at widest point. Several paired eyes are present on many radioles. The collar has two lobes. Body segments number about 100, with nine thoracic setigers and roughly 90 abdominal setigers. Thoracic notosetae all bilimbate; neurosetae with an anterior row of pickax setae and a posterior row of long-shafted uncini. Abdominal setae similar to thoracic setae, but inverted.

Family Serpulidae

Serpulids construct and inhabit characteristic white calcareous tubes (fig. 5.33a). These worms have a brilliantly colored branchial crown composed of bipinnate radioles and an operculum (fig. 5.33b). The thorax has from three to eight segments. The second is usually equipped with a membranous collar. The abdomen has numerous segments. Parapodia are biramous and poorly developed. Notosetae are bilimbate capillaries, and neurosetae are uncini. Abdominal setae are similar to those in thorax, but inverted.

Serpulids can be found on most solid surfaces in the intertidal region, such as rocks, gastropods, and piers. These animals are suspension feeders like sabellids. (ref: Fauvel, 1927; Rioja, 1942a, b; Day, 1967).

Dexiospira marioni Caullery and Mesnil

This serpulid has three thoracic segments and from 15 to 20 abdominal segments. From one to two simple setae are present in the abdominal parapodia. The tube is dextrally coiled in a regular fashion, and has a median ridge or keel. This species has been recorded from Guaymas, La Paz, and Panamá. (ref: Bush, 1904; Rioja, 1942b).

Eupomatus sp.

Small red worms with symmetrical bodies that are 1.5 to 2 mm long and .2 to .5 mm wide. The operculum has two disks (fig. 5.33c). The lower disk has a crenulate margin with 18 crenulations while the upper disk has seven spines. Each spine of the upper disk has a recurved distal tooth and a lower, medially placed lateral process on its inner surface (fig. 5.33d, e). There are seven thoracic setigers and over 60 abdominal segments present. The white, sinistral tubes are tightly coiled and measure 1 mm in diameter.

Eupomatus sp. is to be found on the undersurface of rocks in the intertidal zones of Puerto Peñasco. The reader is referred to Bush (1904), for additional information on the genus *Eupomatus* in the Gulf of California.

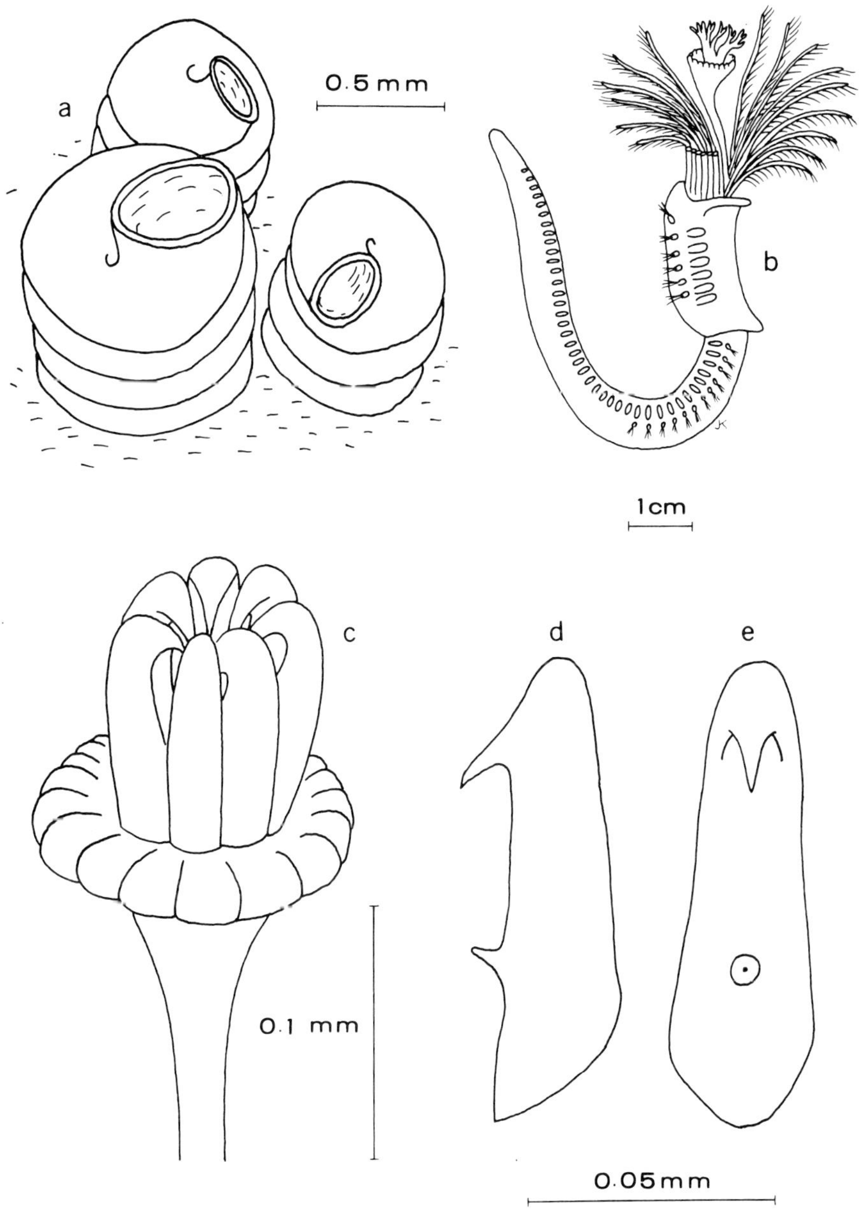

Fig. 5.33 Serpulidae. (a) Tubes of *Spirorbis.* *(b)* Entire view of *Hydroides norvegica* (modified from McIntosh, 1923). (c) Operculum of *Eupomatus* sp. (d) Lateral aspect of the upper opercular spines of *Eupomatus* sp. (e) Inner aspect of the upper opercular spine of *Eupomatus* sp.

Mollusca (Snails, shellfish, etc.)

by Roy S. Houston*

The most well-known group of invertebrates in the Gulf of California is the Mollusca. The richness and diversity of this group is exemplified by more than 3,300 species described from the Panamic region, of which a majority occur in the Gulf.

It is virtually impossible to treat this phylum fully in the present handbook. It was initially felt that this group should be excluded entirely, because of their comprehensive coverage in Myra Keen's *Sea Shells of Tropical West America* (1971, Stanford University Press). Upon reconsideration, it was decided that the molluscs are too important a group to be left out entirely. In keeping with the title of this handbook, only the more common species of molluscs are included. Keys have been written for the prosobranch and pelecypod families.

If the reader wishes to pursue this subject further, he may refer to Dr. Keen's work, which is the most recent and authoritative work on molluscs of this region. Keen (1963), Olsson (1961), Morris (1966), and Marcus and Marcus (1967), are also useful in the identification of members of this phylum in the Gulf of California.

Taxonomy: The taxonomy of genera and species is at this time based on shell and radulae characteristics. The higher taxa, however, are separated by differences in internal anatomy, in addition to differences in shell structure. As mentioned earlier, familial keys to the pelecypods and prosobranchs have been included. However, keys to the Polyplacophora (chitons) and opisthobranchs have not, and identification of specimens from these groups should be made by using the illustrations and species descriptions. The family and species descriptions and photographs that follow are taken from Keen (1971), and we are indebted to her and to the Stanford University Press for permission to use this material.

The classes of molluscs discussed in this chapter are: Gastropoda (snails); Pelecypoda (clams, oysters); Amphineura (chitons); Scaphopoda (tusk shells); and the Cephalopoda (octopus, squid). The gastro-

*Department of Biological Sciences, University of Arizona.

pods usually have an external, coiled, univalve shell. Exceptions are the sea slugs, which have an internal reduced shell or have lost it completely, and the limpets, which have an uncoiled external shell. The pelecypods exhibit an external shell with two valves, while the shell of the amphineurans is divided into eight dorsal plates. The scaphopods are characterized by having a tubular shell which is open at both ends and shaped like an elephant's tusk. The shell of cephalopods is absent, or reduced and internal, and the foot is modified into tentacles or arms.

Preservation: If a mollusc is to be used for dissection or histological study, it is imperative that the soft parts be relaxed before fixation. If this is not done, the animal will become contracted and severely distorted when placed in a fixative. If only the shell is desired, the animal may be boiled for 5 minutes, and the soft parts removed from the shell with a dissecting needle.

There are numerous ways to narcotize molluscs. The following methods are most widely used because they are quite successful, and the chemicals are easily obtained. These procedures are taken in part from Russell (1963).

Narcotizing Agents:

1. Alcohol, 70% added slowly to sea water containing the animal.
   a) *Chromodoris*
   b) *Natica*
2. 3% alcoholized sea water solution.
   a) Polyplacophora
   b) Prosobranchiata
   c) Opisthobranchiata
   d) Scaphopoda
3. Chloral hydrate — crystals dropped directly into water containing the animal, or animals placed directly into a 10 to 15% solution.
   a) *Aplysia*
   b) Polyplacophora
   c) Octopoda
   d) Scaphopoda
   e) *Berthellina*
4. Chloroform — added slowly
   a) Pelecypoda
5. Formalin, 10%, added slowly to sea water containing animal until it has relaxed.
   a) Nudibranchiata
6. Magnesium sulfate, 20 to 30% solution — added slowly to sea water containing animal until it has relaxed.
   a) Nudibranchiata
   b) Pelecypoda

7. Propylene phenoxytol — shake 5 ml of propylene phenoxytol into 20 ml of sea water. Add this to a small quantity of sea water containing the animal (do not let the emulsion directly touch the animal).
   a) Molluscs with siphons
8. Fresh water — drown the animal in boiling water.
   a) Pulmonata

Fixative:

1. Bouin's solution — best general fixative for all molluscs. Place the animal in Bouin's fluid for 24 to 48 hours.

Storage:

1. 70% alcohol for all molluscs.

KEY TO THE COMMON FAMILIES OF PELECYPODA IN THE GULF OF CALIFORNIA

1 Pallial sinus present 2
-Pallial sinus absent 4
2(1) Generally 3 cardinal hinge teeth; shell ovate to heart-shaped; beaks in front of midline .... ***Veneridae***
-Two cardinal hinge teeth; shell not ovate to heart-shaped; beaks more or less at midline 3
3(2) Shell triagonal, wedge-shaped .... ***Donacidae***
-Shell rectangular, flattened .... ***Solecurtidae***
4(1) Hinge teeth present 5
-Hinge teeth absent 8
5(4) Hinge with toxidont teeth 6
-Hinge without toxidont teeth 7
6(5) Teeth in more or less straight row; shell trapezoidal .... ***Arcidae***
-Teeth in curved row; shell subcircular .... ***Glycymerididae***
7(5) Shell heart-shaped; adductor scars equal; 2 cardinal hinge teeth .... ***Cardiidae***
-Shell pear-shaped to cylindrical; adductor scars unequal; crenulatelike teeth at anterior end .... ***Mytilidae***
8(4) Shell fan-shaped; two adductor scars .... ***Pinnidae***
-Shell not fan-shaped; one central adductor scar 9
9(8) Shell with regular outline 10
-Shell with irregular outline .... ***Anomiidae***
10(9) Radial ribbing irregular; no ears projecting from hinge ... ***Ostreidae***
-Radial ribbing regular; ears projecting from hinge 11
11(10) Shell obliquely symmetrical .... ***Limidae***
-Shell not obliquely symmetrical 12
12(11) Shell with strong radial spinose ribs; ears small .... ***Spondylidae***
-Shell smooth or with deep radial and concentric ribs, not spinose; ears large .... ***Pectinidae***

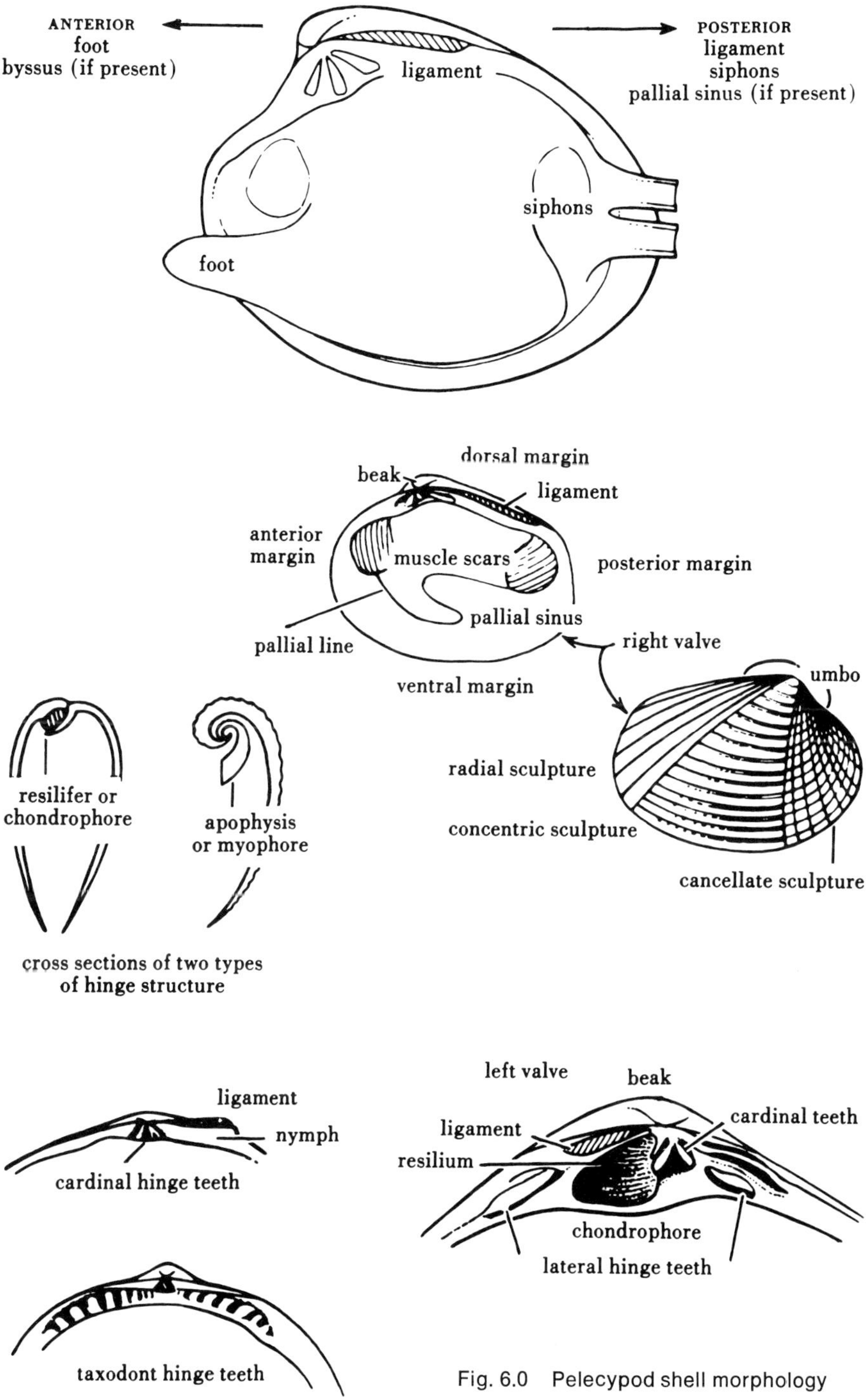

Fig. 6.0 Pelecypod shell morphology

CLASS PELECYPODA

SUBCLASS PTERIOMORPHIA

Order Arcoida

Family Arcidae (ark shells)

The hinge is long, and the ligament is distributed in grooves or pits above it. The two muscle scars are equal in size, and connected by an entire pallial line. The shells of many species are covered by a fibrous or velvety periostracum. In some species, the animal is anchored to the substrate by a byssus that protrudes through a gape on the ventral margin of the shell. Many of the arks may be found intertidally on mud flats or under rocks.

Anadara formosa (**Sowerby**)

The shell is elongate, the posterior dorsal area rather broad. There are about 35 to 38 flat-topped ribs, the posterior ones wider, the anterior ones finely nodulous and grooved down the middle. A large specimen measures: length, 121 mm; height, 58 mm; diameter, 64 mm. This is not a common species, though it is known to occur from Isla de Cedros to Paita, Peru. It has been dredged in depths of 6 to 45 fathoms.

Anadara grandis (**Broderip and Sowerby**)

On all counts this is the largest of the Panamic province arks. The white shell, with a smooth, dark periostracum and about 26 ribs, thickens with age until it becomes impressively massive. A single valve 145 mm (6 inches) long, in the Stanford Collection, weighs 1.5 pounds. Length, 96 mm; height, 79 mm; diameter, 75 mm. The size and the relative abundance of the species make it a staple food clam, especially in the southern part of its range. Specimens may be taken in extreme low tide on sand bars from Bahía Magdalena, Baja California, through the Gulf of California and south to Sechura Bay, Peru.

Arca pacifica (**Sowerby**)

The shell is irregularly ribbed and variable in shape, mostly with an expanded posterior portion, notched near the posterior-dorsal margin. Narrow, rounded V-shaped brown bands ornament the whitish shell. A large specimen of *A. pacifica* measures: length, 128 mm; height, 72 mm; diameter, 80 mm. Specimens may be found adhering to the under sides of rocks in the intertidal zones, or dredged in depths to 75 fathoms from Laguna Ojo de Liebre (Scammon Lagoon), Baja California, through the Gulf of California south to Paita, Peru.

Fig. 6.1
Anadara formosa

Fig. 6.2
Anadara grandis

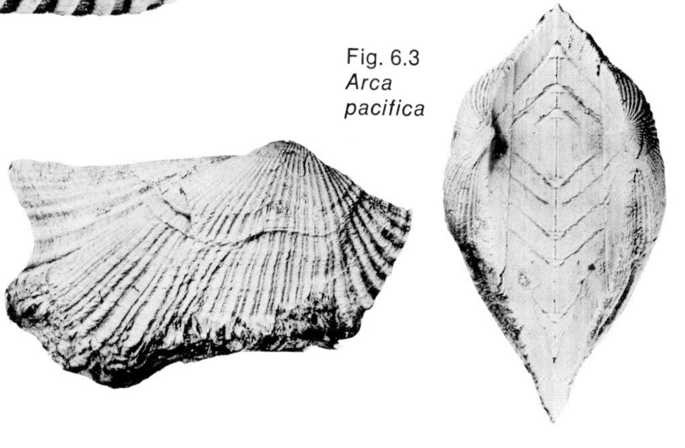

Fig. 6.3
Arca pacifica

Noetia reversa (Sowerby)

The shell is of medium size, inflated, equivalve, with about 36 ribs. The beaks are large, prominent, posterior, and curved backward. Length, 37 mm; height, 34 mm; diameter, 34 mm. Although the habitat has been reported as intertidal, buried in mud, the most recent records are from dredgings, in depths of 12 to 40 fathoms, from Bahía Concepción to Peru.

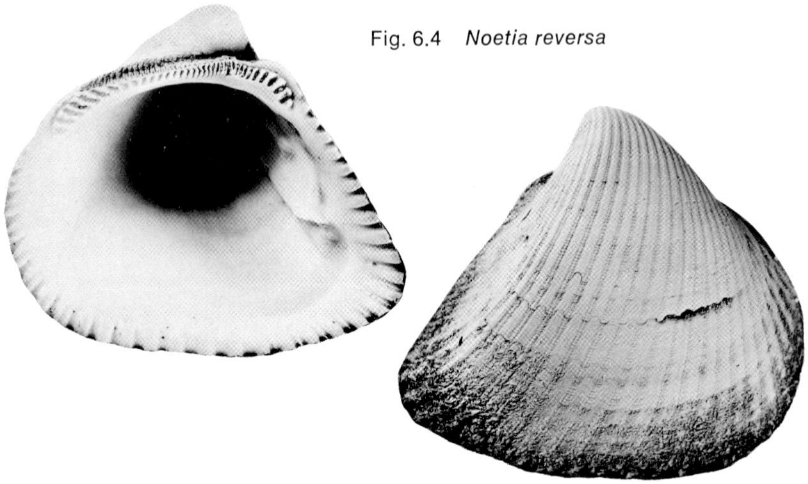

Fig. 6.4 *Noetia reversa*

Family Glycymeridae (bittersweet shells)

Like the ark shells, these have a taxodont hinge; that is, the hinge edge of the shell is modified into a row of similar teeth. The teeth in Glycymeridae are silghtly chevron-shaped or bent, but the shell itself is porcelaneous and solid, not like that of the Nuculidae. The hinge plate in Glycymeridae is arched, and the shell may be so symmetrical that one can tell which is front and which is posterior only by looking at the animal itself. The foot would protrude at the anterior margin, and the siphons at the posterior part. In most shells the posterior muscle scar has a slight flange around it that gives a clue to the orientation. The ligament is distributed in a series of chevron-shaped grooves below the umbones and may be as much in front of the beaks as behind, whereas in most bivalves the major part of the ligament, if not all of it, is behind the beak. The Glycymeridae are not deep water forms, but, on the other hand, they do not tolerate intertidal conditions well, so that most live specimens must be obtained by dredging in shallow to moderately deep water. Single valves are fairly common in beach drift.

Fig. 6.5
Glycymeris gigantea

Glycymeris gigantea (Reeve)

One of the largest species in the whole family, some specimens attain a height of 4 inches. The nearly smooth surface shows radial striae, and the white ground color is mottled with reddish brown in zigzag patterns. There are about 30 hinge teeth. The species seems to be confined to the Gulf of California area south, from Bahía Magdalena, Baja California, to Acapulco, dredged in depths of 4 to 7 fathoms; not uncommon.

Order Mytiloida

Family Mytilidae (mussels)

Although characteristically the mussels cling by the byssus to the exposed faces of surf-beaten rocks, where the churning water brings a maximum amount of food and oxygen, members of the family have become adapted to a number of other habitats, even to the extreme of a sheltered life in a burrow they have excavated for themselves in rocks or in other shells. The sedentary habits of the mussels, and their tendency to cluster in colonies, have made them a target for several predators, not the least of which is man. Carnivorous gastropods, such as *Thais,* and many kinds of starfish may be expected near these colonies. Mussels form a staple food item in some countries, second only to oysters. Mussel poisoning, from which there have been many fatalities, is caused by the concentration of microscopic one-celled organisms, called dinoflagellates, in their gill cavities at a time when sea water becomes saturated with an abundance of these toxic Protozoa; as, for example, when there is a "red tide." Not only the mussels, but other shellfish, may filter out and concentrate these toxic organisms at times.

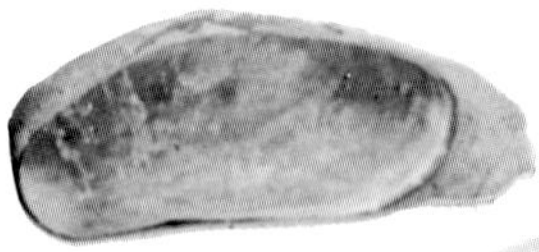

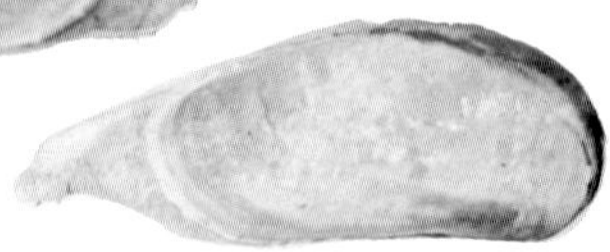

Fig. 6.6
Lithophaga aristata

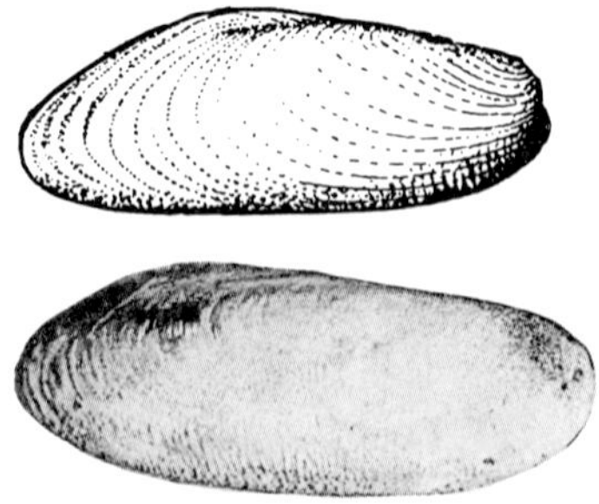

Fig. 6.7
Lithophaga spatiosa

Lithophaga aristata (Dillwyn)

The ends of the incrustation are twisted around each other and look like scissor blades. Length about 25 mm; width, 10 mm. Most specimens are to be found boring into other molluscs, rather than into stones, in the intertidal area and down to depths of as much as 165 fathoms. The species is widely distributed, from its type locality in west Africa to the West Indies, the Mediterranean, the Red Sea, Australia, and Japan, and from southern California to Peru; common.

Lithophaga spatiosa (Carpenter)

This is the largest of the Panamic *Lithophagae*. The incrustation is thin, equal over all the surface, and tends to be arranged in rows of pustules. The mussel has been found boring in valves of *Pinctada* and *Ostrea* from shore down to 15 fathoms. San Felipe, at the head of the Gulf to Ecuador; not uncommon.

Modiolus capax (Conrad)

The bright orange-brown of the shell, its tumid shape, and the serrate peristracal hairs all serve to mark this species. Length, 81 mm; width, 40 mm; diameter, 36 mm. Specimens may be found intertidally on rocks or boulders, or dredged in mud to 46 m. Santa Cruz, California, through the Gulf, to Paita, Peru.

Family Pinnidae (pen shells)

The pinnas, or pen shells, are slender, fragile, and so brittle that in life they must be protected. Accordingly, the mollusc that forms this shell buries itself in the mud of some quiet bay, with the pointed end of the shell down, anchored there by a tuft of fine fibers (the byssus) secreted by glands of the foot. The gaping posterior end of the shell protrudes just above the surface, where, at a very low tide, the observant collector may discover specimens.

Pinna rugosa Sowerby

Spinose on the outside, the shells have about eight rows of somewhat tubular spines, though in old specimens these may be worn down or almost obsolete. An unusually large specimen in the Stanford Collection measures 23 inches. Small specimens may be found near shore, but the larger ones live in deeper water. From the tip of Baja California south to Panamá, on mud bars.

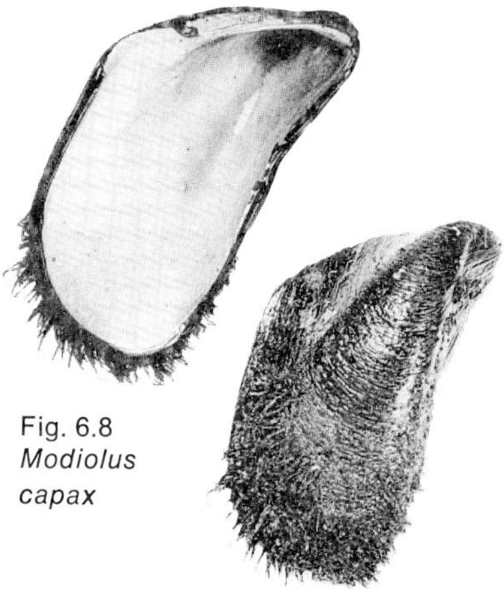

Fig. 6.8 *Modiolus capax*

Fig. 6.9 *Pinna rugosa*

Order Pterioida

Family Ostreidae (oysters)

The oysters reveal some relationship to the pectens (scallops) by the central ligament, which is much larger than in the pectens, and it is not bounded by crura or ridges. There is also a single, nearly central adductor muscle scar. Here the resemblance to pectens ends, for the oysters do not have "ears" on either side of the hinges, and do not have regular radial ribbing. The ribs tend to divide with growth, and cementation of the shell to some object brings about distortions of shape and surface. The presence or absence of denticles along the margins of the shell is important to watch for, as this feature does not seem to be subject to variation. Color and shape of the ligamentary area, and the size, color, and placement of the adductor muscle scar are also significant features.

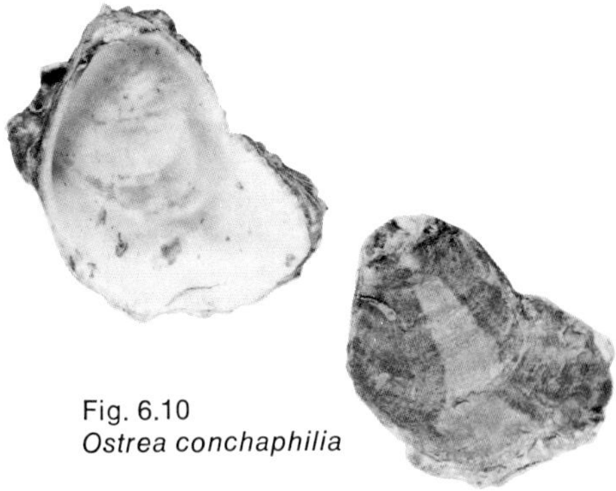

Fig. 6.10
Ostrea conchaphilia

Fig. 6.11
Leptopecten tumbezensis

Ostrea conchaphilia **Carpenter**

Typically, this shell is thin, nearly circular, and flat, an inch to 2 inches in diameter. The outside is radially striped with dark brown, purple or even orange; the inside is white or greenish, with denticles along the margin for about a third of the distance from the hinge to the ventral edge. The ligament area is narrow. Southern California through the Gulf to Panamá, intertidally to a depth of at least 15 fathoms.

Family Pectinidae (scallops)

Commonly known as scallops, the numerous members of this family have a worldwide distribution in both warm and cold seas. Many of them are confined to deep water, and many are never found between tides. Pectens are characterized by a shell that is more or less circular in outline, with a straight, toothless hinge (although strong ridges, or *crura,* may be developed parallel to the hinge line) and two projections called ears on each valve along the hinge. There is a small, but strongly developed, triangular ligament. Adults have only one muscle scar in either valve, although there are two such scars in the young stages, as in the majority of bivalves. The shell structure is highly porcelaneous, tending to become glassy.

Contrary to the usual pelecypod habit, most of the pectens are free-swimming and move about by "flapping" the two valves together and propelling themselves by a jet of water expelled at intervals. Also, unlike most bivalves, the scallops have numerous eyes or *ocelli* along the edge of the mantle that are sensitive to light.

Leptopecten tumbezensis **(Orbigny)**

The left valve varies from a light reddish brown to almost black, sprinkled with tiny light-colored dots, and many specimens also have larger white blotches or rays. The right valve is white to yellowish, with a tendency toward dark rays along the posterior margin. The 14 to 15 ribs are separated by interspaces of almost equal width. About 32 mm across. The Gulf of California to Peru, offshore in depths of 12 to 40 fathoms; not uncommon.

Family Spondylidae (spiny oysters)

The Spondylidae share with the Pectinidae the centrally placed ligament on the hinge, but differ by the lack of a byssal gape in the right valve and by the irregularity of shell caused by attachment to some surface.

Spondylus princeps Broderip

One of the most colorful bivalves of the Panamic Province, the shell varies from rose-red through orange to white, with festoons of strong, curved spines. The shell is about 4 to 6 inches in diameter. The hinge ridges are so interlocked in many specimens that the valves cannot be separated without breaking the teeth. The Gulf of California to Ecuador; not uncommon.

Fig. 6.12
Spondylus princeps

Fig. 6.13 *Lima pacifica*

Fig. 6.14
Anomia adamas

Family Limidae (file shells)

The limas or file shells resemble the pectens in having ears, but the shell is obliquely asymmetrical, with the anterior end the longer and the valves gaping widely. Although able to swim, the animals are less active than most of the pectens, tending to bury themselves on the sea floor in a nest constructed of bits of rubble cemented together by mucus. The mantle edge is fringed, and the living animal is a beautiful object to watch in an aquarium when it is fully expanded or moving about.

Lima pacifica **Orbigny**

The shell is long and narrow, of a dull texture and ivory white color. Length, 19 mm; height, 30 mm; diameter, 9 mm. The northern end of the Gulf of California to Peru; under rocks at low tides; not uncommon.

Family Anomiidae (jingle shells)

Like the oysters, the anomias or jingle shells are attached, usually to other shells, but in a different way. Instead of cementing the shell to some surface, the animal produces a byssal plug of tough, chitinous material that projects through an opening in the lower valve and acts as a firm anchor. Actually, the lower valve is not truly perforate, as it is usually described. It grows in a half-moon shape around the projecting byssus, one side faster than the other, until in the adult stage an almost symmetrical "window" is developed. On a smooth surface, such as the inside of a dead bivalve shell, the jingle may have a glassy texture, with a lustrous or nacreous sheen within. On a rough surface, however, the texture is much coarser, and the irregularities of the substrate may be perfectly mirrored in both valves of the jingle, though such markings may be oblique to or even at right angles to the normal direction of sculpture that one would expect, depending upon the position in which the jingle has become attached to its host shell. Added to the resultant variability of shape is considerable variation in color.

Anomia adamas Gray

Color of the shell varies from lustrous white to bright orange. About 40 mm across. The Gulf of California to Central America; not uncommon.

SUBCLASS HETERODONTA

Order Veneroida

Family Cardiidae (cockles)

The members of this family are commonly known as heart (Greek, *kardia*) cockles, or simply as cockles. The color patterns and sculpture, being more constant than in most bivalves, offer a ready means of species separation. The shell is equivalve, but rarely equilateral, with porcelaneous texture. The internal margin is serrate to crenulate in all but the smoothest species. The two adductor muscle impressions are of about the same size and shape, joined by an indistinct, smoothly curved, pallial line. The siphons are short and the foot well-developed and long, capable — as it obtains leverage in the sand and contracts — of moving the animal about in short leaps. The periostracum is conspicuous in some species, thin and transparent in others.

Trachycardium panamense (Sowerby)

The shell is massive and may attain a height of 90 mm. The ribs are triangular in outline, with fine nodes along the sharp crest, especially on the umbones, and the rather wide spaces between ribs are flat. The shell is white under a greenish, persistent periostracum, except for spots of brown on the umbones. The species is a common one in mud, from low-tide line to a depth of several fathoms, ranging from Laguna San Ignacio, Baja California, throughout the Gulf of California, but despite its name, apparently no farther south than Costa Rica; common.

Fig. 6.15 *Trachycardium panamense*

Family Veneridae (Venus clams)

From the standpoint of color, beauty of form, and diversity, the Venus clams are the most advanced of the bivalves. The shell is porcelaneous, whether large or small, and likely to have a polished, highly finished appearance. The shape is mostly ovate or cordate (heart-shaped), inequilateral, with the beaks well in front of the midline. Sculpture may be radial or concentric, or both. The ligament is external, on a special platform or nymph. Additional nymphs, resembling teeth, may complicate the already complex hinge, which consists of three cardinal teeth in each valve, one or more of which may be grooved or bifid. Some genera have an anterior lateral hinge tooth in the left valve, fitting into a lateral socket in the right valve. The pallial line has a more or less deep sinus, reflecting the habit of burrowing below the surface of the sea floor.

Tivela planulata (Broderip and Sowerby)

The tightly closed trigonal, flattened shell with radiating color bands in shades of light to dark brown makes up the distinctive form. Length, 55 mm; height, 43 mm; diameter, 23 mm. The Gulf of California to Ecuador.

Fig. 6.16
Tivela planulata

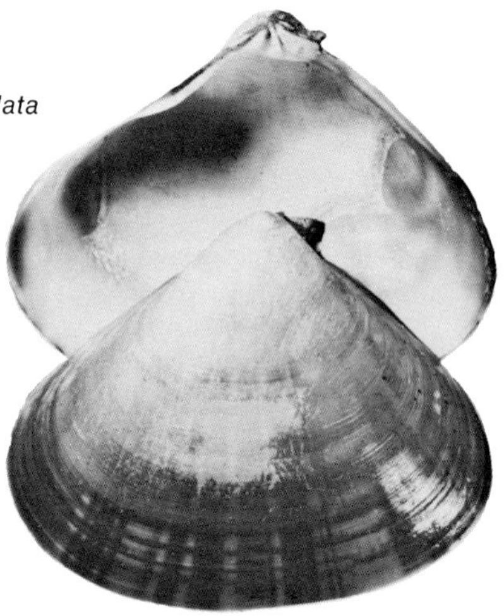

Pitar lupanaria (Lesson)

Like the thorny oyster, *Spondylus,* this has spines so long and slender one wonders how the clam moves about without breaking them. The shell is white, tinted with violet, with violet spots at the bases of the

Fig. 6.17
Pitar lupanaria

spines. Length (not including spines), 53 mm; height, 43 mm; diameter, 29 mm. Bahía Ballenas, west coast of Baja California, through the Gulf of California, and south to Negritos, Peru; common on sandy beaches, especially after storms, and offshore to depths of 13 fathoms.

Dosinia dunkeri (Philippi)

The surface is covered with fine and very regular concentric ribs. The pallial sinus is angular, and points toward the middle or upper part of the anterior adductor muscle scar. Length, 56 mm; height, 55 mm; diameter, 31 mm. Bahía Magdalena, west coast of Baja California, through the Gulf of California, and south to Zorritos, Peru; common on mud flats and offshore to depths of 30 fathoms.

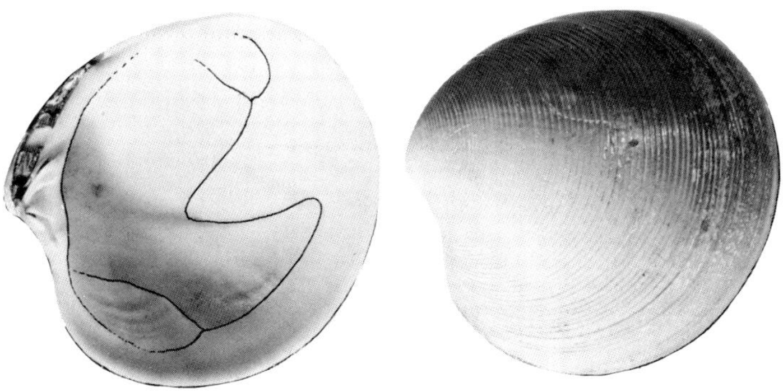

Fig. 6.18 *Dosinia dunkeri*

Fig. 6.19 *Chione amathusia*

Chione amathusia (Philippi)

The sculpture is fine, with interribs between the primary radial ribs, and with the concentric lamellae raised into prickly scales. The shells are mostly a light brown in color. Length, 40 mm; height, 36 mm; diameter, 26 mm. The Gulf of California to Mancora, Peru, mostly offishore in depths to 40 fathoms.

Protothaca grata (Say)

Description of the color patterns of this variable form would be a task, for no two specimens are quite alike. The ground color is an off-white, on which are laid intricate designs in brown and purplish black; or wide areas may be a solid color. Within, the shell is white to wholly purple, the margin weakly crenulate. Length, 40 mm; height, 34 mm; diameter, 23 mm. Cabo Colnett, west coast of Baja California, through the Gulf of California and south to Chile, on mud flats and offshore in depths to 215 fathoms; common.

Family Donacidae (bean clams)

The members of this family are solid little shells, mostly trigonal in shape, valves tightly closing, with a moderately large pallial sinus. The hinge has two cardinal teeth, and one or more lateral teeth on either side. The ligament is external, on a small platform.

Donax punctatostriatus Hanley

Probably the most abundant of the Panamic Province donacids, this is a sturdy, fair-sized shell, easily recognized, in spite of its color variants and variations in size and shape, by the pits in the interspaces between ribs. The majority of specimens are bright brown, violet-stained within.

Length, 45 mm; height 28 mm; diameter, 15 mm. Laguna San Ignacio, west coast of Baja California, through the Gulf of California, and south to Negritos, Peru.

Family Solecurtidae (gari shells)

The shells are rectangular to cylindrical, and more than twice as long as they are high. The hinge is much the same as in the Psammobiidae.

Tagelus affinis (C. B. Adams)

The shell is white under a yellowish brown periostracum. It is one of the smaller species of *Tagelus*: length, 55 mm; height, 21 mm. In the typical form, the pallial sinus extends just beyond a line drawn vertically through the beaks, but in the northern part of the range, specimens with a pallial sinus 60 percent the length of the shell are common. The Gulf of California to Panamá, on mud flats or offshore to depths of 40 fathoms.

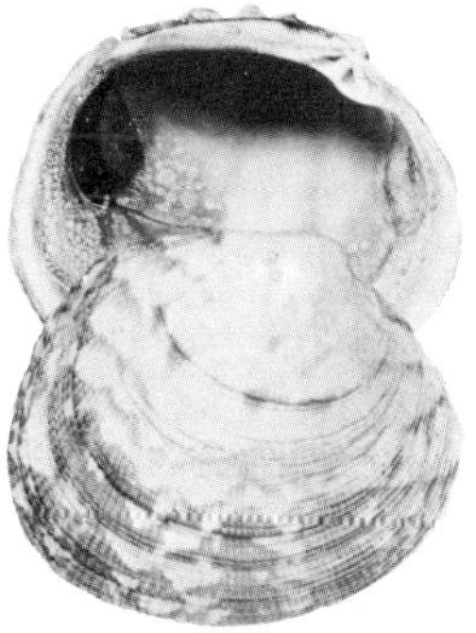

Fig. 6.20 *Protothaca grata*

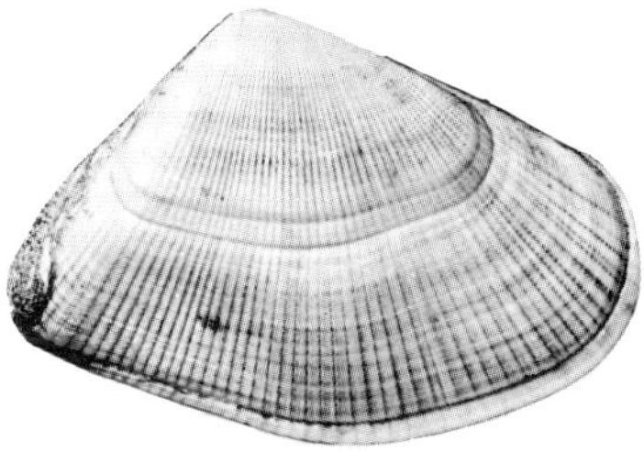

Fiq. 6.21 *Donax punctatostriatus*

Fig. 6.22 *Tagelus affinis*

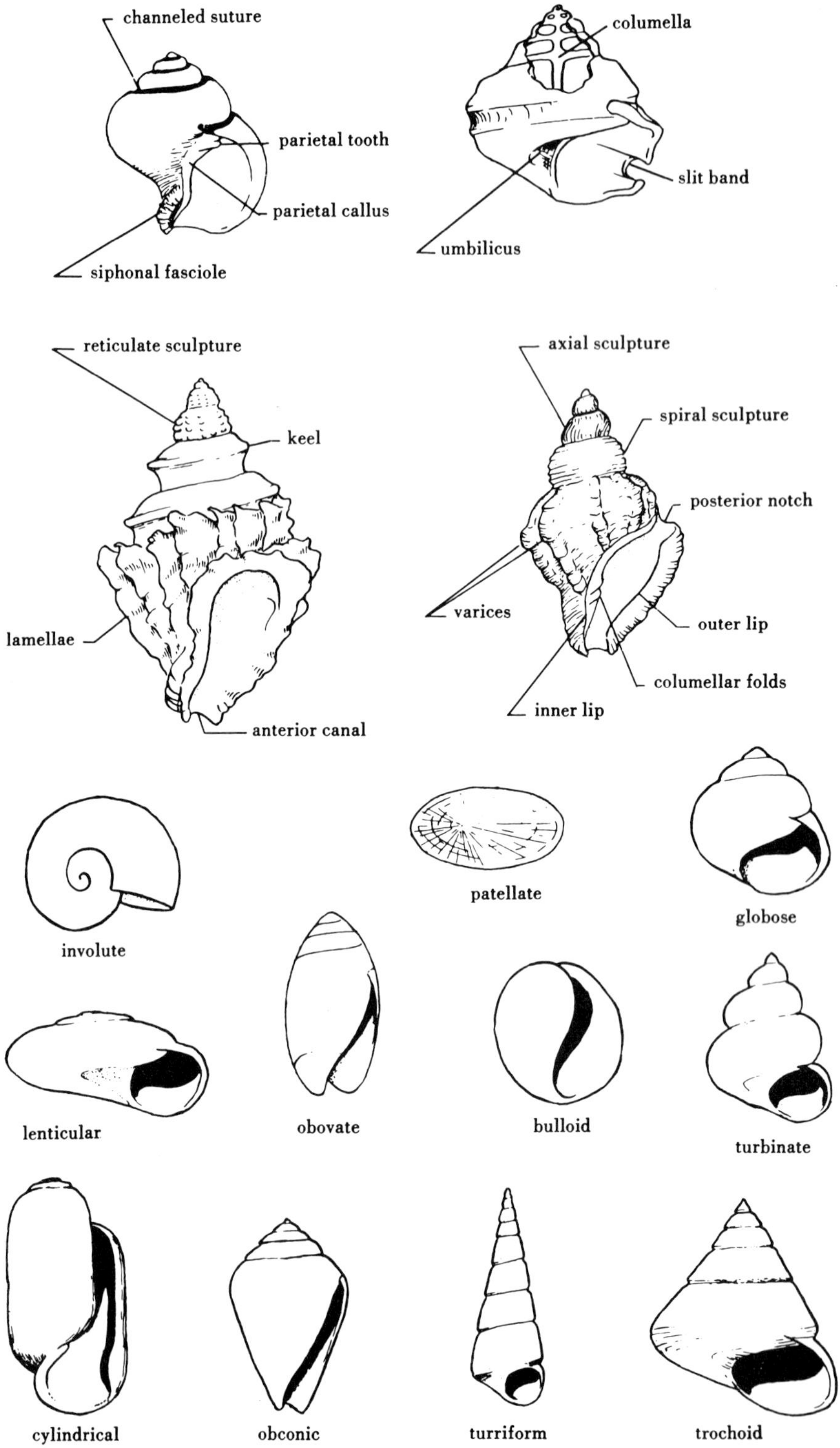
channeled suture
parietal tooth
parietal callus
siphonal fasciole
columella
slit band
umbilicus
reticulate sculpture
keel
lamellae
anterior canal
axial sculpture
spiral sculpture
posterior notch
varices
outer lip
columellar folds
inner lip
involute
patellate
globose
lenticular
obovate
bulloid
turbinate
cylindrical
obconic
turriform
trochoid

KEY TO THE COMMON FAMILIES OF PROSOBRANCH GASTROPODS OF THE GULF OF CALIFORNIA

1 Shell patellate 2
-Shell not patellate 4
2(1) Shell with internal cuplike or diaphragmlike processes .... ***Calyptraeidae***
-Shell without internal processes 3
3(2) Shell with opening at apex of shell .... ***Fissurellidae***
-Shell without opening at apex of shell .... ***Acmaeidae***
4(1) Aperture greater than one-half the length of the shell 5
-Aperture less than one-half the length of the shell 12
5(4) Operculum present; shell not glazed 6
-Operculum not present; shell glazed 22
6(5) Chitinous operculum 7
-Calcareous operculum 10
7(6) Shell pear-shaped .... ***Melongenidae***
-Shell not pear-shaped 8
8(7) Outer lip expanded with a notch at each end .... ***Strombidae***
-Outer lip not expanded, notches absent 9
9(8) Shell conical; shell smooth .... ***Conidae***
-Shell not conical; shell reticulate, knobby .... ***Thaididae***
10(6) Umbilicus present 11
-Umbilicus absent .... ***Neritidae***
11(10) Shell turbinate; aperture round .... ***Turbinidae***
-Shell globose; aperture angular .... ***Naticidae***
12(4) Shell top-shaped .... ***Trochidae***
-Shell not top-shaped 13
13(12) Operculum solid, scaly .... ***Littorinidae***
-Operculum not solid, horny 14
14(13) Shell turriform 15
-Shell not turriform 18
15(14) Apical region of shell turriform, lower region loosely coiled and wormlike .... ***Vermitidae***
-Shell tightly coiled throughout 16
16(15) Shell with a short, strongly developed anterior canal .... ***Cerithiidae***
-Shell without, or with a weakly developed, anterior canal 17
17(16) Heavy axial sculpture .... ***Potamididae***
-Light or no axial sculpture .... ***Turritellidae***
18(14) Shell fusiform 19
-Shell not fusiform 20
19(18) Outer lip thickened at center, notch absent .... ***Columbellidae***
-Outer lip not thickened, notch present .... ***Turridae***
20(18) Shell with varices and occasionally with spines .... ***Muricidae***
-Shell without varices or spines 21
21(20) Spire strongly shouldered and tapered; no callus .... ***Buccinidae***
-Spire weakly shouldered and pointed; callus present .... ***Nassariidae***
22(5) Shell oval, with aperture toothed on both lips .... ***Cypraeidae***
-Shell cylindrical, without teeth .... ***Olividae***

Fig. 6.23 Prosobranch shell morphology

CLASS GASTROPODA
SUBCLASS PROSOBRANCHIATA
Order Archaeogastropoda
Family Fissurellidae (keyhole limpets)

The fissurellas, or keyhole limpets, can be divided into three subfamily groups on the basis of the position of the anal opening, the shape of the callus margin on the inside of the opening, and the shape of the muscle scars.

Diodora inaequalis (Sowerby)

The riblets of the outer surface are visible on fresh specimens as white radiating lines within. The shell is yellowish or grayish, rayed with dull black, and the sides are arched slightly upward. Most specimens have the hole very near the front margin, though there is some variation in its placement. Length, 27 mm; width, 16 mm; height, 8 mm. Laguna San Ignacio, west coast of Baja California, through the Gulf of California and south to the Galapagos Islands; common, under rocks.

Family Acmaeidae (limpets)

The limpets of this family have carried the development of a conical shell so far that they have lost the spiral, even in the young stages. The main difference in the animal from that of the family Patellidae is the presence of a single feather-shaped gill. The shells are never iridescent within. The muscle scar is horseshoe-shaped, open in front.

Limpets are to be found on rocks and seaweed, mostly in the intertidal zone, and many of them have a more or less permanent home base to which they return after feeding. They are entirely herbivorous.

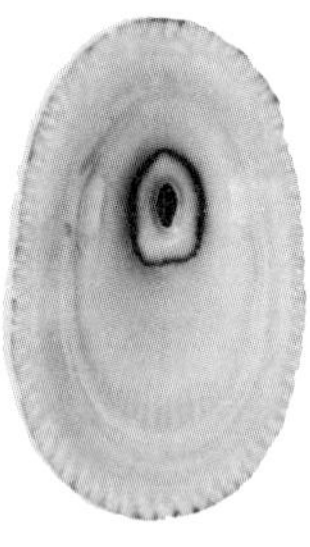
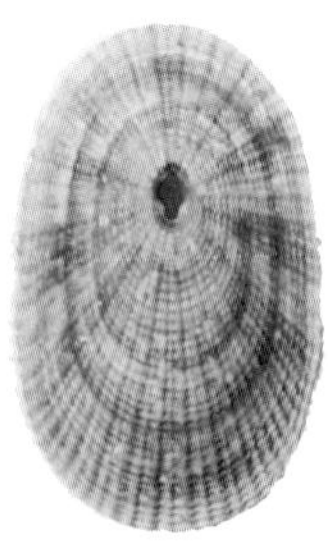

Fig. 6.24 *Diodora inaequalis*

Fig. 6.25 *Collisella atrata*

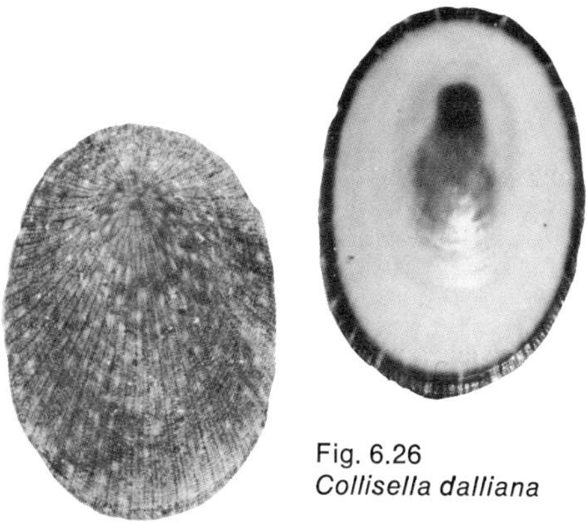

Fig. 6.26
Collisella dalliana

Fig. 6.27
Collisella stanfordiana

Collisella atrata Carpenter

On a whitish ground are black stripes and lines, with solid black on the eroded apex. The inside is white, with a yellowish gray and black-checkered border and brownish clouding of the central area. Length, 31 mm; width, 24 mm; height, 14 mm. The young are flatter, with sharp, close, unequal ribs that crenulate the margin. Bahía Magdalena, west coast of Baja California, through the Gulf of California to Acapulco; on rocks in high tide pools.

Collisella dalliana (Pilsbry)

The shell of Dall's limpet is low, its dorsal margin arching smoothly to the apex, the brown surface sculptured with fine and somewhat beaded riblets that are flecked with radially arranged white spots. The inside is bluish white, with the outer color pattern showing through the thin shell. Length, 46 mm; width, 32 mm; height, 6.5 mm. Restricted to the Gulf of California, especially in the northern part, on rocks at about the midtide line.

Collisella stanfordiana (Berry)

This has been confused with the preceding species, but is thinner, more depressed, with the apex nearer the anterior margin. Length, 27 mm; width, 21 mm; height, 4 mm. Sonoran coast of Mexico, on rocks at midtide level; locally common.

Family Trochidae (pearly top shells)

The top shells are conical, with a round, (entire) aperture, the inner layer being nacreous. The operculum is chitinous, round, and multispiral.

Tegula rugosa (A. Adams)

The rough turban shell is heavy and dull gray, variegated with brown, black, or red streaks; the whorls are roughly and irregularly sculptured, with oblique folds and some spiral threads. The umbilicus is deep and white within. Diameter, 27 mm; height, 26 mm. The species seems to be confined to the Gulf of California, on rocks.

Family Turbinidae (turbins)

Strong-shelled, solid and few-whorled, variously sculptured, but with a calcareous operculum; the aperture nacreous within.

Turbo fluctuosus Wood

The spiral ribs vary in number and in prominence; some may even be nodose. The shell is non-umbilicate, with a wide white columella. The operculum is granular, with a deep central pit and a marginal band of granular ribs and deep grooves. Diameter, 65 mm; height 58 mm. Isla de Cedros, west coast of Baja California, through the Gulf to Paita, Peru, among stones, intertidally; very common.

Family Neritidae (nerites)

The shells are low-spired, nearly globular in shape, and, apparently to take full advantage of the inner space, the animal resorbs the walls of the inner whorls, which is unusual in the Gastropoda. The outer lip may be thickened near the aperture, or toothed within. The inner lip may not only be toothed, but may be spread out as a mass of callus, or even as a sort of deck, so that the shell comes to look like a slipper limpet. The operculum is calcareous and oddly shaped, but with a little clawlike process, or apophysis, that fits under the edge of the columella, and enables a tight fit when the animal withdraws into the shell. This effectively seals in moisture during periods when the animal is out of the water — as at the time of a low tide — and permits it to live in the upper splash zone of the shore. The marine forms are solid-shelled, the freshwater ones thin-shelled. None are nacreous.

Fig. 6.28 *Tegula rugosa*

Fig. 6,29 *Turbo fluctuosus*

Fig. 6.30 *Nerita scabricosta*

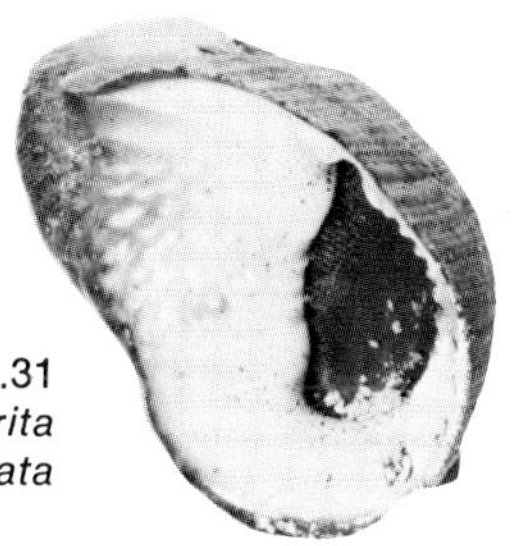

Fig. 6.31 *Nerita funiculata*

Nerita scabricosta Lamarck

This is one of the largest species of the genus. The shell is a dark gray, with spiral ribs that are rough-surfaced and somewhat irregularly spaced. It is the common form of the Gulf of California. An average specimen of *N. scabricosta* measures: diameter, 38 mm; height, 45 mm. The outer coast of Baja California through the Gulf to Ecuador, intertidally on rocks.

Nerita funiculata Menke

Smaller than *N. scabricosta,* this has irregular wrinkles and pustules on the callus area. Diameter, 20 mm; height, 15 mm. Baja California through the Gulf and southward to Peru, on rocks, intertidally.

Order Mesogastropoda

Family Littorinidae (periwinkles)

The littorines, or periwinkles, are among the commonest marine snails of the world, and few are the areas that have none clinging to the rocks between midtide and hightide, often so high they receive only an occasional splash of sea water. They are true inhabitants of the littoral — the seashore — and, like the nerites, have probably provided the ancestral stock for some of the present-day land snails. The operculum is not calcareous as in the nerites, but is a solid little chitinous scale that effectively blocks the aperture and keeps enough moisture inside the shell when the animal is retracted that the gill does not become dry. The littorines mostly have small shells that are porcelaneous in texture. The animals feed upon microscopic plant material that they rasp from the surface of the rocks.

Littorina aspera (Philippi)

The shell is whitish, with axial lineations of brown. The aperture and inner lip are chocolate brown, sculptured with coarse, sharp, spiral ribs. Height, 16 mm; diameter, 10 mm. Bahía Magdalena, west coast of Baja California, through the Gulf, to Panamá; fairly common.

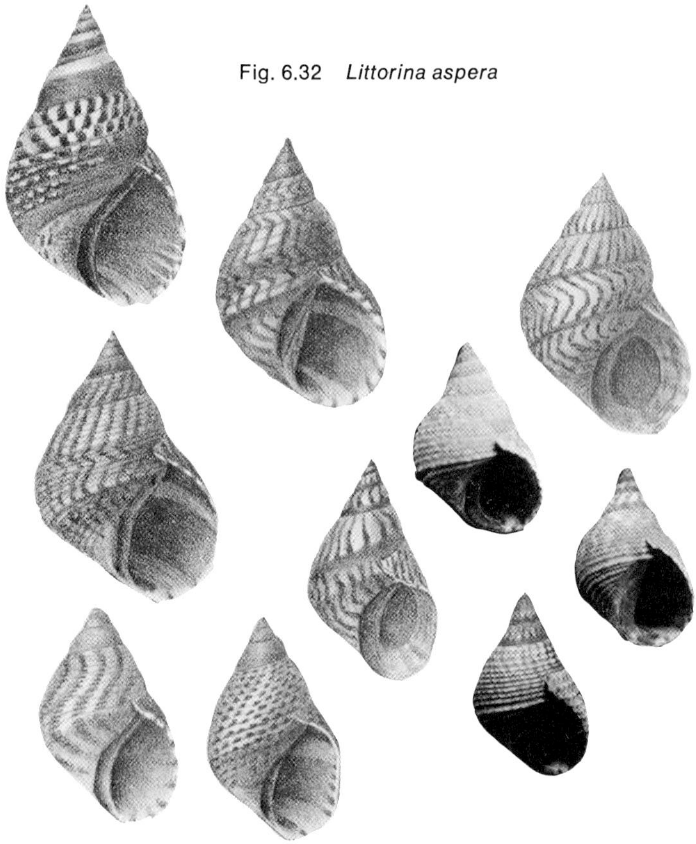

Fig. 6.32 *Littorina aspera*

Family Turritellidae (tower shells)

The tower shells are among the longest and most slender-spired of the gastropods. One wonders how the animal can carry so delicately turned a shell without breaking it, but perhaps that is because animal and shell are safely buried just under the surface of the sea floor, where the animal feeds on detritus — fine bits of seaweed and other broken fragments — that settle like dust on the ocean floor. Most species live beyond the low tide line.

The shells are somewhat variable — even within the course of growth of one individual — in the expression of the spiral ribs. Axial sculpture is weak. The aperture is round, and closed by a remarkable operculum, a flat or saucer-shaped spiral composed of a spirally coiled chitinous band that is thin and membranous at the edge, supported by flexible chitinous bristles somewhat prolonged beyond the margin. This operculum is so flexible it may be withdrawn well within the shell when the animal retreats.

Turritella leucostoma Valenciennes

A delicate-appearing shell having a light buff coloring, mottled and striped with reddish brown. Each whorl is contracted just above the suture, giving it a characteristic profile. Height, 115 mm; diameter, 20 mm. Isla de Cedros, west coast of Baja California, south through the Gulf of California to Panamá, in depths of as much as 20 fathoms.

Fig. 6.33 *Turritella leucostoma*

Family Vermetidae (worm shells)

The worm gastropods are among the most puzzling of shells, and authors have confused them at every classificatory level — from species through genus and family to phylum, for many tube-building annelid worms have been described as vermetid gastropods, and vice versa. The distinction between the vermetids and the worms is clear-cut: the gastropods have a three-layered shell that is glossy within, and that begins with a tightly and spirally coiled embryonic shell, whereas the tube-building worms have a two-layered shell that is dull-surfaced within, and begins with a single tubular chamber. The Vermetidae also have been confused with the Vermicularias (here treated as a subfamily of Turritellidae, though ranked by some authors as a family), but the nuclear whorls are very different. In the Vermetidae, the nuclear whorls, which the little snail has when it emerges from the egg capsule, are two or three in number, with an aperture twisted forward and upward. Immediately upon emergence from the protection of the parent's tubular shell, the young vermetid finds a suitable place for starting a free life and attaches the shell to the substrate, beginning to coil its adult whorls at a right angle to the nuclear whorls. In changing the plane of coiling, the Vermetidae show some relationship to the Architectonicidae. The Vermicularias continue the regular spiral Turritella-like coiling and gradually become lax or unwound.

Serpulorbis margaritaceus (Chenu, *ex* Rousseau MS)

The brownish tubes are either regularly wound into a flat spiral, or may be nearly straight, but they are always firmly attached to rocks or to other shells. Small scales roughen the spiral ribs. Diameter of a large coil 70 mm, of an average coil, 30 mm; diameter of tube, 10 to 13 mm. Gulf of California to southern Mexico; common.

Family Cerithiidae (horn shells)

Marine forms, small to large in size.

Cerithium stercusmuscarum Valenciennes

The bluish gray to brownish shell is finely speckled with white, and has a well-marked spiral row of pointed tubercles below the suture. Height, 25 mm; diameter, 11 mm. Baja California to Peru; common, on sand flats and in estuaries.

Family Potamididae (horn shells)

Shells brownish to horn-colored, with anterior canal less strongly developed than in Cerithiidae. Habitat: brackish water bays.

Cerithidea mazatlanica Carpenter

The shell is dark brown to black, sculptured with a network of axial and spiral ribs that rise into rounded beads where they cross. Height, 27 mm; diameter, 9 mm. Range, the Gulf of California, where astronomical numbers occur.

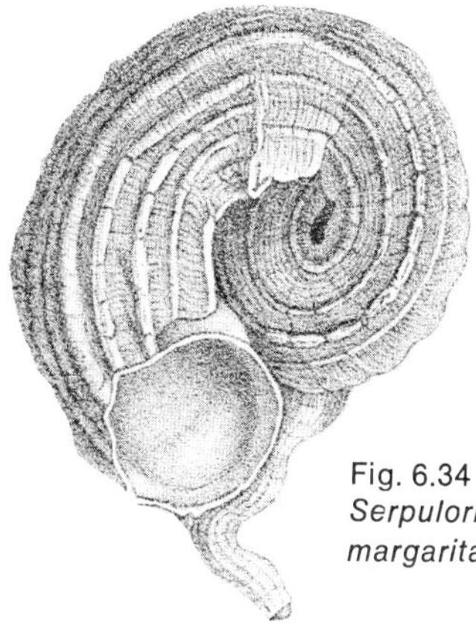

Fig. 6.34 *Serpulorbis margaritaceus*

Fig. 6.35 *Cerithium stercusmuscarum*

Fig. 6.36 *Cerithidea mazatlanica*

Family Strombidae (strombs)

The molluscs of this family have a narrow, peculiarly arched foot, the greatly elongate posterior portion of which (metapodium) bears at its end the relatively small operculum. The foot facilitates an unusual mode of locomotion for a gastropod. Instead of the normal gliding motion, there is a scrambling or leaping, with the pointed operculum used as a sort of claw. Many collectors have been startled on picking up the shell to find that the animal inside is adept at kicking in an attempt to escape. The development of the eyes is also noteworthy — situated at the ends of long, stout stalks, they are more complex and apparently keener than in any other gastropod group. In food habits, the strombs seem to be herbivorous, feeding on bottom deposits and finely cropped algae.

Strombus gracilior Sowerby

This species has a high spire armed with blunt spines. The color is a yellowish brown with a lighter central band, the aperture white, edged with orange brown. The periostracum is thin, horn-colored, or varnished in appearance. Length, 75 mm (3 inches); width, 50 mm (2 inches). The Gulf of California to Peru, on sand flats and in muddy lagoons, especially in mangrove areas.

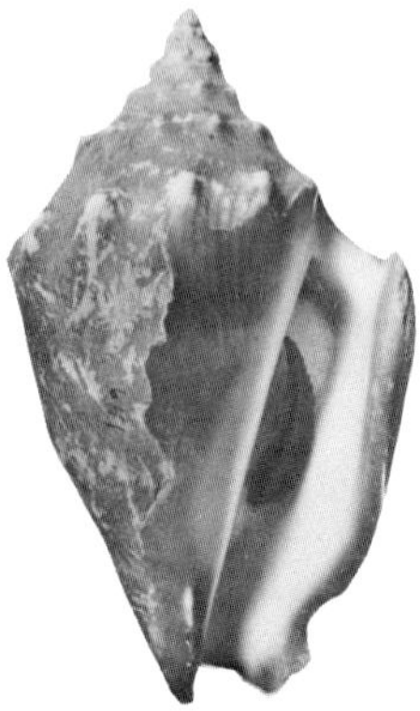

Fig. 6.37 *Strombus gracilior*

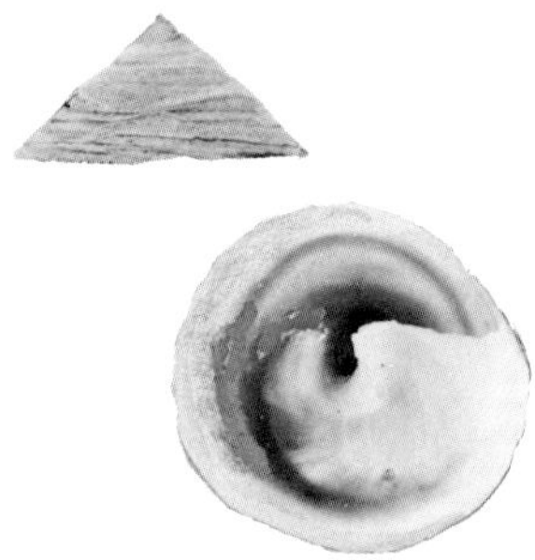

Fig. 6.38 *Calyptraea mamillaris*

Family Calyptraeidae (cup-and-saucer limpets)

Calyptraea mamillaris Broderip

The shell is white and smooth, with a flush of purple or brown on the spire. Diameter, 35 mm; height, 15 mm. Bahía Magdalena, outer side of Baja California, through the Gulf and southward to Paita, Peru, offshore in depths to 20 fathoms; not uncommon on mud flats, attached to rocks and other shells.

Crucibulum spinosum (Sowerby)

The shell is yellowish white, with or without curved purplish to brownish rays. The apex is recurved, its tip to one side, like a little tailed stocking cap. Oblique wrinkles roughen the surface, overlain by more or less well-developed rows of spines, some of which are even tubular. The internal cup is white, and attached along most of one side. Diameter, 20 mm; height, 7.5 mm. California, and southward through the Gulf to Tomé, Chile, on stones and dead shells, intertidally, and offshore in depths to 30 fathoms; common.

Fig. 6.39
Crucibulum spinosum

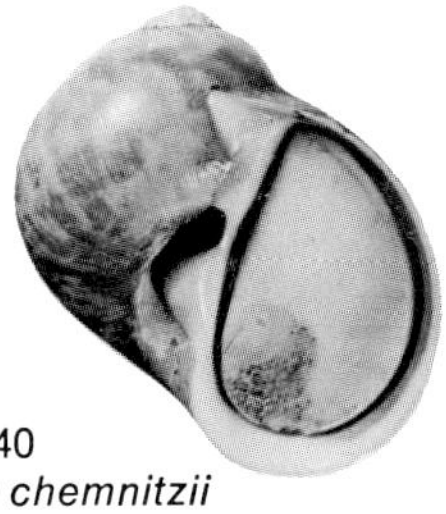

Fig. 6.40
Natica chemnitzii

Family Naticidae (moon snails)

Moon snails have a short spire and large body whorl; aperture angular above, broadly rounded below, with a more or less callus on the inner lip, operculum horny.

***Natica chemnitzii* Pfeiffer**

The shell is grayish blue or grayish yellow, with four to five bands of arrow-shaped markings that are either brownish or nearly white, and the interior is dark brown, banded. The operculum is white and smooth, except for a roughened area near the nucleus. The variant called *N. unifasciata* has a white band on the upper part of the whorls, and is more common in the southern part of the range. Height, 33 mm; diameter, 31 mm. Bahía Magdalena, west coast of Baja California, through the Gulf and south to northern Peru; intertidally on mud flats; common.

Family Cypraeidae (cowries)

The shell of an adult cowry differs greatly in appearance from that of a young specimen. Immature shells are fragile, somewhat polished, with an elongate body whorl, short spire, and sharp lip. The color pattern shows a tendency to vague spiral bands. As the shell nears maturity, the outer lip turns abruptly in toward the columella, and both lips begin to show signs of teeth. The shell also becomes stronger and heavier, and the color pattern begins to take on the adult character. Fully mature specimens have a narrow aperture with well-developed teeth (in most species), a distinctive color pattern, and a fairly sturdy shell. A heavy layer of enamel is deposited by the mantle upon the surface of the shell, often leading the uninitiated to assume that the shell has been varnished or polished. This layer of enamel almost or entirely conceals the spire of the shell. In some species the enamel is heavily deposited about the sides of the shell (known as the margins), forming callus deposits. The margins may be heavily marked with so-called "lateral spots." The term "base" is commonly used to describe the flattened part of the adult shell where the aperture and teeth are located, while the opposite side is known as the "dorsum." A periostracum is completely lacking.

Cypraea annettae Dall

The dorsal surface of Annette's cowry is heavily spotted with chestnut-brown on a pale bluish white base, and shows a faint brown banding, the sides well-rounded and spotted. The base varies from flesh color to a rich brown, the teeth being lighter or almost white. A purple interior shows clearly through the rather broad aperture. Young specimens are lighter in color, weakly spotted, with five brown bands. Length, 40 mm; width, 23 mm. The outer coast of Baja California to and throughout the Gulf; fairly common under rocks.

Order Neogastropoda

Family Muricidae (rock shells)

The rock shells reach a climax among the gastropods in complexity of spines, from symmetrical, slender ranks, as in the Venus-comb to intricately frilled varices. Varices — periodic thickenings of the outer lip — are characteristic of the family, the shape being determined by the convolutions of the edge of the mantle. The operculum is horny, oval, mostly with an apical nucleus.

These are carnivorous snails, many of them being classed as "drill shells," but they prefer to chip the edges of clam shells in getting at the juicy meat rather than to drill in the manner that *Natica* or *Polinices* do. The holes an ambitious muricid makes are rougher, less even, and less finely beveled than those of the moon snails. Many of the muricid species have developed a tooth on the outer lip that seems to be useful as a lever in holding or wedging apart the valves of mussels or clams.

Hexaplex erythrostomus (Swainson)

The pink-mouthed murex is probably the most abundant member of the family in the Panamian Province. The white shell is dull-surfaced on the outside, but polished and bright pink within. Now and then one finds a specimen in which the pink is reduced to a faint trace around the edge of the lip. About 4 inches in length.

Muricanthus nigritus (Philippi)

Black murexes prey especially upon clams, chipping the edges of the valves in order to insert the proboscis and suck out the rich juice. *M. nigritus* seems to be confined to the Gulf of California, where it is fairly common.

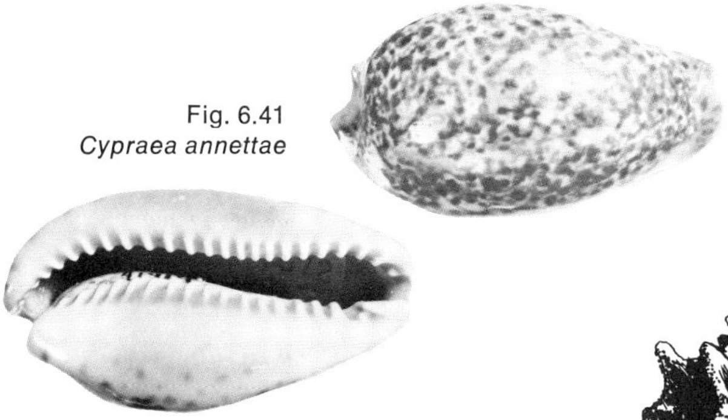

Fig. 6.41
Cypraea annettae

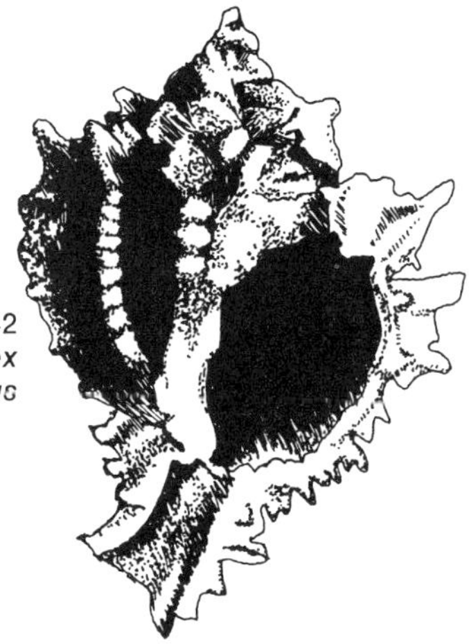

Fig. 6.42
Hexaplex erythrostomus

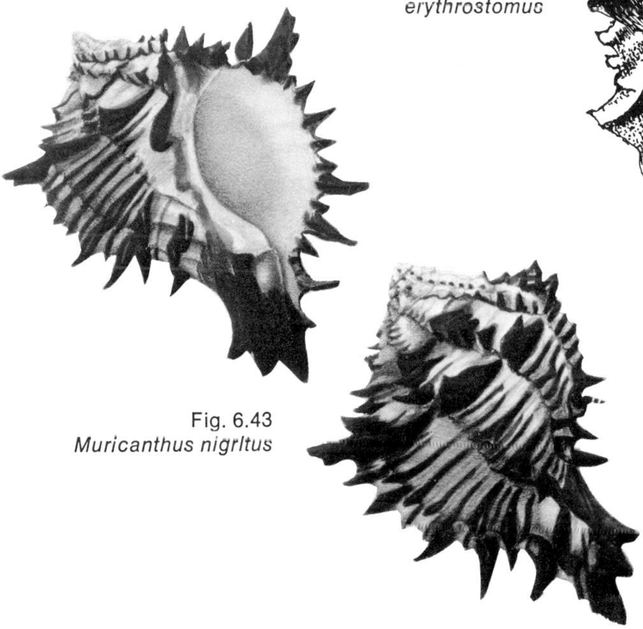

Fig. 6.43
Muricanthus nigritus

Family Thaididae (dye shells)

The thaids are active predators, creeping about over the rocks in pursuit of their prey — clams, mussels, barnacles, and so on. They are not so varied in form as the Muricidae or so variable as the Coralliophilidae. The shells are of medium size, rather solid, lacking varices, but with knobby or slightly spinose sculpture. The operculum is chitinous, with a heavy polished ridge along the inside of the outer edge, and the nucleus is at about the middle of this ridge.

Thais biserialis (Blainville)

The exterior of the shell is dark gray, and may vary from nearly smooth to roughly corded, the color pattern varying from plain to mottled with light and dark brown. Mature specimens may attain a length of more than 75 mm (3 inches), with a diameter of 50 mm. Isla de Cedros, west coast of Baja California, through the Gulf, and south to Chile. On rocks, intertidally, not uncommon.

Acanthina angelica I. Oldroyd

The species somewhat resembles the Baja Californian *A. lugubris,* but shells are smaller and more slender, the spire higher, and the coloration less subdued, the aperture being white within. Length, 39 mm; diameter, 24 mm. The species seems to be restricted to the Gulf of California; not uncommon.

Morula ferruginosa (Reeve)

This species is somewhat similar to *Muricopsis zeteki* in shape and coloring, but it is smaller, and the apertural teeth distinguish it. The shell is dark gray to black, obscurely banded with white, the nodules black; within, the aperture is bluish to brownish. Length, 20 mm; diameter, 10 mm. Bahía Magdalena, west coast of Baja California, through the Gulf, and south along the Sororan coast of Mexico at least as far as Guaymas, intertidally under rocks; fairly common.

Family Buccinidae (whelks)

The buccinids are mostly strongly sculptured, with a somewhat tapering spire, fairly large body whorl, and short canal. They are carnivorous, but are more scavengers than they are active predators, content to feast on dead fish and other scraps. The operculum is ovate, thin, and horny, its nucleus apical.

Cantharus gemmatus (Reeve)

A brownish shell under an olive-green periostracum, the spiral ribs touched with dark brown at their nodose junctions with the axial ribs. The aperture is white. About 35 mm in length. Mazatlán, Mexico, to Ecuador, on rocks between tides; not uncommon.

Fig. 6.44
Thais biserialis

Fig. 6.45
Acanthina angelica

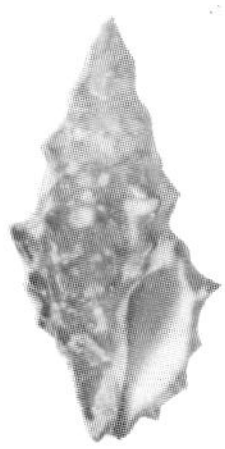

Fig. 6.46
Morula ferruginosa

Fig. 6.47
Cantharus gemmatus

Family Columbellidae (dove shells)

Dove shells are mostly small, though rarely microscopic in size, with boldly marked, smooth, or ribbed shells. The animals are not as timid as many molluscs, and may be seen plowing about on the floor of a tide pool at low tide — especially in the tropics — searching for the minute organisms on which they feed. Some have developed the picturesque habit of spinning an elastic thread of mucus, attached to a bit of seaweed or even to the surface film of a quiet tide pool, from which the little owner of the shell dangles and swings. As the animals live in abundance in the intertidal area, the shells may be cast ashore by the tides. They thus form a considerable part of the fine beach drift on the shores of the world, for these are adaptable molluscs. Not a few species, or even genera, have moved out of the tropics into temperate or boreal seas.

Columbella fuscata (Sowerby)

The chestnut-brown shell is dotted and irregularly spotted with white. Just below the suture is a band of triangular white markings. The shell is covered with a thin, smooth, light olive periostracum, and the aperture is lavender-colored. Length, 18 mm; diameter, 11 mm. Bahía Magdalena, west coast of Baja California, throughout the Gulf, and south to Peru, under rocks between tides; common.

Fig. 6.48 *Columbella fuscata*

Fig. 6.50 *Anachis coronata*

Fig. 6.49 *Columbella strombiformis*

Fig. 6.51 *Anachis sanfelipensis*

Fig. 6.52 *Anachis varia*

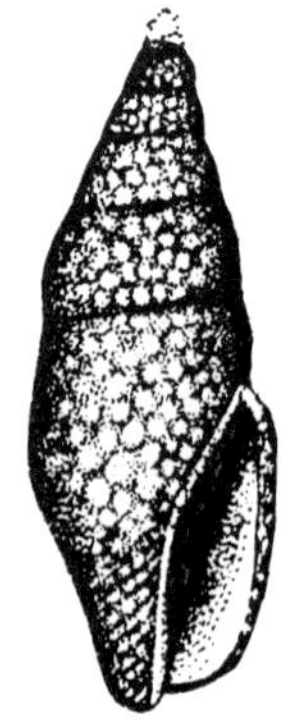

Fig. 6.53 *Mitrella guttata*

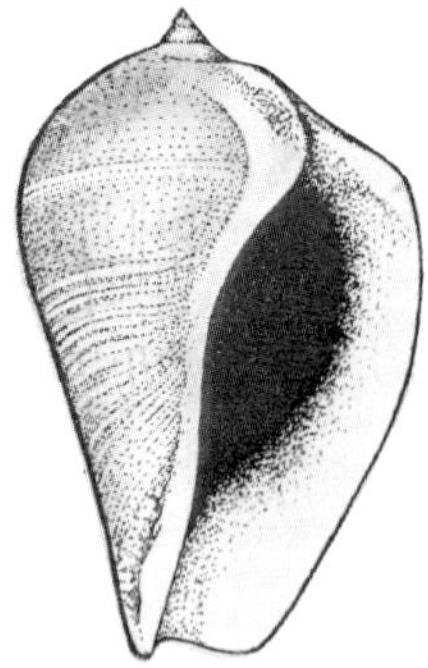

Fig. 6.54 *Melongena patula*

Columbella strombiformis (Lamarck)

Largest of the west American pyrenes, this has a swollen body whorl, with a ridge at the upper part that forms a sort of spout at the aperture. Under a shaggy olive periostracum, the shell is dark reddish brown, spotted and marked with zigzag lines and triangular blotches of white. The aperture is white or tinted with orange. About 30 mm in length. The central part of the Gulf of California, south to Peru, under rocks between tides; not uncommon.

Anachis coronata (Sowerby)

The shell is yellowish white, mottled or striped with brown. Length, 14 mm; diameter, 6 mm. The outer coast of Baja California through the Gulf and south to Panamá; not uncommon.

Anachis sanfelipensis Lowe

A flesh-colored shell, with blotches of light brown between the axial ribs, and wavy axial lines of brown on the base. Length, 17 mm; diameter, 6.5 mm. The species is restricted in range to the northern end of the Gulf of California, in the lower tidal zone.

Anachis varia (Sowerby)

This brown shell has a white band at the suture and on the periphery. About 19 mm in length and 9 mm in diameter. Sonora, Mexico, to Panamá; not uncommon.

Mitrella guttata (Gmelin)

The shells are dark brown, with rows and bands of circular light spots. The tip or apex of the shell is broken off in most specimens. About 14 mm long and 4 mm wide. Bahía Magdalena, west coast of Baja California, through the Gulf, and south to Panamá, under stones between tides; common.

Family Melongenidae (conchs)

Melongena patula (Broderip and Sowerby)

The bright brown shell is banded with pale yellow or white. The aperture may be yellowish to pinkish. Young shells are more spindle-shaped, but the smooth columella gives the clue that they are not fasciolarids or fusinids. A large adult shell may be as much as 10 inches long, but most are about 6 to 7. The northern part of the Gulf of California to Panamá; fairly common on sand and mud flats.

Family Nassariidae (basket whelks)

The nassas or basket shells are mostly scavengers, greatly attracted by the smell of decaying flesh. Plowing along the sea floor or a tidal mud flat on the oddly divided foot, they leave a characteristic track that may lead to a small mound of sand under which the snail is buried. The Nassariidae are at their best in subtropical to tropical waters, and the variations they exhibit have been the despair of many a student.

Nassarius iodes (**Dall**)

Callus well-developed, white; shell light colored, with a violet-brown band in the middle of the whorl, and feebly sculptured. Length, 8 mm; diameter, 5 mm. Gulf of California, especially the northern part; common.

Nassarius moestus (**Hinds**)

Callus thin, margins white; last whorl smooth near aperture. Length, 10 mm; diameter, 6 mm. Northern end of the Gulf of California to Panamá; not uncommon.

Nassarius tiarula (**Kiener**)

A buff to brown shell, with a fillet of dark brown just below or between the nodes of the shoulder. Length, 16 mm; diameter, 10 mm. Throughout the entire Gulf of California, and south to Panamá; fairly common.

Family Olividae (olives)

Like the cowries, the olives are brilliantly polished shells, and for the same reason — two wide lobes of the soft parts fold back over the shell when the animal is moving about, and the living tissue protects the delicate glazed surface of the shell from being dulled. The mantle also has an unusual threadlike extension that lies along the suture, the function of which is unknown. Most olives like sandy shores, where they plow about just beneath the surface at low tide, leaving a characteristic trail.

Oliva spicata (**Röding**)

Most variable of the west American olives, this one is basically a buff-olive color, flecked with small, irregular dots of brown. The spire is fairly high. About 45 mm in length, 23 mm in diameter. The range is from the northern end of the Gulf of California to Panamá; common.

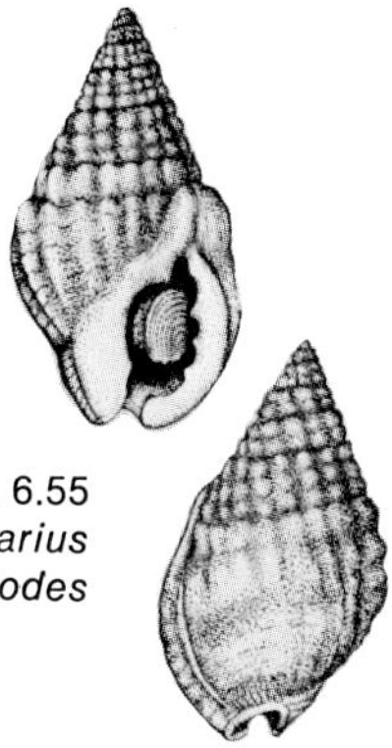

Fig. 6.55 *Nassarius iodes*

Fig. 6.56 *Nassarius moestus*

Fig. 6.57 *Nassarius tiarula*

Fig. 6.58 *Oliva spicata*

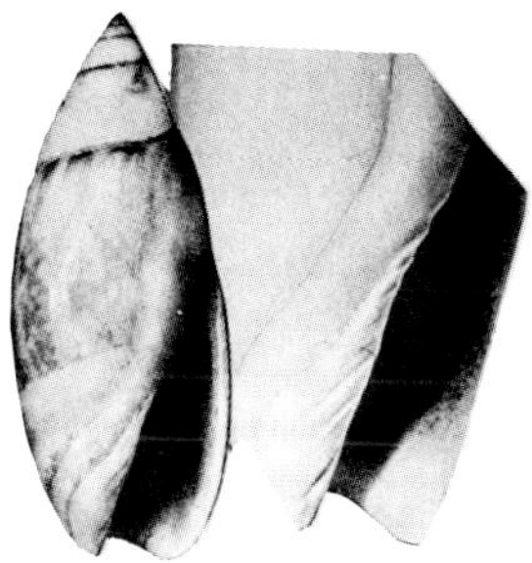

Fig. 6.59 *Olivella dama*

Olivella dama (Wood *ex* Mawe MS)

This is the type species of the genus. It is a stout little shell, white, variably marked with faint zigzags of brownish or grayish color, the apex of the spire and the aperture being violet. About 20 mm in length and 9 mm in diameter.

Family Conidae (cones)

The shell of a typical cone is broad at the top of the body whorl, and tapers gradually to a narrow base. The spire is steplike, or turreted in some forms. Ornamentations with blunt nodes at the shoulder, known as coronations, are frequent, or there may be beads. The aperture is long, narrow, and mostly of uniform width. The outer lip is thin and sharp, and is very easily broken or chipped, as the collector soon learns and as is attested by the frequency with which marked growth flaws occur.

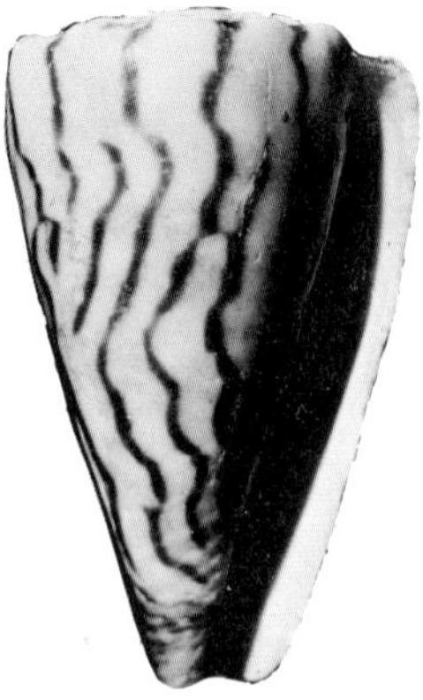

Fig. 6.60
Conus princeps

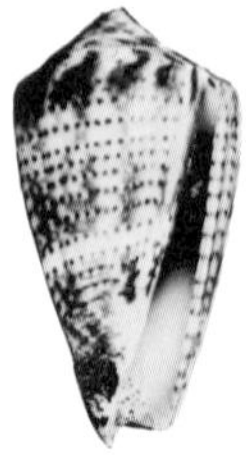

Fig. 6.61
Conus perplexus

Conus princeps Linnaeus

A striking and easily distinguished shell, the princely cone is of an orange or pink color with dark brown, waxy axial stripes that extend up onto the low and well-coronated spire. The aperture is the color of the outside, but without the stripes. Also distinctive is the periostracum, which is heavy and dark with wide-spaced bristly spiral lines. Length, 55 mm; diameter, 33 mm. The head of the Gulf of California to Ecuador, intertidally on rocky ledges; not uncommon.

Conus perplexus Sowerby

The ground color of the puzzling cone is grayish white to pale violet, covered with numerous spiral rows of brown dashes or dots, some of which may be raised into pustules or ridges, especially near the base of the shell. Axial stripes or blotches may appear on spire and body whorl. The aperture is pale bluish violet, deeper within, the periostracum thin, smooth, and light-colored. Length, 26 mm; diameter, 16 mm. Bahía Magdalena, Baja California, throughout the Gulf of California and south to Ecuador, mostly on sandbars but also recorded offshore to depths of 20 fathoms; common.

Family Turridae

The Turridae are at their best in the Panamic province and comprise more named (and unnamed) species than any other family here covered. The hallmark of the family is the anal notch in the posterior margin of the outer lip. The shells may be of various proportions — mostly fusiform in shape — and of various textures and colors.

Crassispira pluto Pilsbry and Lowe

A slender, black shell with a faint yellow band above the shoulder on some specimens, and nodose sculpture at the intersection of the spaced spiral and axial ribs. Length, 19 mm; diameter, 7 mm. Limited to the Gulf of California, where it is fairly common.

Fig. 6.62 *Crassispira pluto*

SUBCLASS OPISTHOBRANCHIATA (No Key)

Order Cephalaspida

Family Aglajidae

Shell internal, forming a flattened spiral coil.

Navanax inermis (Cooper)

Animal dark brown to black, variously spotted and striped. Length, 75 to 225 mm. Diameter of shell about 20 mm, nearly the length of the head shield. Monterey Bay, California, to the head of the Gulf of California.

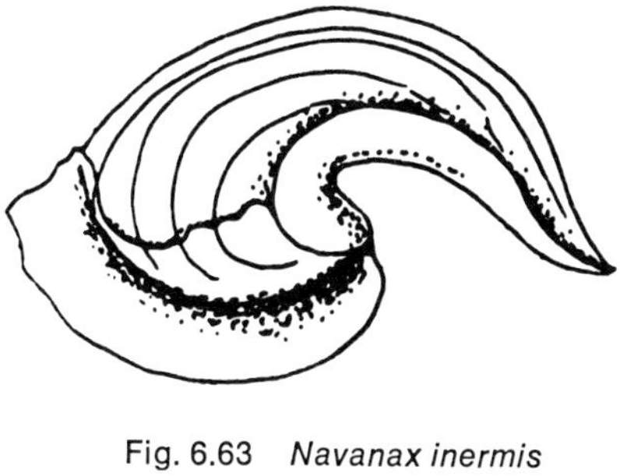

Fig. 6.63 *Navanax inermis*

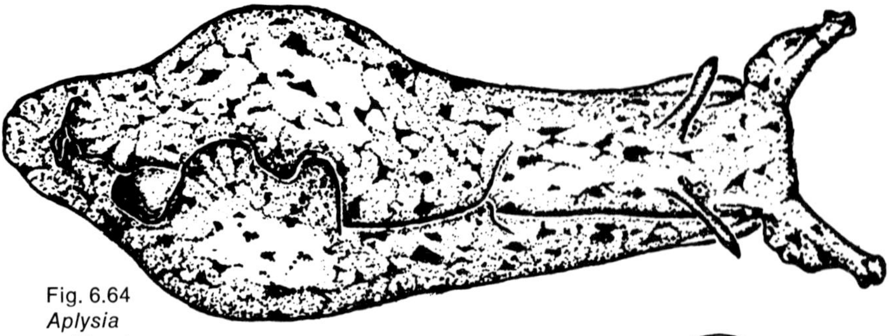

Fig. 6.64
Aplysia californica

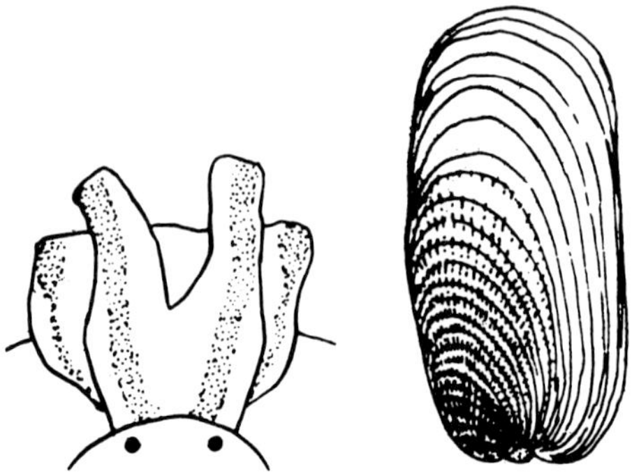

Fig. 6.65
Berthellina ilisima

Order Anaspida

Family Aplysiidae

Popularly called "sea hares," perhaps because the two rhinophores look a little like ears and the animal resembles a crouching rabbit in shape. The shell is a thin, membranous to calcareous, asymmetrical plate, entirely internal. The animals are variously colored or mottled, large (one in California measures up to 30 inches in length and weighs up to 35 pounds), and of herbivorous habits, grazing on seaweeds.

Aplysia californica **Cooper**

Greenish gray with dark streaks and spots and white skin glands. Length, about 130 mm. Humboldt Bay, California, to the northern end of the Gulf of California.

Order Notaspidea

Family Pleurobranchidae

Shell partially covered by the mantle, ear-shaped, thin; shell rarely absent.

Berthellina ilisima Marcus and Marcus

Apricot to orange colored animals; shell flat, light brown, with an iridescent periostracum, the larval coil a white knob; sculpture of coarse growth lines and radiating rows of dots. Length of animal, 60 mm; of shell, 4.5 mm. Throughout the Gulf of California (type locality, Puerto Peñasco, Sonora).

Order Sacoglossa

Family Elysiidae

Body flattened, elongate, with lateral leaflike expansions; foot narrow, tentacles large, sometimes short.

Tridachiella diomedea (Bergh)

Pale yellow to translucent gray, with black velvety triangular spots along the parapodia, which extend the length of the animal and are thrown into six folds. Length to 35 mm. Throughout the Gulf of California and south to Panamá.

Order Nudibranchia

Family Dorididae

Rhinophores retractable into pits; dorsal surface warty, stiffened by spicules in some.

Doris pickensi Marcus and Marcus

White, with a translucent notal border and numerous low, opaque white papillae on the back, the rhinophores, gills, and frontal foot margins yellowish. Length, 25 mm. Puerto Lobos north to Puerto Peñasco, under boulders intertidally.

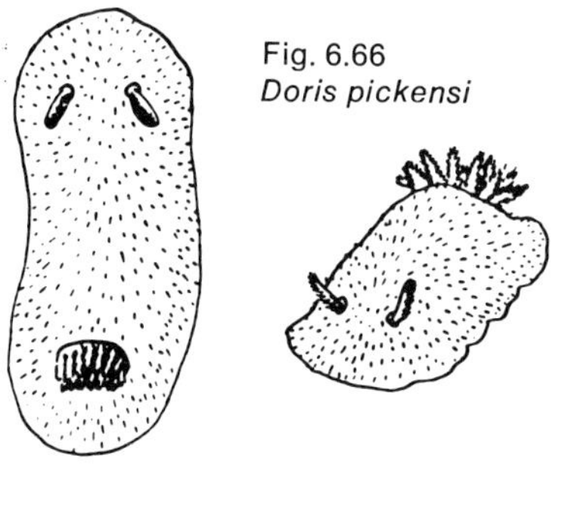

Fig. 6.66
Doris pickensi

Family Chromodorididae

Rudula often with a rudimentary central plate, the lateral teeth short, cuspate.

Chromodoris banksi sonora Marcus and Marcus

The body is white, edged with a thin translucent line and a bright orange line. The back is cream color, interspaced with black to brown dots; first lateral tooth is dentate. Length, 25 mm; width, 11 mm. Puerto Peñasco.

Chromodoris norrisi Farmer

Body white, with dashes of bright orange around the back margins; rhinophores and gills orange-tipped. Over the back are intermingling bright yellow and deep violet dots in alternating patterns. Length, 61 mm; width, 32 mm. Pacific coast of Baja California, through the Gulf of California to Puerto Peñasco.

Chromodoris sedna (Marcus and Marcus)

Body pure white, with three distinct color bands bordering the notum and posterior half of the foot, the inner one opaque white, the middle one vivid orange-red, and the outer one bright yellow. Upper portions of rhinophores and gills deep red. Length, 47 mm; width, 11 mm. Throughout the Gulf of California, south to Santa Cruz, Nayarit, Mexico.

Hypselodoris californiensis (Bergh)

Elongate, the body dark ultramarine blue, with spots or streaks of golden yellow, the back and foot margins pale blue to white. Length, 71 mm; width, 11 mm. Monterey Bay, California, southward along the Baja California coast and throughout the Gulf of California, south to Jalisco, Mexico.

Order Gymnophila

Family Onchidiidae

Animals are oval in outline. The retractible tentacles have eyes at their tips. These molluscs do not necessarily hide under rocks when exposed to the sun.

Onchidella binneyi Stearns

This dark gray slug is the common species of the Gulf of California, occurring in the intertidal zone. In the ventral view, one sees the sole of the foot as being about one-third the width of the animal. Length to 30 mm. Bahía las Cruces, Gulf of California, to Puerto Peñasco.

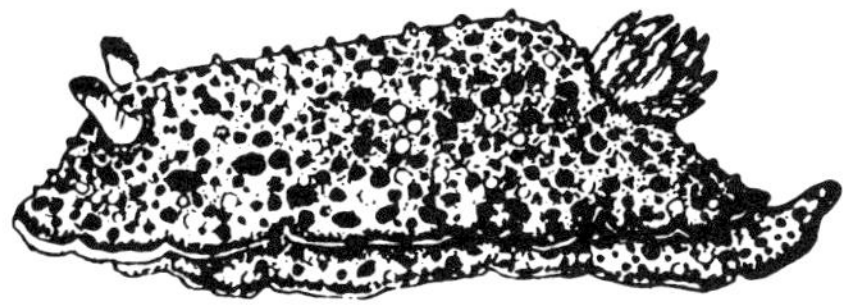

Fig. 6.67
Chromodoris banksi

Fig. 6.68
Hypselodoris californiensis

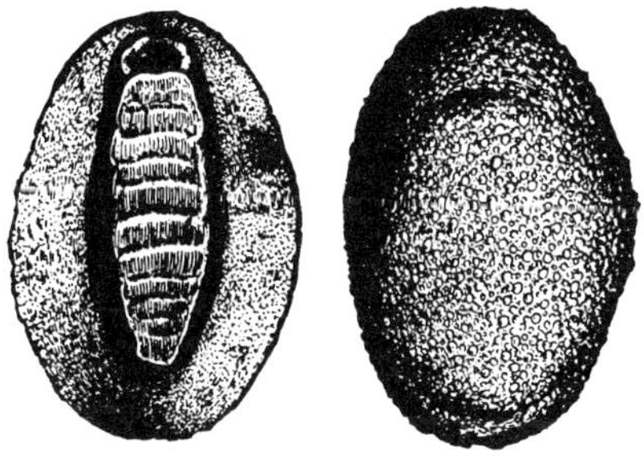

Fig. 6.69
Onchidella binneyi

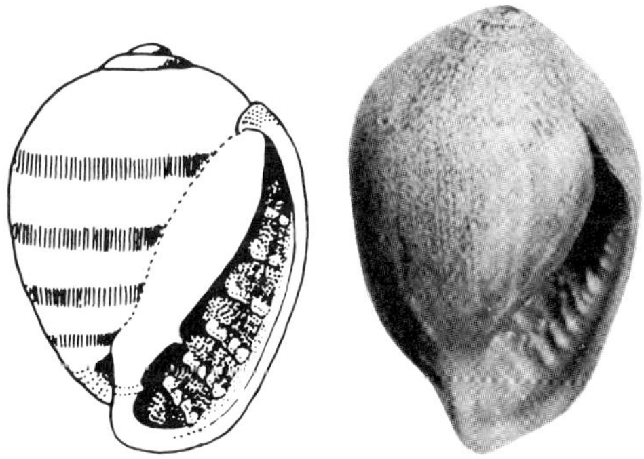

Fig. 6.70
Melampus olivaceus

SUBCLASS PULMONATA (No Key)

Order Basommatophora

Family Melampidae

Shells ovate to conical, the spire with several whorls; not markedly thin; aperture with folds or denticles on one or both lips.

Melampus olivaceus **Carpenter**

The shell is purplish brown, banded with whitish under an olive-green periostracum. The outer lip is sharp, with white ridges within that are toothed at the ends in adults. Length, 16.5 mm; diameter, 9.7 mm. Southern California to Mazatlán, in bays at high tide line, among matted salt grass and drift; common.

Marinula rhoadsi Pilsbry

Pale yellowish, the shell has a dark band at the shoulder, and is suffused with brown below the band. Length, 10 mm; diameter, 5.7 mm. From near the head of the Gulf of California to as far south as Puerto Peñasco.

Family Siphonariidae (false limpets)

Cap-shaped, rather irregularly sculptured shells, mostly dark in color, with a siphonal groove on the right side. Muscle scar continuous at the front, but open at the right. Air breathing inhabitants of the intertidal zone, mainly tropical in distribution.

Siphonaria maura Sowerby

This is a protean form, varying from low- to high-conic, from coarsely to finely ribbed, from uniformly colored to spotted or rayed with white. Length, 22 mm; width, 17.5 mm; height, 7 mm. The range of all three forms is from the Gulf of California to Ecuador, intertidally; fairly common.

CLASS POLYPLACOPHORA (No Key)

Order Chitonida

Family Chitonidae

Insertion teeth deeply cut, or fluted on the outer and bottom surfaces.

Chiton stokesii Broderip

The broadly ovate, blackish brown shell is marbled with white toward the center, its principal distinction being the netlike sculpture of the lateral areas, the cords of the net being irregularly beaded; the surface is usually worn or encrusted, or encrusted in adult specimens. Inside, the shell is smooth, from bluish white to light blue. Length, 80 to 90 mm. Throughout the Gulf to Ecuador, on the upper sides of exposed rocks; not uncommon.

Chiton virgulatus Sowerby

Olive brown, the thin, oblong shell is greenish toward the center, dark bluish green within. It resembles *C. stokesii,* but is narrower, has finer and smoother girdle scales, more numerous radiating riblets, which have a tendency to divide at the ends, and more slits in the insertion plates.

In contrast to *C. stokesii,* the surface is clean at all stages in most specimens. Length, 68 mm; width, 37 mm. Bahía Magdalena, west coast of Baja California, to the northern part of the Gulf of California, under stones; possibly the most common intertidal chiton in the Gulf.

Fig. 6.71 *Marinula rhoadsi*

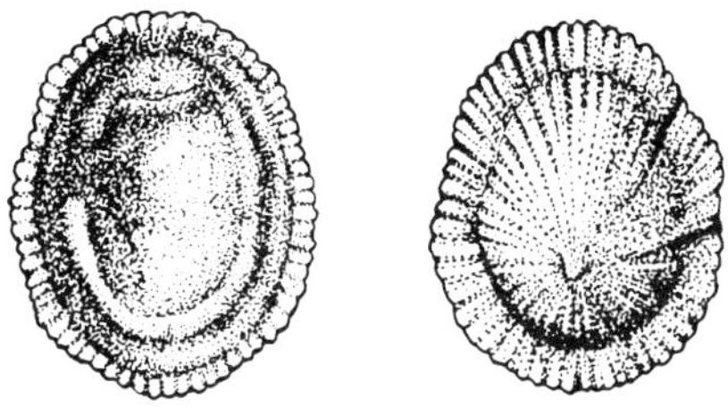

Fig. 6.72 *Siphonaria maura*

Fig. 6.73 *Chiton stokesii*

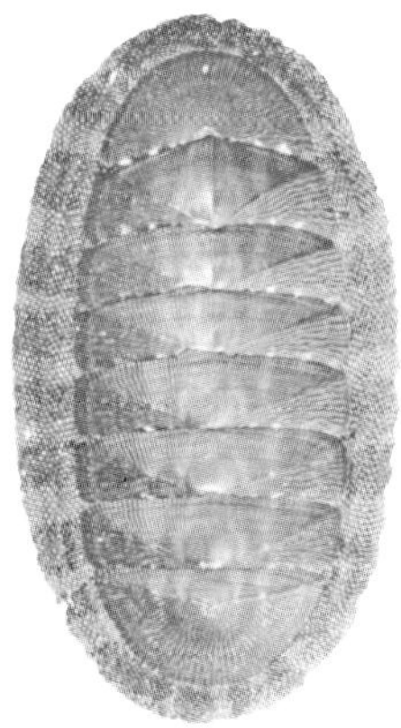

Fig. 6.74 *Chiton virgulatus*

Family Acanthochitonidae

Valves immersed in a smooth or hairy girdle; plates well-divided into areas, the insertion plates sharp and nearly smooth. In most forms, the head valve has five slits.

Acanthochitona exquisita (Pilsbry)

Only a narrow portion of the dark olive valves show, most of the shell being covered with large tufts of green, pink, or bronzy spicules. Inside, the shell is blue. A perfect specimen is an object of great beauty. About 25 mm in length, 15 mm in width. Confined to the Gulf of California, mostly near Bahía Las Animas and La Paz, in sandy mud under stones.

Fig. 6.75 *Acanthochitona exquisita*

Family Ischnochitonidae

All valves with insertion plates, the teeth smooth or nearly so, the posterior valve with a crescentic series of well-developed teeth. This is the largest family of the order, with the most species and the most generic divisions.

Stenoplax conspicua sonorana Berry

This resembles the southern California *S. conspicua,* of which it is considered a subspecies, but is smaller and longer, the valves more sharply ridged, the sculpture coarser. The color is greenish, more rarely brownish, the interior pink with a blue spot at the sinus. The eaves are wide, solid, and a light bluish green. The number of slits are: 13 in the anterior valve, two to four in the medial valves, and 12 in the posterior. The girdle has a plushlike surface of stiff, curved spinelets. Length, 55 mm; width, 21 mm; height, 6.4 mm. The subspecies is restricted to the northern end of the Gulf of California, on reefs and in tide pools; not uncommon.

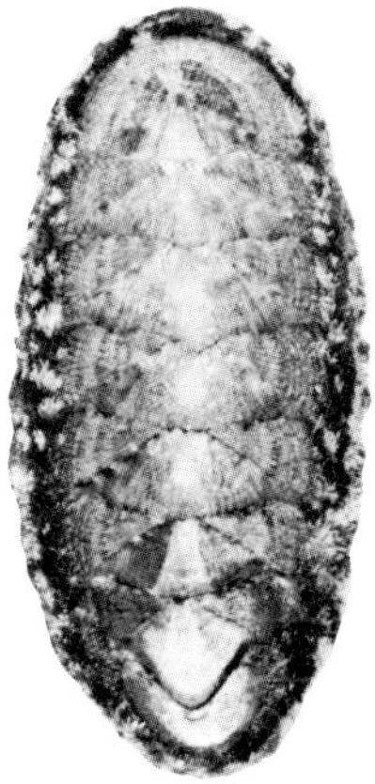

Fig. 6.76
Stenoplax conspicua

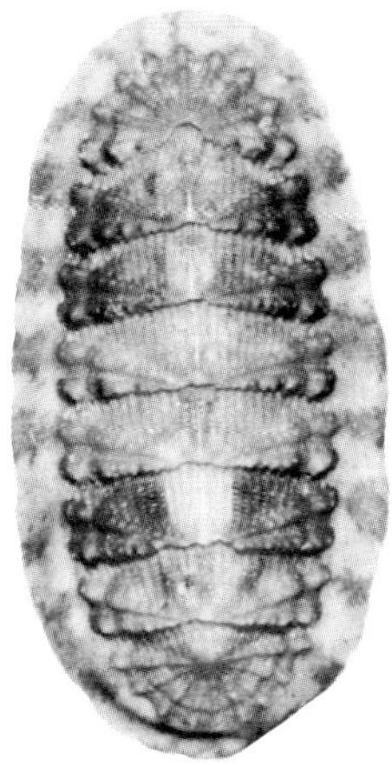

Fig. 6.77
Callistochiton gabbi

Callistochiton gabbi Pilsbry

With a lusterless surface, the shell has light brown central areas and olive brown lateral areas; the interior is light olive with dark markings under the beaks. The sutural plates are well-developed; the anterior valve has nine slits, the posterior valve eight. Length, 14 mm; width, 8.5 mm. The northern part of the Gulf of California to Panamá, under stones; not uncommon.

CLASS SCAPHOPODA (No Key)

Family Dentaliidae

Shell with the greatest diameter at the oral aperture; surface smooth or sculptured; radula with central tooth twice as wide as high; foot of animal conical, not terminating in an expandable disk.

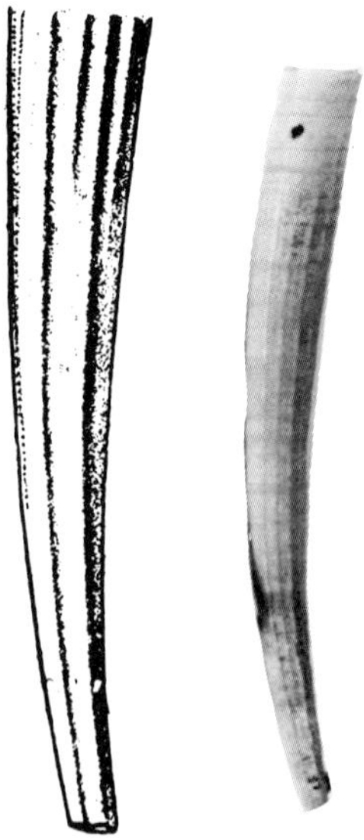

Fig. 6.78
Dentalium oerstedii

Fig. 6.79
Fustiaria splendida

Dentalium oerstedii Morch

Length, 40 mm; diameter, at aperture, 3.5 mm. Puerto Peñasco to Santa Elena Bay, Ecuador, in 4 to 145 m.

Fustiaria splendida (Sowerby)

Length, 45 mm; diameter at aperture, 3.8 mm. Puerto Peñasco to La Libertad, Ecuador, in 2 to 100 m.

CLASS CEPHALOPODA

SUBCLASS COLEOIDEA

Order Octopoda

Family Octopodidae

An ink sac present; bottom-dwelling to bathypelagic forms.

Octopus digueti Perrier and Rochebrune

Northern end of the Gulf of California to Mazatlán, living inside empty shells.

Octopus fitchi Berry

San Felipe to Guaymas. This species has been recorded by Berry and Halstead (1954) as one that will bite readily on being handled, and the saliva may be dangerously toxic.

Arthropoda (Crustaceans)

CLASS CRUSTACEA

General Preservation: Most crustaceans can autotomize their legs, but rarely do so if handled gently. However, if too many crabs or other large crustaceans are put into the same container for any length of time, they tend to fight and will break off appendages (as well as destroy other fragile specimens near them). Most marine crustaceans should be killed by slowly adding fresh water to their sea water container, or simply by allowing the sea water to become putrid. Epsom salts (magnesium sulfate) or menthol crystals may be used to relax the animals if the above methods are undesirable. Knudsen describes various methods of preparing specimens for display and study. Crustaceans always should be preserved in 70 to 90 percent alcohol, and color notes should always be made prior to preservation.

General Taxonomy: The Crustacea are a very large class of arthropods (over 25,000 species) with a traumatic taxonomic life of constant name changings and taxa revisions. Due to its large number of species, many taxonomic categories are used to delimit phyllogenetic affinities within the Crustacea (such as super-, sub-, and infraorder; section; tribe; etc.). The following taxonomic scheme is derived largely from Schmitt *(Crustaceans,* 1965) and Barnes *(Invertebrate Zoology,* 1969), and is followed throughout the sections dealing with the Crustacea. Taxa treated in this handbook are marked with an asterisk.

Class Crustacea
Subclass Cephalocarida
Subclass Branchiopoda
Subclass Ostracoda
Subclass Mystacocarida
Subclass Copepoda
Subclass Branchiura
Subclass Cirripedia
*Order Thoracica
Order Acrothoracia
Order Ascothoracia
Order Apoda
Order Rhizocephala
Subclass Malacostraca
Superorder Leptostraca
Superorder Syncarida
Superorder Pancarida
Superorder Peracarida
Order Spelaeeogriphacea
Order Mysidacea
Order Cumacea
Order Tanaidacea
*Order Amphipoda
*Order Isopoda
Superorder Hoplocarida
*Order Stomatopoda
Superorder Eucarida
Order Euphausiacea
Order Decapoda
Suborder Natantia
*Section Penaeidea
*Section Caridea
Section Stenopodidae
Suborder Reptantia
*Section Macrura
Section Anomura
*Superfamily Thalassinoidea
*Superfamily Galatheoidea
*Superfamily Paguroidea
*Superfamily Hippoidea
*Section Brachyura

Specific taxonomy of the various groups of Crustacea are discussed in their respective sections.

Since much of the literature includes information about the Macrura, Anomura, and Brachyura within single papers, cross-checking literature references for these three groups is recommended.

Crustaceans of the Gulf: Crustaceans are well-represented in the Gulf of California, as they are along almost any seashore of the world. The most conspicuous members of this group, in the intertidal zone, are the barnacles (Cirripedia) and crabs (Brachyura and Anomura). There are many species of amphipods and isopods inhabiting the Gulf littoral, but they are less frequently encountered by the general observer due to their small size. Collecting these small creatures usually requires meticulous examination of sand and mud samples, algae beds, and other microhabitats. The true shrimp (Natantia) are common and abundant, but difficult to catch, resulting in few good, available collections. The mantis shrimp (Stomatopoda) are rarely encountered intertidally, and only eight species are reported as occurring in Gulf waters by W. L. Schmitt (*The Stomatopods of the West Coast of America,* 1940). These curious creatures are most easily obtained by sampling beyond the low tide level with trawls and dredges.

Other Crustacea from the Gulf intertidal, rarely encountered, include: the sand or stone lobsters (Scyllaridae), such as *Evibacus princeps,* which live in the subtidal but may occasionally wash ashore after a bad storm; the ostracods, mysids, copepods, and euphasids, which are largely planktonic and must be captured with a fine mesh net; and the cumaceans, those small, odd crustaceans that live buried in the sandy-mud or mud substrate, usually of estuaries and bays. One occasionally finds a mole crab or sand crab (Hippidea) buried in the sand above or near the water line. These little anomuran crabs are filter feeders, lying just beneath the sand surface and extending their densely fringed antennae into the water as the waves alternately cover and expose them. A pelagic galatheid crab not treated in the keys is *Pleuroncodes planipes* Stimpson, occasionally referred to as the pelagic rock lobster. It has been reported washed ashore in vast numbers throughout its range: Monterey, California (1859, 1960); Isla Espíritu Santo (1940); La Paz (1940); Punta Belcher (1964); and Bahía de los Angeles (no date). I also have found it washed ashore at Bahía Concepción (1971). In the open sea these creatures probably serve as an important food source for turtles and fish. The range of *P. planipes* is given by J. Haig et al. (*The Porcellanidae of the Eastern Pacific,* 1970), as Baja California and the northern Gulf to about 250 km south of Cabo San Lucas, its occurrence on Californian shores being an anomaly. The galatheids are often called "scarlet lobsterettes," a very descriptive name.

KEY TO THE MAJOR GROUPS OF INTERTIDAL CRUSTACEA OF THE GULF OF CALIFORNIA

1 Animal volcano-shaped; enclosed in a series of 4 or 6 heavy, calcareous plates; attached to substrate by a broad base (the barnacles) (figs. 7.1 to 7.3) .... **subclass Cirripedia**

-Animal not volcano-shaped; exoskeleton soft to hard, but not encased in plates as above; free-living, not attached to substrate (crabs, shrimp, etc.) .... **subclass Malacostraca 2**

2(1) Body sticklike and greatly reduced (abdomen vestigial); body usually less than 20 mm long (Caprellids) (fig. 7.4) .... **order Amphipoda (suborder Caprellidea)**

-Body not sticklike and greatly reduced (abdomen not vestigial); body often greater than 20 mm long **3**

3(2) Carapace lacking 4

-Carapace present 5

4(3) Animal compressed laterally (figs. 7.5 to 7.16) ...... **order Amphipoda**

-Animal compressed dorso-ventrally (figs. 7.17 to 7.21) .... **order Isopoda**

5(3) Carapace short, not covering the last 3 to 4 thoracic segments; second maxillipeds modified to form large, mantislike, raptorial claws (the mantis shrimps) (fig. 7.22) .... **order Stomatopoda**

-Carapace completely covers thorax; maxillipeds not as above 6

6(5) Body more or less laterally compressed; first abdominal segment equal to or larger than more posterior segments; pleopods always present and well-developed, used for swimming (shrimp) (figs. 7.23 to 7.29) .... **suborder Natantia**

-Body not laterally compressed, usually dorso-ventrally compressed; first abdominal segment distinctly smaller than more posterior segments; pleopods absent, or if present reduced and not adapted for swimming (crabs, lobsters, and burrowing or mud shrimp) .... **suborder Reptantia** 7

7(6) Abdomen extended and well-developed; large uropods and telson 11

-Abdomen folded under thorax, or reduced and asymmetrical; uropods and telson absent or reduced 8

8(7) Abdomen reduced and tightly flexed beneath thorax; fifth legs not greatly reduced, plainly visible; eyes usually lateral to antennae; carapace fused with epistome (true crabs) (figs. 7.32 to 7.77) .... **section Brachyura**

-Abdomen reduced and folded under thorax, or abdomen asymmetrical, soft and twisted (as in hermit crabs); fifth legs reduced and turned upward, often appearing absent; eyes median to antennae; carapace not fused with epistome (hermit crabs, sand crabs, and porcelain crabs) .... **section Anomura 9**

9(8) Animal occupying a mollusc shell (usually a gastropod shell); abdomen soft, without chitinous exoskeleton; abdomen extended; abdomen asymmetrical (hermit crabs) (fig. 7.91) ... **superfamily Paguroidea**

-Animal not occupying a mollusc shell; abdomen hard, with chitinous exoskeleton; abdomen folded under thorax; abdomen symmetrical 10

10(9) First legs with pincers (chelipeds); body usually crablike; carapace rarely inflated, or highly convex; rostrum often present; rarely found burrowing in sand (porcelain crabs) (figs. 7.78 to 7.90) .... **superfamily Galatheidea**

-First legs without pincers; body not crablike; carapace inflated and highly convex; rostrum never present; found burrowing in sand, often high up on beach, in splash zone (mole or sand crabs) (figs. 7.92 to 7.93) .... **superfamily Hippidea**

11(7) First pair of legs chelate (ghost or mud shrimp) (fig. 7.31) .... **superfamily Thalassinidea**

-First pair of legs not chelate (lobsters) (fig. 7.30) ... **section Macrura**

SUBCLASS CIRRIPEDIA (Barnacles)

Order Thoracica

Preservation: Barnacles may be preserved in buffered 10 percent formalin or 70 percent alcohol. Because of their calcareous nature, alcohol may be preferable; however, alcohol will bleach the colorful plates much more than formalin will. Some collectors prefer to preserve the animals in one of the above for a day or two, and then allow them to air dry. To preserve specimens with the cirri extended, it is necessary to narcotize expanded individuals with chloretone or menthol.

Taxonomy: Most of the New World barnacles were described in Darwin's (1854) classic monograph on the Cirripedia. Since it was written, studies on American barnacles have extended ranges, established numerous subspecies, and isolated new taxonomic characters (largely by use of more powerful microscopes). The taxonomy of this group is based principally on the number and morphology of the calcareous plates, the morphology of the mouth parts, and the mode of attachment to the substrate. The following briefly describes structures a student may need to recognize when using the key to the barnacles.

The armor (shell) of barnacles consists of calcareous *plates* connected by chitinous intervals. Each plate has three parts: a median triangular area, the *parietes;* and two lateral wings, called *radii* when they overlap adjacent plates and *alae* when they are overlapped by adjacent plates. The membranous or calcareous plate upon which the barnacle rests is the *basis*. Barnacles primitively have eight plates. The one farthest from the head is the keel-like *carina* (usually the tallest plate); the one opposite the carina is the *rostrum;* the two *carinolaterals* are on either side of the carina; two *rostrolaterals* are on either side of the rostrum; and last are the middle compartments, referred to as the *laterals*. The sessile barnacles found in the Gulf have a shell wall composed of either six or four plates, depending on the degree of fusion or plate loss. In the genus *Balanus* the rostrolaterals are fused with the rostrum. In the genus *Chthamalus* the carinolaterals are lacking altogether, giving individuals of these two genera only six plates. Further, in *Chthamalus* both end plates of the shell wall (rostrum and carina) are overlapped by the adjacent compartmental plates, while in *Balanus* one of the end plates (rostrum) overlaps its adjacent plates. In the genus *Tetraclita* the rostrolaterals are fused with the rostrum and the carinolaterals are absent, giving it four plates.

Barnacles of the Gulf: The vast majority of the barnacles of the Gulf of California are sessile and belong to the order Thoracica, suborder Balanomorpha. These are the only barnacles treated in the following key, although a piece of wood or pumice will occasionally wash ashore covered with goose barnacles *(Lepas).*

Steinbeck and Ricketts reported 12 species of barnacles from the Gulf and concluded that the majority of the Gulf fauna in this class was of Panamic origin, with a few unique species and a few species showing affinities with the Californian Province. Combining the range extensions of Steinbeck and Ricketts with the geographic distributions given by Pilsbry (1916), who reported four species from the Gulf, we find that approximately 44 percent of the Gulf species (or subspecies) may be associated with the Panamic fauna, 22 percent show affinities to the Californian fauna, and 33 percent are apparently unique to the Gulf. These figures are approximations, since many of the barnacles are nearly cosmopolitan, being carried freely on ship hulls to ports around the world, thereby making zoogeographic studies of this group extremely difficult. There are, in fact, many subspecies known only from ship bottoms!

Dr. Arnold Ross of the San Diego Museum of Natural History has informed me (personal correspondence) that the Gulf of California barnacle fauna isn't well known yet, and cannot be properly investigated until the Pacific and Caribbean fauna is better known. He states that there are probably six or eight species of *Tetraclita* in the Pacific and Caribbean regions that are presently considered subspecies of *Tetraclita squamosa* (which is Indo-Pacific, not Eastern Pacific).

Barnacles are found attached to rocks, wharf pilings, crustacean exoskeletons, and other hard surfaces, from the splash zone (*Tetraclita squamosa* and *Chthamalus fissus*) to the midintertidal (most *Balanus* and *Chthamalus*), but rarely in the low intertidal.

KEY TO THE COMMON INTERTIDAL BARNACLES OF THE GULF OF CALIFORNIA

1 Individual plates appear fused and indistinguishable externally; plates with deep, vertical striations; shell composed of 4 plates (visible internally) (fig. 7.2) .... ***Tetraclita squamosa***

-Individual plates visible externally; plates smooth, or with weak striations; shell composed of 6 plates 2

2(1) Shell red, purple, or blue, with vertical stripes 3

-Shell white or dull gray, without colorful stripes 5

3(2) Orifice triangular, with 3 well-defined sides .... ***Balanus trigonus***

-Orifice not triangular, but round, oval, or teardrop-shaped 4

4(3) Each plate with 3 visible sections, a median and 2 lateral (the parietes and alae) (fig. 7.1) .... ***Balanus tintinnabulum***

-Each plate with only the median and one lateral triangle visible, the lateral alae being hidden under the adjacent plates .... ***Balanus amphitrite***

5(2) Both end plates of shell wall (rostrum and carina) overlapped at sides by adjacent plates; calcareous walls not porous; opercular plates darker than wall 6

-Only one end plate of shell wall (carina) overlapped by adjacent plates; calcareous walls porous; opercular plates usually same color as wall 7

6(5) Base star-shaped; plates ivory white (although they may be corroded or encrusted); shape depressed conic ... ***Chthamalus anisopoma***
-Base rounded; plates white to dingy gray; shape convex conic or irregular (fig. 7.3) .... ***Chthamalus fissus***
7(5) Orifice triangular, with distinct angles .... ***Balanus trigonus***
-Orifice rounded or oval, not triangular .... ***Balanus concavus***

SYSTEMATICS AND SPECIES DESCRIPTIONS

Cirripedia

Suborder Balanomorpha

Family Balanidae

Balanus tintinnabulum (Linnaeus)

This species of sessile barnacle is one of the best known in the world. It has been found in nearly every ocean port, and has been studied and described numerous times in the last two to three centuries. This barnacle is, as are most barnacles, freely transported around the globe on ship bottoms, making zoogeographical studies nearly impossible. A numerous species, it has many subspecies, several of them known only from ship bottoms. In his monograph on sessile barnacles of America, Pilsbry (1916) listed two subspecies from the Panamic region, *B. tintinnabulum cocopoma* and *B. tintinnabulum peninsularis.*

B. tintinnabulum is easily recognized by its large size (4 to 6 cm in diameter) and its red, or purple, and white vertical striations. The parietes are permeated by pores, as are the radii, which are well-developed. The

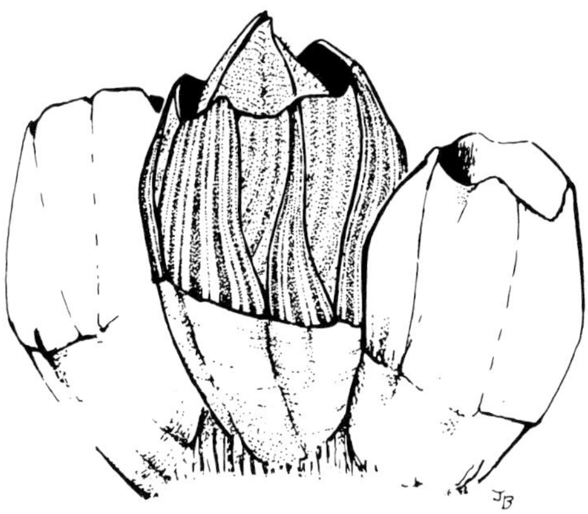

Fig. 7.1 *Balanus tintinnabulum* (from Allen, 1969)

parietes and alae of each plate are exposed and easily visible. The shell base is nearly circular, while the valves form a conical to elongate structure. This species is a common inhabitant of boat bottoms, wharf pilings, and rocks in the midintertidal zone. It has been reported on the west coast of America from California to Peru. Steinbeck and Ricketts reported the Californian subspecies, *Balanus tintinnabulum californicus,* from the Gulf of California, as well as *B. tintinnabulum peninsularis.*

Balanus amphitrite Darwin

This is another widely distributed and well-known species of sessile barnacle, containing numerous subspecies. In this species, the base is calcareous and porous, but the parietes are not permeated by pores The parietes are purplish blue with white lines, and the radii are a drab white. The base diameter may reach 25 mm, but it is usually much less. Pilsbry (1916) lists one subspecies in the Gulf, *B. amphitrite inexpectatus.* It reaches 15 mm in diameter, and has a diamond-shaped orifice. The species occurs in the midintertidal on rocks and shells, especially oyster shells. It superficially resembles *B. tintinnabulum;* however, it may be distinguished by its hidden alae, the parietes being the only exposed section of the plate. It is found in tropical, subtropical, and warm temperate seas of the world.

Balanus concavus Bronn

The parietes are weakly ribbed (occasionally smooth), with pink to purple stripes. This barnacle is strongly conical, with a thin, flat base, and a maximum diameter of around 35 mm. It seems to prefer attaching itself to parts of other animals, such as shells and crab legs, and carapaces. Steinbeck and Ricketts reported it from the Sonora side of the Gulf of California. *B. concavus* ranges from southern California to Peru.

Balanus trigonus Darwin

This species has ribbed parietes, mottled with purplish red. The orifice is broad and trigonal. The shell is conical, generally depressed, with irregularly branching or regular longitudinal ribs. The radii are pale pink or white. The basal diameter is 10 to 25 mm. *B. trigonus* may be recognized easily by its triangular orifice and ribbed and colored walls. The scutum has numerous rows of tiny pits. The angles of the triangular orifice in many specimens are broken to form a pentagonal orifice.

This is another barnacle common on ship bottoms, often associated with other species. In Japan, it is a common inhabitant of crab carapaces. It typically forms large clusters of individuals, but it may also be solitary. This *Balanus* is worldwide in tropical, subtropical, and temperate seas.

Fig. 7.2 *Tetraclita squamosa* (from Allen, 1969)

Tetraclita squamosa (**Bruguiere**)

A widely distributed and well known species, with numerous subspecies. It is easily recognized by its deeply eroded, longitudinal striations. *T. squamosa* is large and conical, with the base diameter reaching 30 mm. It has only four valves, instead of the usual six. The valve junctions are usually indistinguishable externally, so the entire animal appears to be of a single plate.

The species is found in the high intertidal, on exposed rocks, usually well above the *Balanus* zone, and is probably the most commonly encountered barnacle in the Gulf. It is worldwide in tropical, subtropical, and temperate seas. Students familiar with the solitary California subspecies will be impressed by the gregariousness of the Gulf form, which often entirely covers large rocks and reefs, to a thickness of several individuals in the high intertidal. Because of its heavy striations, it is often called the thatched-roof barnacle. It is dull brick red to gray or white in color, and very porous.

Family Chthamalidae

Chthamalus fissus **Darwin**

This and the following species have both end plates (rostrum and carina) overlapped laterally by the adjacent plates. The lateral plates are without deep ribs. The walls are not porous. This species is white to dingy gray, with a round base. It is a small barnacle (7 to 8 mm in diameter) that usually occurs in great colonies on rocks and pilings from the splash

Fig. 7.3 *Chthamalus fissus* (from Allen, 1969)

zone to the midintertidal. It occurs from California to the Gulf of California, and with *Tetraclita* is one of the most commonly encountered barnacles of the Gulf.

Chthamalus anisopoma Pilsbry

A barnacle that superficially resembles *C. fissus,* but may be distinguished by the strongly ribbed parietes, long elliptical orifice (oval in *C. fissus*), and star-shaped base. The plates are usually ivory white. The species occurs in the high and midintertidal throughout the Gulf.

SUBCLASS MALACOSTRACA
Superorder Peracarida
Order Amphipoda

Preservation: See general preservation, Crustacea.

Taxonomy: The taxonomy of this group is based principally on the external anatomy, in particular that of the various appendages. Due to the overwhelming number of amphipods, both in individuals and species, and their generally small size, the taxonomy and identification of this group has become largely a matter for specialists. For this reason, Dr. J. Laurens Barnard, one of the world's foremost amphipod experts, was called upon. He graciously agreed to write the following on amphipod identification, as well as the taxonomic key.

"Identification requires patience and careful microscopic technique. The following key is designed so that only a few operations are required. The key fits mainly the upper half of the Gulf and deals with the rocky intertidal gammarideans only. To properly use the key the following items will be needed: high quality compound and dissecting microscopes, glass slides and cover slips, glycerine, needles, fine jeweler's forceps, and depression slides. If depression slides are not available to mount the thick body parts, use flat slides with pins or sand grains to hold the cover slip above the animal parts. Do not attempt, at first, to identify one isolated amphipod. Rather, collect several hundred and sort out many similarly appearing ones. This will give you a *series* to work with, allowing comparisons and making identification easier.

"Cut off the last three body segments in one piece and place it in a drop of glycerine on a glass slide and cover with a clean cover slip. Be sure that the top of the last three segments points upwards. Examine this with a compound microscope, locating *uropod 3* and the *telson*. The structure of these plus general observations of the body under the dissecting scope will provide most of the data to use the key.

"The mount of the posterior end of the body is the most important observation. These last three body segments are known collectively as the *urosome* and bear three pairs of *uropods* and a single telson. Identify uropods 1 to 2 so as to eliminate them from your search for uropod 3 and the telson. Usually uropod 3 is large and easily seen, but in three genera *(Heterophlias, Corophium,* and *Podocerus)* it is very small, and one may have to overturn the urosome and observe it from a ventral view. In this case, it will appear as a small flap below the telson. If no definite third uropod is found, discard the specimen, as all amphipods in this key have a third uropod. Several genera, especially *Pontogeneia,* tend to lose the third uropod during collecting and transporting of specimens. The telson is always symmetrical but is occasionally damaged, one half being lost. This can be confusing and misleading, as the damaged telson then appears to be uncleft (formed of only one lobe).

"You may need to examine *antenna 1* to find the *accessory* (secondary) *flagellum.* This lies on the medial and distal end of article 3 of the first antenna (antenna 1). It may be large and visible under the dissecting scope, or it may be small, making it necessary to mount the antenna on a glycerine slide and search it out. If the specimen has broken antennae, do not attempt identification. The *rostrum* may be seen by depressing the antennae with a pin while holding the amphipod with forceps. In some cases you will need to observe the rostrum from the top view *(Heterophlias),* but in others, when the *mandibular palp* needs to be seen, it can be observed under the dissecting scope by searching beneath antenna 2 where it joins the head. The palp is a long, flexible setose appendage. Do not confuse this with the *maxillipeds,* those larger mouth parts covering the entire mouth field.

"If one is to deal with the amphipods effectively, one must become a serious student of their taxonomy. If one is making a biological survey of communities and a casual amphipod is collected (a few specimens only), it is scarcely worthwhile trying to identify these, or even have an expert deal with them. Amphipods are so abundant and diverse that a lone record or two of a few broken specimens is virtually worthless. Time rather should be spent dealing with vast quantities of specimens. When submitting specimens to an expert, one should ensure that they be well-preserved, in clean preservative and placed in a small glass vial, stoppered with cotton, this small vial then placed in a larger vial or jar and stoppered. A paper label should be enclosed in the large vial but not in the small vial with the amphipods. Large paper labels ruin specimens by rubbing off appendages. The label should be small and bear a code number. In the letter sent to the expert, the full collection data should be keyed to the code numbers in the vials. Do not enclose large labels in the vials as these are a nuisance to the expert. Always be certain to obtain permission from the specialist before sending him the material.

"The following key is carried only to genus, as species have not yet been fully studied in the Gulf. For a double check of your identifications, you may also utilize the keys to genera in Barnard (1969b). Most of the literature of the Gulf amphipods may be found in Barnard (1969a), but identifications to species remains difficult because of poor descriptions, misidentifications, and sparsity of illustrations in the literature.

Sex in Amphipods: Much of the variability in specimens of amphipods lies in sexual differences. In the following Gulf genera, males and females have strong distinctions in gnathopods 1 to 2, (first two pairs of thoracic appendages): *Podocerus, Ericthonius, Photis, Ampithoe, Lembos, Gammaropsis, Hyale,* and *Elasmopus*. In the other genera, gnathopods 1 to 2 are similar between the sexes. The advice that one deal with many specimens aims partly at this problem of recognizing males and females. One simply studies them intensively to find gross differences in gnathopods (male gnathopod 2 larger than in female in above genera, except *Lembos,* which has gnathopod 1 dominant and thus male gnathopod 1 is enlarged). Adult females can also be recognized by having *brood plates* attached on the inner sides of coxae 2 to 6. The basal article of each thoracic appendage is called a *coxa,* and is often enlarged and platelike. It will bear both a gill and a brood plate in females. The brood plate will often be setose, so that female coxae 2 to 6 will each have two pairs of appendages attached inside (besides having the rest of the leg attached to it). Gills and brood plates are very difficult to sort out without actually dissecting off a leg (preferably leg 3) and examining it individually. When this is done, be sure the basal coxa is also removed, as this is where the extra appendages are attached. If brood plates are absent, one may presume one has either a juvenile or a male. Males may be confirmed by examining the ventral surface of thoracic segment 7 to find two tiny *penial processes* arising near the inner edge of coxa 7 on each side. Do not confuse the penial processes with the gills. If the gill is touched it usually falls off, but the penial process is firmly anchored. The penial processes are so small that they are usually not worth looking at and *for using the following key it will not be necessary to distinguish between males and females.* In many populations males will be quite rare."

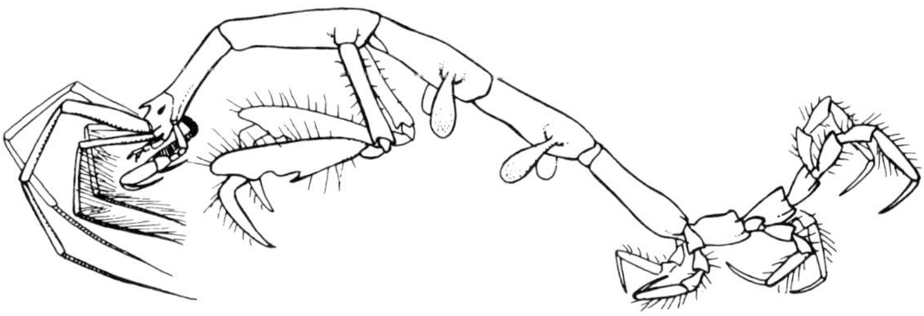

Fig. 7.4 A caprellid amphipod (from Allen, 1969)

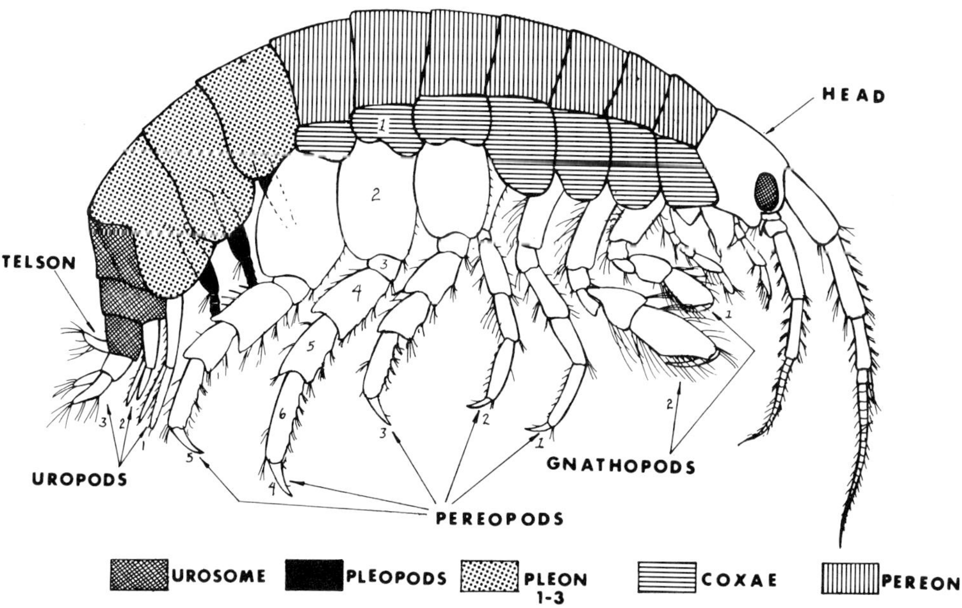

Fig. 7.5 Amphipod morphology (modified from Barnard, 1969)

KEY TO THE COMMON GENERA OF GAMMARIDEAN AMPHIPODS OF THE GULF OF CALIFORNIA

by J. Laurens Barnard*

1 Telson puffy, fleshy or thick; telson uncleft or weakly hollowed dorsally **2**
-Some part of telson thin, lamellar; telson cleft or uncleft **9**

2(1) Uropod 3 lacking ramus, composed only of setose or fuzzy peduncle **3**
-Uropod 3 with 1 to 2 rami **5**

3(2) Body with appearance of an isopod, broad and dorsoventrally flattened but with heavy middorsal ridge (ridge either full length or near posterior part of body); abdomen partially flexed under thorax **4**
-Body with appearance of tanaid, cylindrical but slightly flattened and dorsally smooth (rostrum short or thin, gnathopods thin and heavily setose, coxae 1 to 4, small); abdomen not flexed under thorax (fig. 7.6) .... ***Corophium***

4(3) Rostrum large, a flat rectangular projection between antennae (dorsal view); gnathopods 1 to 2 appearing as thin pereopods; poorly setose (body appearing like *Stegosaurus* dinosaur) (fig. 7.7) .... ***Heterophlias***
-Rostrum weak or absent; gnathopods 1 to 2 well developed; heavily setose; gnathopod 2 larger than 1 (fig. 7.8) .... ***Podocerus***

5(2) Uropod 3 with one ramus; male gnathopod 2 complexly chelate; body thin .... ***Ericthonius***
-Uropod 3 with two rami; male gnathopod 2 not chelate; body usually thick and of ordinary gammaridean plan **6**

6(5) Inner ramus of uropod 3 half or less as long as outer ramus (fig. 7.9) .... ***Photis***
-Rami of uropod 3 projecting equally **7**

7(6) Rami of uropod 3 ordinary, as long as or longer than peduncle, thin, lanceolate, and poorly setose apically **8**
-Rami of uropod 3 fleshy, thick, shorter than peduncle, with inner ramus heavily setose apically and the outer ramus with 2 large reverted hooks apically (spines) (fig. 7.10) .... ***Ampithoe***

8(7) Gnathopod 1 larger than gnathopod 2 (check carefully, as this is true also of females, though the gnathopods are small and at first glance appear identical in size) .... ***Lembos***
-Gnathopod 2 larger than gnathopod 1 ...... ***Gammaropsis* (includes *Eurystheus* and *Megamphopus*)**

9(1) Telson uncleft **10**
-Telson cleft (with a slit down the middle) **12**

10(9) Gnathopod 1 of special carpochelate form (figs. 7.11 to 7.12) .... ***Leucothoe***
-Gnathopod 1 not of special carpochelate form **11**

11(10) Gnathopod 1 of special propodochelate form (fig. 7.14); coxa 1 large and visible (fig. 7.13) .... ***Synchelidium***
-Gnathopod 1 of ordinary form (figs. 7.15 to 7.16); coxa 1 small and mostly hidden by coxa 2 .... **family Amphilochidae**

12(9) Uropod 3 with one short ramus; telson short, formed of two triangles by a very deep cleft; mandible lacking palp (fig. 7.15) .... ***Hyale***
-Uropod 3 with 2 large rami; telson short or long, but not formed of independent triangles; mandible with palp **13**

13(12) Head with a large rostrum (between antennae); accessory flagellum absent or with one article; antenna 2 as long as antenna 1 (both often broken, however); gnathopods small in both sexes (fig. 7.16) .... ***Pontogeneia***
-Rostrum small or absent; accessory flagellum with 2 to 5 articles; antenna 2 much shorter than antenna 1 (rarely broken); male gnathopod 2 enlarged .... ***Elasmopus***

*Smithsonian Institution, Washington, D.C.

SYSTEMATICS AND GENERA DESCRIPTIONS

Amphipoda
from Barnard (1969b)

Family Corophiidae

Corophium Latreille

Accessory flagellum absent; article 3 of antenna 1 shorter than article 1, flagellum longer than article 3; antenna 2 equal to or longer than antenna 1, stout; gnathopods small, gnathopod 1 subchelate, gnathopod 2 simple, extremely setose; uropod 2 biramous; uropod 3 flattened.

Fig. 7.6
Corophium sp.
(from Barnard, 1969)

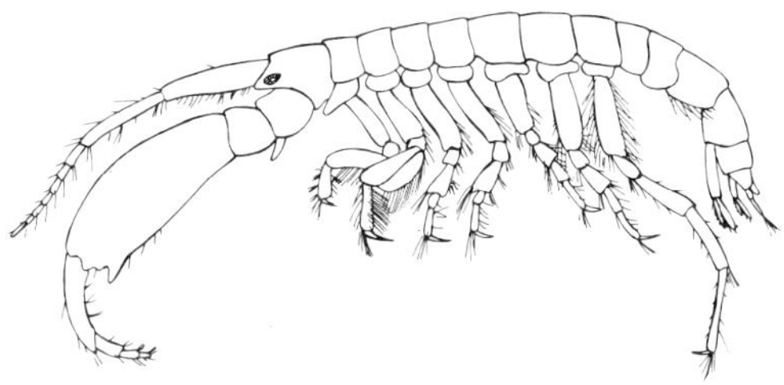

Ericthonius Milne Edwards

Accessory flagellum absent; article 3 of antenna 1 longer than article 1, flagellum subequal to peduncle; antenna 2 slender, equal to antenna 1; male gnathopod 2 larger than 1; uropod 2 biramous; uropod 3 with its single, medium length, and uncinate ramus shorter than peduncle.

Family Phliantidae

Heterophilias Shoemaker

Maxilla 1 lacking palp; maxillipedal palp 4-articulate; gnathopods simple; inner ramus of pleopod 3 half as long as outer; uropod 3 lacking rami. The only species of this genus so far published is *H. seclusus* Shoemaker, but others will no doubt be found, or designed to fit this genus eventually.

Fig. 7.7
Heterophilias sp.
(from Barnard, 1969)

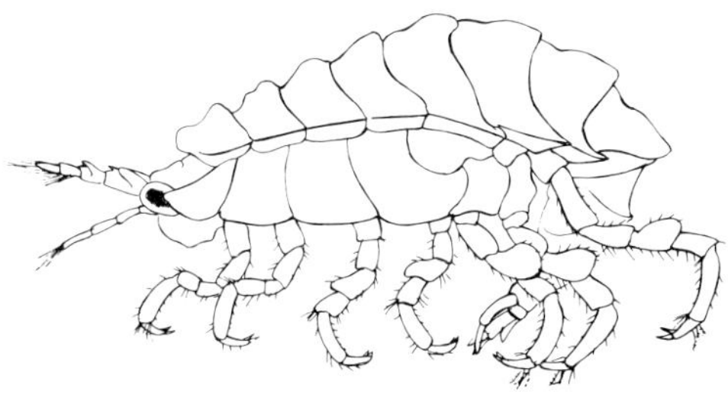

Family Podoceridae

Podocerus Leach

Accessory flagellum present; antenna 1 shorter than 2; pleon with six segments and three pairs of uropods; uropod 2 with two rami; uropod 3 lacking rami.

Family Isaeidae

Photis Kroyer

Accessory flagellum absent, or formed of a vestigial article or scale; article 3 of antenna 1 equal to or longer than article 1; gnathopods subchelate; uropod 3 biramous, outer ramus equal to or shorter than peduncle, inner ramus much shorter than outer ramus.

Gammaropsis Liljeborg

Accessory flagellum 3- or more articulate; article 3 of antenna 1 equal to or longer than article 1; gnathopods subchelate; uropod 3 biramous, rami equal to each other, variable in length, generally equal to or longer than peduncle.

Family Ampithoidae

Ampithoe Leach

Antenna 1 lacking accessory flagellum; mandible with palp; gnathopods large, subchelate, gnathopod 2 equal to or larger than 1; article 6 of pereopods 3 to 5 scarcely widened apically, rarely prehensile; outer ramus of uropod 3 with two hooks.

Family Aoridae

Lembos Bate

Article 3 of antenna 1 shorter than article 1, accessory flagellum present; male gnathopod 1 subchelate, with articles 4 to 5 lacking teeth, article 6 not narrower but usually longer than 5, female gnathopod 1 subchelate; male gnathopod 2 subchelate, with article 5 longest, poorly setose on anterior edge; rami of uropod 3 equal, longer than peduncle.

Family Leucothoidae

Leucothoe Leach

Coxa 2 at least as long as broad, rounded ventrally and anteriorly, coxa 1 not concealed; mandibular palp 3-articulate; outer plate of maxilliped reaching less than halfway along palp article 1.

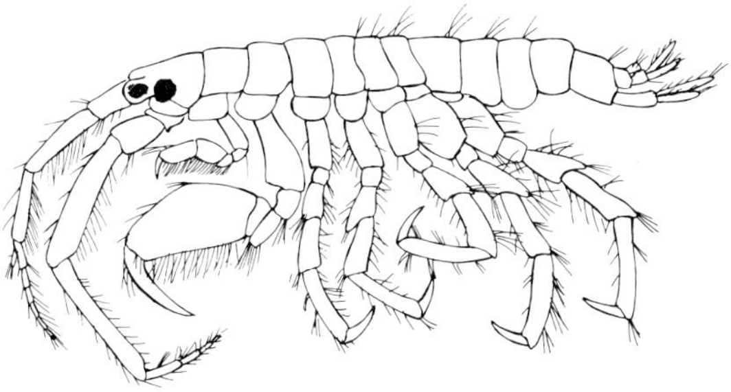

Fig. 7.8
Podocerus sp.
(from Barnard, 1969)

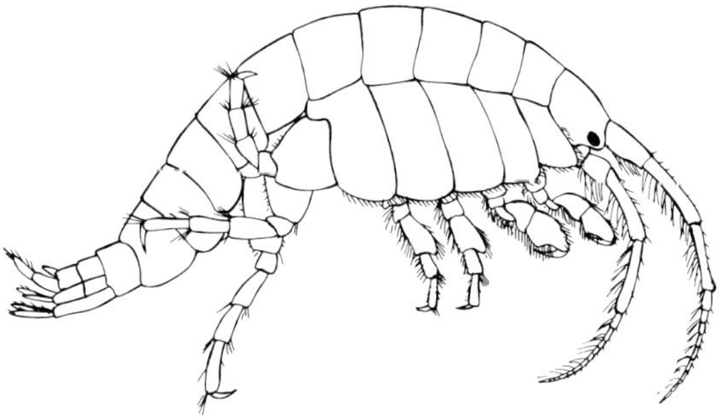

Fig. 7.9
Photis sp.
(from Barnard, 1969)

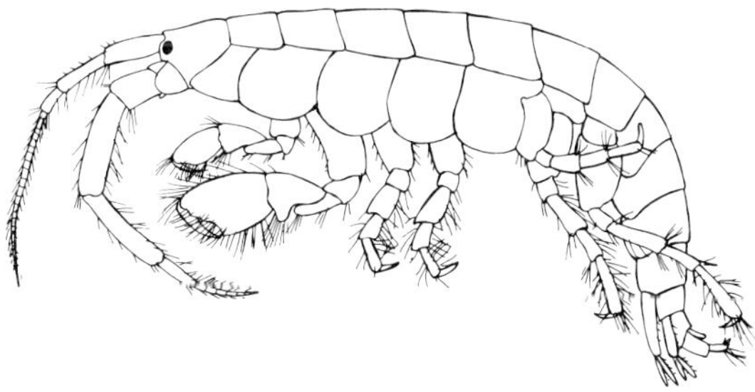

Fig. 7.10
Ampithoe sp.
(from Barnard, 1969)

Fig. 7.11
Leucothoe sp.
(from Barnard, 1969)

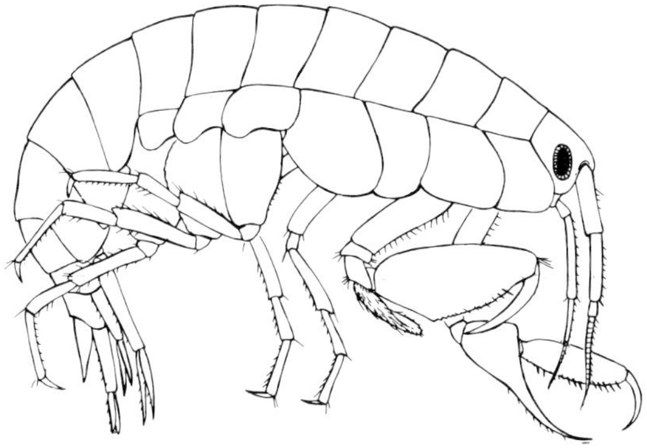

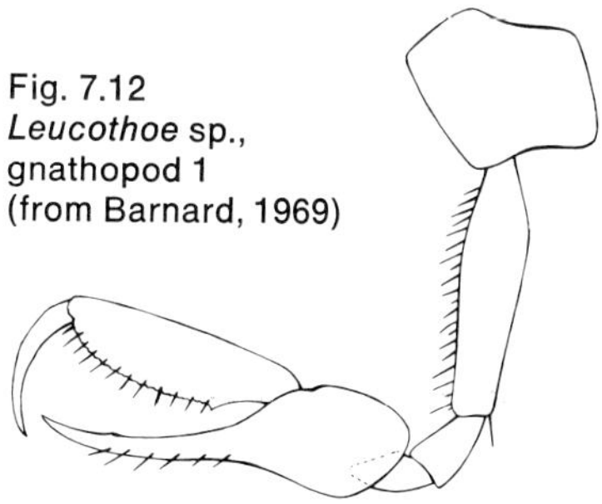

Fig. 7.12
Leucothoe sp.,
gnathopod 1
(from Barnard, 1969)

Family Oedicerotidae

Synchelidium Sars

Cutting edge of mandible projecting, toothed, molar small, conical, with apical spine, unringed; inner lobes of lower lip flat but separate; gnathopods dissimilar, gnathopod 1 subchelate, stout, article 5 forming a lobe guarding article 6, gnathopod 2 slender, chelate, article 5 unproduced; uropod 2 reaching end of uropod 3.

Family Amphilochidae

A large family (17 genera) of amphipods, represented in the Gulf by a number of species. Accessory flagellum absent; coxa 1 very small, partially hidden by following coxae; rostrum conspicuous; coxae 3 to 4 enlarged, either overlapping or, when immensely enlarged, with contiguous margins abutting; uropod 2 shortened; uropod 3 biramous, peduncle elongate; telson entire in all but one genus, often elongate and triangular, also short and linguiform.

Family Talitroidea

Hyale Rathke

Article 4 of maxillipedal palp unguiform; gnathopods subchelate in both sexes, male gnathopod 2 larger than 1, article 5 not projecting between articles 4 and 6, female gnathopod 2 like gnathopod 1; uropod 3 lacking inner ramus; telson cleft.

Family Eusiridae

Pontogeneia Boeck

Accessory flagellum absent; gnathopods not eusirid, article 5 not lobate, slightly longer than article 6; pereopods 3 to 5 with articles 4 to 6 each not longer than article 2, article 4 slender; outer lobes of lower lip not broadly separated.

Family Gammaridae

Elasmopus Costa

Accessory flagellum three or more, occassionally 2-articulate; lower lip with inner lobes; gnathopods normal; uropod 3 variable in length, rami equal, rectangular, outer 1-articulate; telson deeply cleft; urosome occasionally with dorsal teeth; palp article 3 of mandible falcate, stout.

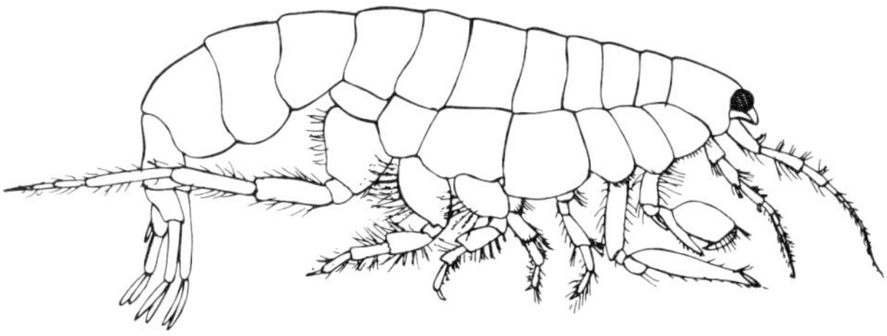

Fig. 7.13 *Synchelidium* sp.
(from Barnard, 1969)

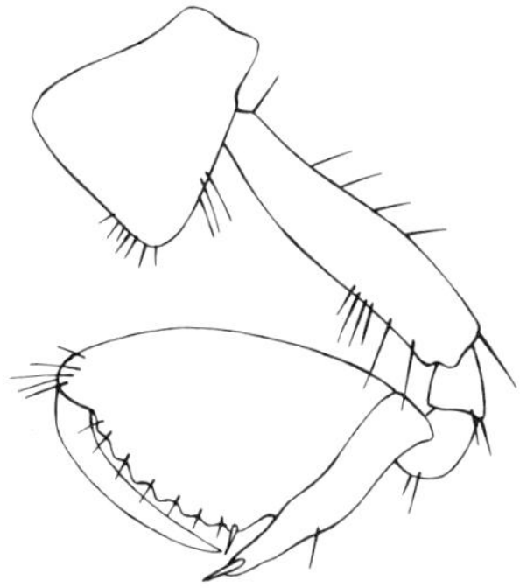

Fig. 7.14 *Synchelidium* sp.,
gnathopod 1 (from Barnard, 1969)

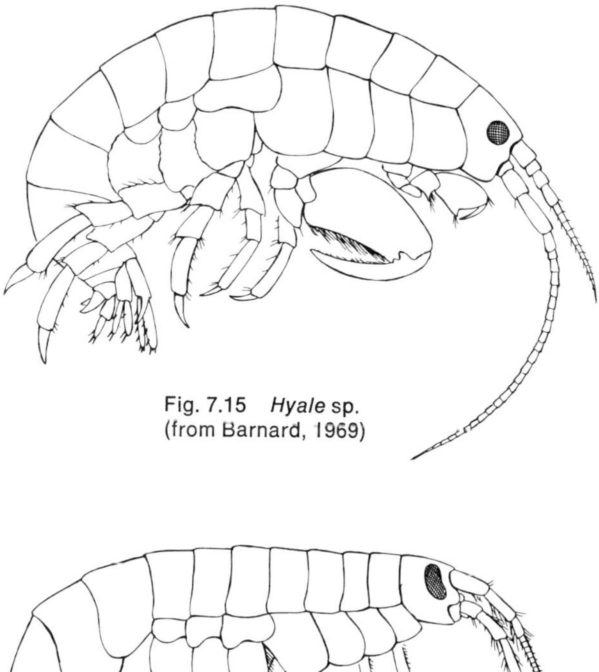

Fig. 7.15 *Hyale* sp.
(from Barnard, 1969)

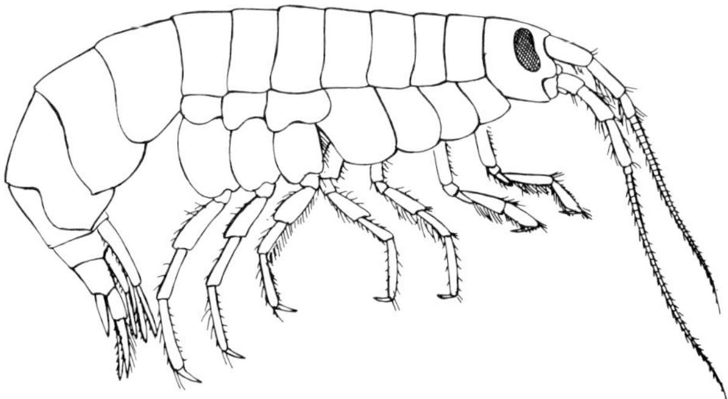

Fig. 7.16 *Pontogeneia* sp. (from Barnard, 1969)

Order Isopoda

Preservation: See general preservation, Crustacea.

Taxonomy: The isopods form an order of the superorder Peracardia, which is that group of crustaceans without a carapace, or with a rudimentary one. The isopods generally are considered to be phylogenetically the most advanced group of the Peracardia (the Peracardia also include the amphipods, mysids, cumaceans, and tanaidaceans), and are distinguished from all other peracaridean crustaceans by their sessile eyes, dorsoventrally compressed body, uniramous thoracic appendages, abdominal heart, 5 pairs of leaflike pleopods, single pair of uropods, and 7, usually distinct, peraeonal somites with sub-similar nonchelate peraeopods (Menzies and Frankenberg, 1966). The taxonomy of the Isopoda is based largely on the external morphology of the mouth parts, legs, pleon, and other structures.

The isopod body may be divided into three regions: the *cephalon* (or head), the *peraeon* (or thorax), and the *pleon* (or abdomen). The peraeon usually consists of seven free segments, each with a pair of legs, the *peraeopods,* that may be prehensile, meaning modified for clinging to a host, as in the parasitic species (in which case they are hooked and curve inward near the tip), or that may be ambulatory, that is, used for walking or swimming. The pleon primitively consists of seven segments, but most modern forms have these fused into one to six free segments. The appendages of the pleon are called *pleopods,* which are usually small, lamellate structures on the ventral surface used for respiration and swimming. The last segment of the pleon is the *pleotelson* (or simply *telson).* The pleotelson usually bears the flattened *uropods,* which are formed from the terminal pair of pleonal appendages. Each uropod bears an inner and outer ramus, the *endopod* and *exopod* respectively. The head bears two pairs of antennae, the innermost of which is designated as *antenna 1,* the outermost as *antenna 2.*

The classification used in this handbook follows that of Schultz (1969), Richardson (1905), and Van Name (1936).

Isopods of the Gulf: The isopod Crustacea of the Gulf of California are not well-known. Steinbeck and Ricketts reported 12 species from this region, of which five range from the Gulf northward and none have strictly Panamic affinities. Many of the Steinbeck and Ricketts identifications remain questionable. Ohmart (1964) reported eight species from the upper Gulf, in the region of Puerto Peñasco. Schultz (1969) reported only four species with distributions in the Gulf, *Meinertia gilberti, Meinertia gaudichaudii, Rocinela aries,* and *Rocinela tuberculosa.* He didn't include the beach-inhabiting terrestrial forms (Oniscoidea), which would add at least three other species: *Ligia occidentalis, Ligia exotica,* and *Tylos latreilli.* The author's own work on Isopoda has just begun at the time this manuscript goes to press, but he has already turned up about 20 species from the Gulf. Unfortunately, they have not been properly examined yet and so will receive only the briefest mention in the following discussion. Needless to say, a thorough study of the isopod fauna of the Gulf of California will add significantly to our knowledge of these creatures in the Eastern Pacific.

KEY TO THE COMMON GENERA OF MARINE AND SHORE ISOPODS OF THE GULF OF CALIFORNIA

Note: The following key includes all seven suborders of free-living isopods known to occur in the Northern Hemisphere, whether known from the Gulf of California or not. It thus may be used to identify an isopod to suborder, regardless of where it was collected, north of the equator. This has been done because this particular group of crustaceans is so poorly known from the Gulf that no one can be certain what species may yet be found there.

1 Five pair walking legs; mandibles of males and juveniles large, conspicuous and anteriorly directed; pleotelson usually T-shaped or triangulate .... **S.O. *Gnathiidea***

-Seven pair walking legs; mandibles not as above; pleotelson not as above **2**

2(1) Uropods lateral, operculate, or absent **3**

-Uropods terminal **5**

3(2) Uropods operculate **7**

-Uropods lateral or absent **4**

4(3) Body elongate, length at least 7 times width; uropods partially or largely folded over and above pleotelson (fig. 7.17) .... **S.O. *Anthuridea* Genus *Mesanthura***

-Body not elongate, length never 7 times width; uropods never folded up over pleotelson (figs. 7.18-7.19) .... **S.O. *Flabellifera* 8**

5(2) Length less than 2 mm; blind; body length about 7 times width; none yet reported from Gulf .... **S.O. *Microcerberidea***

-Length usually greater than 3 mm; blind or with eyes; body length less than 7 times width **6**

6(5) Terrestrial forms, pleopoda fitted for air breathing; first pair antennae rudimentary and inconspicuous; 3 or more free pleonal segments (fig. 7.21) .... **S.O. *Onisoidea* (in part) Genus *Ligia***

-Aquatic forms, pleopoda not fitted for air breathing; first pair antennae not rudimentary; all pleonal segments may be free, or all but 1 or 2 may be fused .... **S.O. *Asellota***

7(3) First pair of antennae rudimentary and inconspicuous; terrestrial forms with pleopoda fitted for air breathing .... **S.O. *Onisoidea* (in part) Genus *Tylos***

-First pair of antennae large or small, but never rudimentary or inconspicuous; aquatic forms with pleopoda not fitted for air breathing (fig. 7.20) .... **S.O. *Valvifera* Genus *Erichsonella***

8(4) Pleon composed of 3 or less visible segments (plus pleotelson) **9**
-Pleon composed of 4 to 5 visible segments (plus pleotelson) **10**
9(8) Exopod and endopod about the same length; exopod capable of folding under endopod .... **Genus *'Dynamenella'***
-Exopod and endopod about the same length in females, but exopod of males greatly elongate; exopod of male not capable of folding under endopod .... **Genus *Paracerceis***
10(8) Peraeopods of single individual both prehensile and ambulatory .... **Genus *Rocinela***
-Peraeopods either all prehensile or all ambulatory **11**
11(10) Peraeopods prehensile; animal often parasitic on fish **12**
-Peraeopods ambulatory; animal free-living **13**
12(11) Basal segments of antennae 1 dilated and touching at base .... **Genus *Meinertina***
-Basal segments of antennae 1 compressed and not touching (fig. 7.19) .... **Genus *Lironeca***
13(11) Telson and uropods with a fringe of long, terminal hair .... **Genus *Eurydice***
-Telson and uropods without long, terminal hair (fig. 7.18) .... **Genus *Cirolana***

SYSTEMATICS AND GENERIC DESCRIPTIONS

Isopoda

Suborder Anthuridea

The anthurids are found mostly on the bottom or on wharf pilings, often living in a burrow of their own construction or in one of another marine invertebrate. The bodies of these isopods are extremely elongate. Males are rare in this group, and may be distinguished by their larger and/or more numerous eyes and their long, hairy, second antennae.

Family Anthuridae

Mesanthura K. H.Barnard

This genus is distinguished by the following set of characteristics: mouth parts for chewing and biting; paired statocysts on telson; maxilliped with five articles; pleopod 1 operculiform; pleon sutures indistinct dorsally; body pigmented; and telson not sculptured. I have found only one

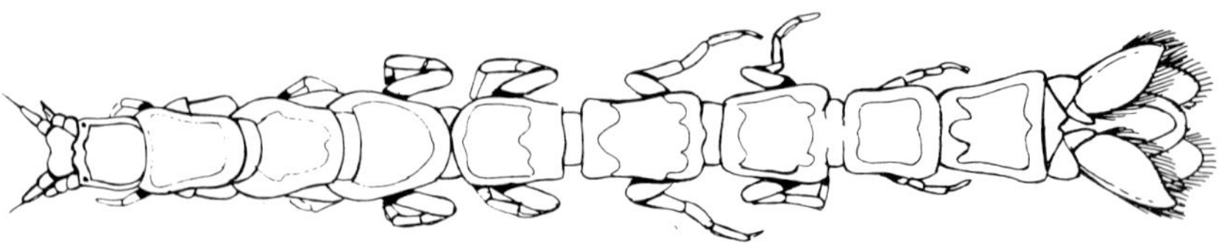

Fig. 7.17 *Mesanthura* sp.

species of this genus (indeed of the entire suborder) in the Gulf. It is possibly *Mesanthura occidentalis* Menzies and Barnard, which has previously been known only from southern California to Bahía de San Quintín, Baja California. My specimen was collected from algal mats in the intertidal of Puerto Peñasco. To my knowledge, members of this suborder have not been recorded yet in the literature from the Gulf of California.

Suborder Flabellifera

A large group of isopods, most of which are either free-living shore dwellers or parasites on fish. Some members of this group have both ambulatory and prehensile peraeopods on a single individual, while others have the legs strictly one type or the other.

Family Cirolanidae

Eurydice Leach

The pleon of these isopods has six distinct segments, the peduncle of the second antennae has four articles, and the basal article of the peduncle of the first antennae is extended straight in front at right angles to the remainder of the antennae, the opposing basal articles meeting at the midline of the cephalon. The uropods are provided with plumose setae, and are inserted on the ventral side of the pleon. Species of this genus are differentiated by the patterns of fringing setae and spines along the posterior margin of the pleotelson. Most of them are planktonic. Ohmart (1964) reported an unidentified species from Puerto Peñasco, and Steinbeck and Ricketts reported *Eurydice caudata* Richardson from the plankton of Punta San Marcial, Bahía Concepción, San Lucas Cove, and Bahía San Carlos. Schultz (1969) reported two species from California: *E. caudata* from Catalina Island, and *E. branchuropus* Menzies and Barnard from Point Loma. I have not yet found any members of this genus in the Gulf.

Cirolana Leach

The genus *Cirolana* may have more species than any other North American genus. It is characterized by the presence of five free pleonal segments and the pleotelson, the uniformity of the ambulatory legs, and the lateral uropods. Members of this genus do not have the first peduncular article of the first antennae projecting outward at right angles to the remainder of the antennae (as is the case in the genus *Eurydice,* which they closely resemble). The uropods are fanlike, lateral, and indurated. These isopods are common and usually abundant in most intertidal

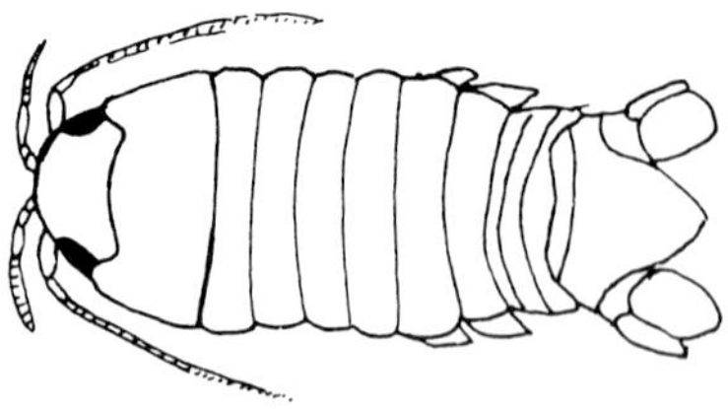

Fig. 7.18 *Cirolana* sp.

habitats of temperate and tropical shores. Ohmart (1964) reported *C. mayana* Ives from Puerto Peñasco, and Steinbeck and Ricketts reported *C. harfordi* (Lockington) from Isla Tiburón. Schultz (1969) mentions four species from California, two of which are known from the west coast of Baja California (*C. harfordi* and *C. diminuta* Menzies). No others appear to have been recorded from the Gulf of California, although I have collected one species of this genus at Bahía Cholla in the northern Gulf. My specimen closely resembles, but *is not, Cirolana diminuta.*

Family Sphaeromatidae

Paracerceis Hansen

Members of this genus have the apex of the pleotelson incised in the males, and formed into a ventral, tubular channel in the females. The exopod of the males is about twice the length of the endopod, while in the females they are about the same size. The dorsal surface of the pleon usually bears various types of ornamentation. This is a difficult genus to work with, as there is a good deal of intraspecies variation. Although none have been reported in the literature from the Gulf, I have collected at least two species from the chambers of the white sponge *Leucetta* sp. at Guaymas: *Paracerceis sculpta* Holmes, and a new species. I have collected *P. sculpta* also at Puerto Peñasco. Identifications of isopods of this genus were made by Ernest W. Iverson, California Academy of Sciences.

'*Dynamenella*'

This genus is placed in apostrophes, because reliable taxonomic characters for distinguishing species of the genera *Dynamenella* and *Dynamenopsis* in North American forms are not known. Steinbeck and Ricketts reported an unidentified species of *Dynamenella* from the Gulf, and Schultz (1969) reported four species of *Dynamenella* from the west coast of North America in general, including *D. dianae* Menzies from Bahía de San Quintín, and *D. glabra* (Richardson) and *D. benedicti* (Richardson) from Monterey Bay, California. I have collected members of this genus from Puerto Peñasco and Guaymas.

Family Cymothoidae

Meinertia Stebbing

Species of this genus are large parasites on fish. They are usually found attached to the gills or mouth cavity of their host. The heads of these isopods are more or less immersed, or set, in the first peraeonal segment. The eyes are distinct, and the first pair of antennae are dilated, contiguous at the base. Schultz reported two species from the west coast of México, both from Mazatlán; *M. gilberti* Richardson (from *Mugil hospes,* a mullet) and *M. gaudichaudii* (Milne-Edwards) (from *Thynnus* sp. and *Sarda chilensis,* members of the tuna group). I have not collected any of this species in the Gulf, although I strongly suspect they occur at least as far north as Guaymas and probably Bahía Kino, as probably do the specific hosts the Mazatlán forms were recorded from.

Lironeca Leach

These are moderately large isopods, parasitic on fish, and usually found clinging to the gills of their host. The body is oval, more or less twisted to one side, the cephalon usually deeply immersed in the first peraeonal segment. Schultz (1969) reported four species from the west coast of America: *L. vulgaris* Stimpson (Washington to Baja California), *L. panamaensis* Schioedte and Meinert (México to Panamá), *L. californica* Schioedte and Meinert (central and southern California), and *L. convexa* Richardson (southern California to Ecuador). Although none have been recorded from the Gulf of California, I have found what appears to be specimens of *Lironeca panamaensis* clinging to the gills of the common Gulf mullet (*Mugil cephalus*) at Puerto Peñasco. Little work has been done on the external isopod parasites of fish, but were more collections available and the group's natural history better known it would probably be found that most of the isopod parasites have distributions closely paralleling those of their hosts.

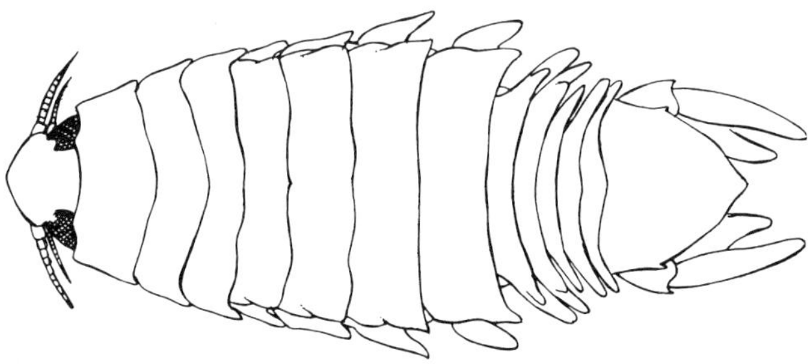

Fig. 7.19 *Lironeca* sp.

Family Aegidae

Rocinela Leach

Members of this genus have the body depressed, with the pleon not much narrower than the peraeon. The eyes are well-developed and the maxillipedal palp is of two articles. This genus includes both parasitic and free-living species. Steinbeck and Ricketts reported *R. aries* Schioedte and Meinert from the intertidal of Bahía San Francisquito. Schultz (1969) reported two species from the Gulf of California, *R. aries* Schioedte and Meinert and *R. tuberculosa* Richardson, and a third, *R. laticuada* Hansen, from the west coast of México.

Suborder Valvifera

Schultz (1969) states the following characters are always present in members of this suborder from North America: the uropods form valves which fold over the branchial chamber as a cover; the flagellum of antenna 1 is always formed of a single fused segment; the mandible has no palp; the pleon always has several fused segments; and the frontal margin of the cephalon rarely has a long rostrum.

Family Idoteidae

Erichsonella Benedict

This genus is distinguished by the following set of characters: second antennae with a single flagellum; palp of maxillilpeds composed of four articles; pleon consists of a single somite. Apparently no members of this genus have yet been recorded from the Gulf. Schultz (1969) reported *E. crenulata* and *E. pseudoculata* from southern California. I have collected two individuals of this genus from the intertidal region of Puerto Peñasco that may represent an undescribed species.

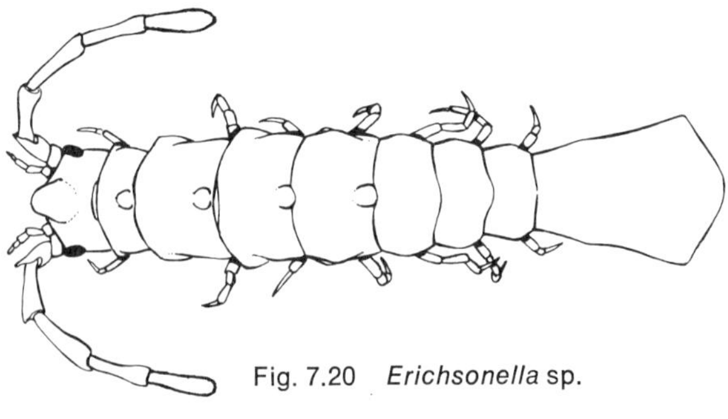

Fig. 7.20 *Erichsonella* sp.

Suborder Onisoidea

This suborder of isopods represents all of the terrestrial species, many of which are shore forms inhabiting the splash or spray zones. The legs are ambulatory in nature. The uropods are terminal (although situated ventrally in the genus *Tylos*), and composed of a peduncle and two branches, the branches being uniarticulate. The pleopoda are fitted for breathing air, the outer opercular plate of the first two pairs (and sometimes all five pairs) containing air cavities or tracheae. The abdomen is composed of six well-defined segments, and the first pair of antennae are rudimentary and inconspicuous.

Family Ligidae

Ligia Fabricius

Ligia is a large genus of terrestrial isopods whose semi-marine forms usually occur in the very high tide or splash zones of the intertidal, but do not enter the water. There are two common species of *Ligia* on the west coast of North America, *Ligia pallasii* (Brandt) and *Ligia occidentalis* (Dana). The former has the peduncle of the uropods about as broad as long, while the latter has the peduncle several times longer than broad. *L. pallasii* has not been reported from the Gulf, while *L. occidentalis* is probably the most common isopod of the intertidal region throughout the Sea of Cortez. *L. occidentalis* is the large (to 35 mm), dark brown or purple to black isopod that is seen scampering over the rocks and pilings above the high tide line. It usually is found in great numbers, but is difficult to capture because it runs so fast to the seclusion of rock crevices. It would be difficult indeed to find an area on the Gulf coast with large rocks or wharf pilings that does not harbor a large colony of these animals. Manton (1951) stated *L. occidentalis* is able to take up to 16 steps a second with

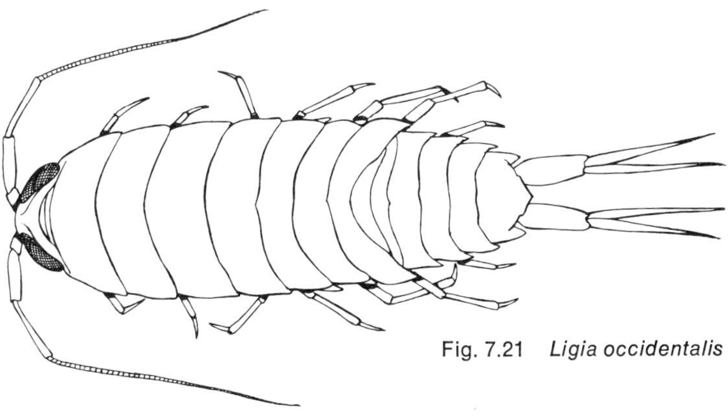

Fig. 7.21 *Ligia occidentalis*

each leg (and it has 14 legs). Steinbeck and Ricketts reported *Ligia exotica* (Roux), a tropical species, from the intertidal of Puerto Escondido and Bahía San Carlos, and Mulaik (1960) reported it from La Paz and Topolobampo. I have collected *L. occidentalis* along all rocky coastlines from Puerto Peñasco to Guaymas, and I have collected *L. exotica* on the rocks of Mazatlán.

Family Tylidae

Tylos Latreille

This genus occupies about the same habitat in the Gulf as does *Ligia,* but it is found usually on a pure sand or small pebble substrate, as opposed to the large rocks and boulders *Ligia* seems to prefer. Both, however, are scavengers in the spray zone and high intertidal when the tide is at ebb. Species of *Tylos,* unlike all other shore isopods, resemble the terrestrial pill-bugs, or common "sow-bugs." They are found feeding on cast-up seaweed and dead animals. Ohmart (1964) reported *Tylos latreilli* Audouin and Savigny from the sand and tidepools of Puerto Peñasco, which also is a locality given by Mulaik (1960). However, Van Name (1936) reported this species as restricted to the Atlantic Ocean, and specimens I have collected in the Puerto Peñasco region more closely resemble *Tylos punctatus.* These two species, as well as several others, are all very similar, and the exact taxonomic relationship of the various species of *Tylos* is not well-known.

Superorder Hoplocarida
Order Stomatopoda

These shrimplike crustaceans are called "mantis shrimps," because the second thoracic appendages (maxillipeds) are greatly enlarged, the terminal segment being turned back to fit into the one above it, like a knife fits into its handle. This arrangement gives the creature the general appearance of a seagoing praying mantis. There are six walking legs, on the last three thoracic segments. The abdomen is flattened dorso-ventrally and bears six pairs of flattened swimming appendages that actually serve the dual purpose of locomotion and respiratory gills. Adult mantis shrimp live in burrows or crevices among rocks or on muddy bottoms offshore. They are seldom found intertidally, although they are commonly dredged up in fishing trawls. The only way I have ever been able to collect them in the Gulf is by digging through shrimp boat catches. Dr. Raymond B. Manning of the Smithsonian Institution has been kind enough to identify the specimens I have accumulated from shrimping hauls in the northern

Gulf. These are listed below, but, since none have been found in the intertidal zone, I have not included a key or species descriptions:

Squilla bigelowi Schmitt
Squilla mantoidea Bigelow
Squilla tiburonensis Schmitt

In addition, Schmitt (1940) listed the following Panamic forms: *Squilla swetti* Schmitt from Bahía de Petatlán, Mexico; *Squilla aculeata* Bigelow from Sinaloa, Mexico, and south to Chile; *Squilla hancocki* Schmitt from Bahía de Petatlán, and Tangola-Tangola, Guererro, Mexico; *Squilla biformis* (Bigelow) from La Paz; *Squilla panamensis* Bigelow from Bahía de Petatlán, Mexico, south to Colombia; *Squilla parva* Bigelow from Tangola Tangola and Bahía de Petatlán, Guererro, Mexico, to Ecuador; *Pseudosquilla lessonii* (Guerin) from California to Chile, but not recorded from the Gulf of California; *Pseudosquilla veleronsis* Schmitt from Bahía de los Angeles, Gulf of California, to Oaxaca; *Lysiosquilla desaussurei* Stimpson from Mazatlán; *Lysiosquilla digueti* Coutiere from La Paz (as a commensal in the burrow of a *Balanoglossus*); *Lysiosquilla mccullochae* Schmitt from Isla San Francisco, Gulf of California; *Gonodactylus oerstedii* Hansen from the Gulf of California to Ecuador; and *Gonodactylus stanschi* Schmitt from Isla Angel de la Guarda to Bahía Tangola-Tangola, Mexico, including Isla Isabela and Islas Tres Marias.

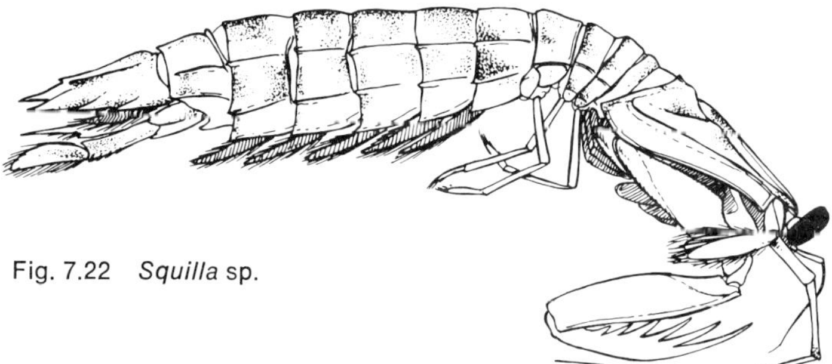

Fig. 7.22 *Squilla* sp.

Superorder Eucarida
Order Decapoda
Suborder Natantia

Preservation: See general crustacean section.

Taxonomy: The taxonomy of the true shrimp is based primarily on external features of the carapace and various appendages. The classification used in this handbook follows that of Schmitt (1921, 1935) and Holthuis

(1951, 1952), and the key is based to a large extent on these monographs. Because of the large number of poorly known species of shrimp from this region, the key is to genus only. For extensive keys to the species of the true shrimp, the serious student is referred to the above publications and to other references in this section, especially Holthuis (1947 to 1955). Mention of the hepatic spine in the key refers to the short spine on the sides of the carapace, about one-third of the distance back from the head. Note that the third maxilliped is often very long, and can easily be mistaken for the first walking leg.

Natantia of the Gulf: Very little work has been done on the intertidal shrimp of the Gulf of California, although the Mexican government does encourage research on the commercial shrimp of the genus *Penaeus*. Members of this group are, however, rarely found in the intertidal, although immature individuals are occasionally collected in bays and estuaries. Castro (1966) stated that (in regards to the species he studied) the shrimp of the Gulf of California have strong Californian affinities. Of the ten species of known specific status he discussed, five were Californian forms, two Panamic, one Indo-Pacific, and one was previously known only from the west Atlantic. Steinbeck and Ricketts reported 13 species of shrimp from the Gulf, and found the fauna to be well-distributed between Panamic and Californian forms.

The shrimp most often encountered in the intertidal zone of the Gulf are the pistol shrimp (snapping shrimp) of the genera *Alpheus* and *Synalpheus*, and the red-striped shrimp *Lysmata californica*.

Shrimp are most easily captured by the use of a net in large tidepools, or by overturning large stones and grabbing them by hand. The subtidal shrimp can be retrieved from the shrimp trawls, which are actually small otter trawls and are good sources of many invertebrates. Shrimp fleets operate out of Puerto Peñasco, Topolobampo, Guaymas, Mazatlán, and other ports.

KEY TO THE GENERA OF INTERTIDAL NATANTIA OF THE GULF OF CALIFORNIA

1 Lateral portions of second abdominal pleura expanded to overlap the first abdominal pleura; mainly intertidal **2**
-Lateral portions of second abdominal pleura not as above; mainly subtidal **9**

2(1) Wrists of second legs undivided **3**
-Wrists of second legs divided (multiarticulate) **6**

3(2) Posterior margin of telson with 3 pairs of spines; pleurobranch absent from 3rd maxilliped **4**
-Posterior margin of telson with 2 pairs of spines; pleurobranch present on 3rd maxilliped (fig. 7.23) .... ***Palaemon***

4(3) Hepatic spine present 5
-Hepatic spine absent (fig. 7.25) .... ***Pontonia***
5(4) Supraorbital spine present; mandible with a palp .... ***Palaemonella***
-Supraorbital spine absent; mandible without a palp (fig. 7.24) .... ***Periclimenes***
6(2) First legs much larger than others; eyes covered by carapace; rostrum small or wanting 7
-First legs not much larger than others; eyes not covered by carapace; rostrum prominent 8
7(6) Large cheliped laterally compressed; no pterygostomian spine; front of carapace with or without lateral spines (fig. 7.26) .... ***Alpheus***
-Large cheliped not laterally compressed, but rounded; pterygostomian spine present; front of carapace with lateral spines (fig. 7.27) .... ***Synalpheus***
8(6) Carpus of second pair of legs with 3 segments .... ***Hippolyte***
-Carpus of second pair of legs with 7 or more segments (fig. 7.28) .... ***Lysmata***
9(1) Carapace and abdominal segments with a dorsal keel; sides of body with a large, anterior pigment spot; no foliaceous appendage on the inner margin of the base of the antennules .... ***Sicyonia***
-Carapace and abdomen without a dorsal keel; sides of body without a large pigment spot; inner margin of basal joint of antennules with a foliaceous appendage (fig. 7.29) .... ***Penaeus***

SYSTEMATICS AND SPECIES DESCRIPTIONS

Natantia

Section (or Tribe) Caridea

Family Palaemonidae

Subfamily Palaemoninae

Genus *Palaemon*

This genus is characterized by the following set of characters: rostrum well-developed, toothed; carapace smooth, with antennal and branchiostegal spines; telson with two pairs of dorsal and two pairs of posterior spines; posterior margin of telson pointed; two feathered setae between the inner posterior spines. Holthuis (1952) reported five species from the west coast of the Americas, of which only one, *Palaemon ritteri* Holmes, is known to occur in the Gulf. This species ranges from southern California to Peru, including the Gulf of California. Steinbeck and Ricketts recorded it from the midrift islands, and Holthuis (1952) reported it from Bahía de Agua Verde, Isla San Francisco, and Cabo San Lucas. *P. ritteri* is a brightly colored tidepool shrimp resembling the "broken-back" shrimp of California shores. The rostrum is long and bears seven to eight dorsal teeth, one or two of which are behind the orbit, and three prominent ventral teeth. The length of this species is about 20 to 40 mm. The shrimp are translucent, with longitudinal reddish brown

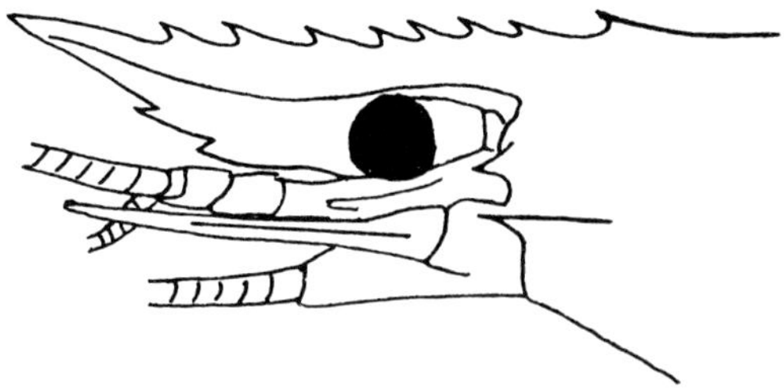

Fig. 7.23 *Palaemon ritteri* (lateral view of anterior region).

Fig. 7.24 *Periclimenes infraspinis* (lateral view of anterior region).

Fig. 7.25 *Pontonia pinnae* (lateral view of anterior region).

lines along the thorax and abdomen. There is a red spot, and occasionally two smaller black spots, on each uropod. I have collected this species throughout the Gulf, usually in tidepools, where they tend to aggregate under large rocks.

Subfamily Pontoniinae

Genus *Periclimenes*

A large genus, distinguished by the following combination of characters: rostrum compressed and provided with teeth; carapace smooth, may be provided with supraorbital, antennal, and hepatic spines; abdomen generally with all the segments broadly rounded (pleurae); carpus of first leg not segmented, second legs stronger than first. Holthuis (1951) reported two species from the Gulf of California. The first, *Periclimenes infraspinis* (Rathbun), formerly *Urocaris infraspinis,* has the rostrum straight and reaching about to the beginning or end of the third segment of the antennular peduncle. The upper margin of the rostrum bears five to eight teeth, the lower margin one to two small teeth in the extreme distal part. The sixth abdominal segment is twice as long as the fifth, and somewhat longer than the telson. The telson has two pairs of dorsal spines, and the posterior margin bears three pairs. According to Chace (1937), the color in life is semitranslucent, pale brown. This species ranges from southern California to Costa Rica and the Galapagos, as well as being reported from the lower Gulf of California. The second species, *Periclimenes lucasi* Chace, ranges from Baja California to Panamá, and has been reported from the extreme southern portion of Baja California, around Cabo San Lucas. This species has the rostrum failing to reach the end of the antennular peduncle. The upper margin of the rostrum is convex and bears eight to nine teeth, the lower margin one to three teeth near the apex. The sixth abdominal segment is about the same length as the telson. The sure way to distinguish these two species is to examine the dactylus (last segment) of the last three pairs of legs — in *P. lucasi* the dactyli are simple, while in *P. infraspinis* they are biunguiculate.

Genus *Pontonia*

This is another large genus of shrimp, with three species reported from the Gulf by Holthuis (1951). The genus is distinguished by the following set of characters: body dorsoventrally depressed; rostrum depressed, when teeth present, very small and few in number (1 to 2); carapace smooth, hepatic and supraorbital spines never present; carpus of first legs unsegmented; last three legs with bifid dactylus. The first of the three species reported from the Gulf is *Pontonia pinnae* Lockington, characterized by its strongly depressed and flattened rostrum, and the sixth abdominal pleura which ends in a strong spine, reaching over the base of the uropods. This shrimp is reported as a common commensal on the hatchet clam, *Pinna.* I have found it free-living in tidepools at Puerto Peñasco where the hatchet clam is known to exist. The second species is *Pontonia longispina* Holthuis, characterized by its depressed and somewhat down-curved rostrum, and large, well-developed dorsal spines on the telson (the dorsal spines of *P. pinnae* are very small and inconspicuous). The last species, *Pontonia margarita* Smith, also has the down-curved rostrum, but may be differentiated from *P. longispina* by examining the dactyli of the last three pair of legs. These are slender, with the posterior margin about straight, in *P. longispina,* but very broad, with the posterior margin distinctly convex, in *P. margarita.* Holthuis (1951) lists the distribution of *P. pinnae* as the Gulf of California to Panamá; *P. longispina* as the Gulf of California (Isla Angel de la Guarda and Guaymas); and *P. margarita* as the Gulf of California to Colombia and the Galapagos Islands. *Pontonia margarita* is almost always found living commensally in the pearl oyster, *Pinctada fimbriata* Dunker.

Genus *Palaemonella*

Palaemonella is characterized by the following set of characters: rostrum compressed, well-developed, with teeth; carapace smooth, hepatic and antennal spines present; well-developed postorbital ridge; mandibular palp with one or two joints; carpus of first legs unsegmented. Holthuis (1951) reported *Palaemonella holmesi* (Nobili), formerly *Periclimenes tenuipes,* in shallow coastal waters from southern California to Ecuador, including the Gulf of California. He describes it as follows: rostrum straight or slightly upturned at the tip; six to nine teeth on upper margin (two situated on the carapace behind the orbit), two to four ventral teeth; abdomen with first four pleurae rounded, the fifth ending in a small acute point; posterior margin of the telson with three pairs of spines, the median spine being very long and slender. The size of this species is about 10 to 24 mm, the color pale brown. The only other species of this genus reported by Holthuis from the Americas is *Palaemonella asymmetrica* Holthuis, known only from the Galapagos Islands.

Family Alpheidae (=Crangonidae)

Genus *Alpheus* (=*Crangon*)

These are some of the most common inhabitants of the rocky intertidal of the Gulf. Members of the genus *Alpheus* may be distinguished from the other common genus of snapping shrimp, or pistol shrimp, *Synalpheus,* by their compressed chelae, which in the latter genus are rounded. The length of the large cheliped in this genus often exceeds that of the carapace and abdomen combined. There are at least six or seven species of *Alpheus* known from the Gulf, one of which, *Alpheus clamator* (Lockington) is a common California form, known from the tidepools of southern California and both sides of Baja California. *A. clamator* is reported by Schmitt (1921) as ranging from San Francisco, California, to Cabo San Lucas; *A. malleator* (Dana) is reported by Steinbeck and Ricketts as occurring from Ecuador to the Gulf, where they report it from Bajio Plumo; and *A. lottini* Guerin is reported by Steinbeck and Ricketts (as *A. ventrosus*) as ranging throughout the Indo-Pacific, Hawaiian Islands, and the Gulf of California, where they also reported it from Bajio Plumo. In addition to these, Castro (1966) reported *A. californiensis* Holmes from the northern Gulf. I have collected six species of *Alpheus,* one of which is identical to the description of *A. californiensis* and which is no doubt the same species Castro described. Dr. Fenner Chace, Jr., of the Smithsonian Institution, has identified my specimens of *Alpheus* as follows: from Topolobampo, *A. californiensis* Holmes and *A. normanni* Kingsley; from Guaymas, *Alpheus* sp. (near *A. heterochaelis* Say) and *A. panamensis* Kingsley; and from Puerto Peñasco, *A. paracrinitus* Miers and *Alpheus* sp. (near *A. heterochaelis* Say).

Both the male and female members of this genus have the single enlarged cheliped, and both have the eyes hidden under the opaque carapace. The chelae are carried in an inverted position, with the dactyl or

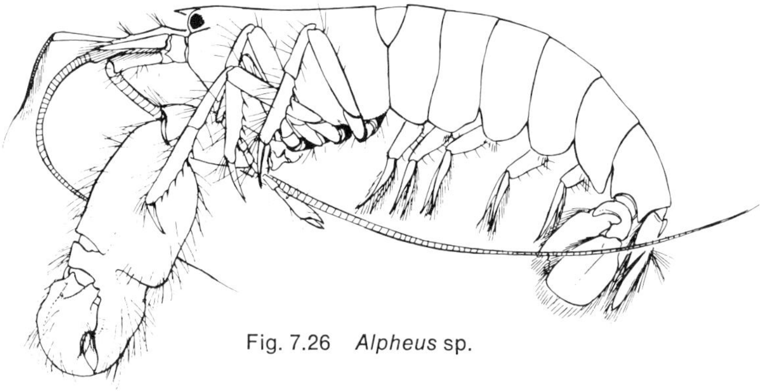

Fig. 7.26 *Alpheus* sp.

movable finger beneath, instead of uppermost as is usually the case. One need not see these curious creatures to be aware of their presence, for the sharp crack of their cheliped can be heard a good distance away. The sound is made by the shrimp's snapping its dactyl against the fixed finger of the large cheliped, the rapid impact being made possible by a trigger device at the joint. It is an odd experience to be walking along a reef at low tide, in the evening when the sea is calm and the air is still, and suddenly hear a chorus of loud reports resounding from the rocks below your feet. A curious phenomenon of these shrimp is that, when the large cheliped is pulled off, the new one will be formed, during the next molt, on the opposite hand. Molting can in fact be induced in these animals by removing the large cheliped. Once it is removed, the shrimp will shed in almost five weeks exactly (MacGinitie, 1968). In the aquarium, these shrimp can be induced to snap by agitating the water near their burrows. A number of functions probably can be assigned to the snapping device, but one that has been observed is the shrimp's ability to stun small fish for subsequent dining

Genus *Synalpheus*

Numerous species of this genus are present in the Gulf intertidal. The most common one in the northern Gulf (and probably occurring throughout the Gulf) is *Synalpheus lockingtoni* Coutiere, characterized by having the rostrum longer than the lateral spines of the anterior carapace margin, and the second joint of the attennules longer than the third. The body of this species is 15 to 30 mm long, and pale green. *S. lockingtoni* is also know from southern California and Baja California. I have collected this and one other species of *Synalpheus* in the Gulf, the other being an undescribed species from Topolobampo. Steinbeck and Ricketts reported at least four additional species of this genus, all from the southern Gulf: *S. digueti* Coutiere, *S. sanlucasi* Coutiere, *S. townsendi mexicanus*

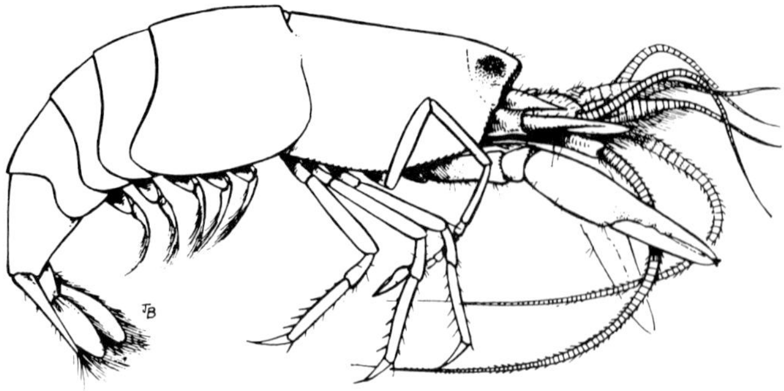

Fig. 7.27 *Betaeus longidactylus* (from Allen, 1969)

Coutiere, and *S. apiocerus sanjosei* Coutiere(?). Coutiere (1909) reported *S. sanlucasi* from Cabo San Lucas, *S. digueti* from the west coast of Baja California to Ecuador, and *S. goodei* from Baja California.

Another alpheid shrimp, the Calfornian *Betaeus longidactylus* Lockington, is reported to occur in the Gulf of California, although I have not collected it there. This species would key out to Synalpheus in this handbook. *B. longidactylus* has an elongated carapace that projects somewhat over the animal's eyes. The legs are red, and there is usually a fringe of yellow setae on the last segment of the abdomen. In adult males, the large chelae are often over half as long as the body, and are usually about equal in size. *B. longidactylus* ranges at least from northern California to the Mexican border, reappearing again in the Gulf. Schmitt (1924) reported it from Bahía de Tepoca.

Family Hippolytidae

Genus *Lysmata*

Probably several species of *Lysmata* are in the Gulf. The most common of these is the red and white striped shrimp common to southern California tidepools, *Lysmata californica* (Stimpson), formerly *Hippolysmata californica* Stimpson. These beautiful shrimp are found clinging to the underside of coarse boulders in the mid- and low intertidal. They may be picked off the rocks with relative ease, but once in the water they are nearly impossible to capture, as they swim backwards with great speed and agility. Linbaugh et al. (1961) stated that these shrimp may be facultative cleaners of fish, aiding in the removal of external parasites, but this behavior has not been seen by the author, either in the field or in the aquarium. These colorful little shrimp make excellent aquarium pets, as they are hardy and attractive, coming out in the evenings to dart about, filtering particles of food out of the water.

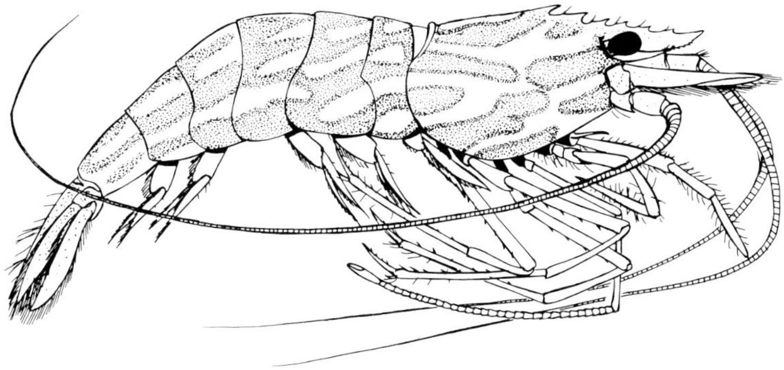

Fig. 7.28 *Lysmata californica*

L. californica is characterized by its red body with longitudinal white stripes (or vice versa), and the six to seven dorsal and three to four ventral teeth of the slender rostrum. The second pair of legs is very long and slender, the merus being divided into about 20 segments. In most individuals, the telson and third maxillipeds are dark mahogany. The body length often exceeds 50 mm. The range of this species is from Point Conception, California, to La Paz, on both sides of Baja California, and on the east side of the Gulf at least as far south as Bahía Kino. A second species, *Lysmata* sp.(?), has been collected at various points in the northern Gulf and probably represents a new species. It is characterized by a small rostrum of variable length. In specimens with a short rostrum (about two-thirds the length of the eye), there is a minute dorsal spine. All individuals have a cluster of about seven fine, ventral bristles on the rostrum. The body is opaque with dark brown spots. Castro (1966) reported a maximum length for this undescribed species of 23 mm in a gravid female. This species may be restricted to the tidepools of the northern Gulf. Dr. Fenner A. Chace has informed me that, based on Castro's description alone, he suspects this *Lysmata* to be in fact a member of the newly proposed genus *Ambidexter*. A third species has been collected only at Guaymas. It is nearly identical to *L. californica,* but lacks the characteristic stripes of the latter.

Genus *Hippolyte*

I have not collected any representatives of this genus in the Gulf. However, Chace (1951) reported *H. californiensis* Holmes as ranging from Bodega Bay, California, to the Gulf of California, and Schmitt (1924) reported it from Isla Patos in the Gulf. This species is green and abounds in the eelgrass beds of northern California. The rostrum has three to five dorsal and three to five ventral teeth.

Section (or Tribe) Penaeidea
Family Penaeidae

Genus *Penaeus*

Members of this genus are the large commercial shrimp collected subtidally by shrimp trawlers. They are rarely found in the intertidal zone. There are three or four principal species that make up the bulk of the Mexican shimp fishery in the Gulf. The first of these is *Penaeus stylirostris* Stimpson. Its rostrum has five to ten dorsal teeth and three to eight ventral teeth. *P. stylirostris* is a Panamic form ranging from the Gulf to Tumbes in northern Peru. It is caught in great numbers out of the ports of Guaymas, Topolobampo, and Mazatlán. The second principal commercial shrimp in the Gulf is *P. californiensis* Holmes. This is a Panamic species, the same one that occurs subtidally as far north as San Francisco Bay, California, and as far south as Peru. *P. californiensis* and *P. stylirostris* are both about 150 to 200 mm long when mature. *P. stylirostris* is white to blue, and commonly called the "white shrimp" or "blue shrimp." *P. californiensis* is light brown to reddish-brown, and is commonly called the "brown shrimp" or "khaki shrimp." *P. californiensis* is the dominant form in the northern Gulf. It has two to three ventral and eight to 11 dorsal teeth on the rostrum. It is closely related to *P. brevirostris* Kingsley, a third species known from the southernmost part of the Gulf. The latter is less common, and may be distinguished on the basis of the length of the gastro-orbital carina, and other minor differences. Schmitt (1921, 1935) treated *P. californiensis* and *P. brevirostris* as the same species, while Burkenroad (1935) separated them as two distinct species. Another species of commercial *Penaeus* is *P. vannamei* Boone, which is caught in the southern Gulf and increases in numbers with the spring and fall rain runoff. It ranges from the northern Gulf to Peru.

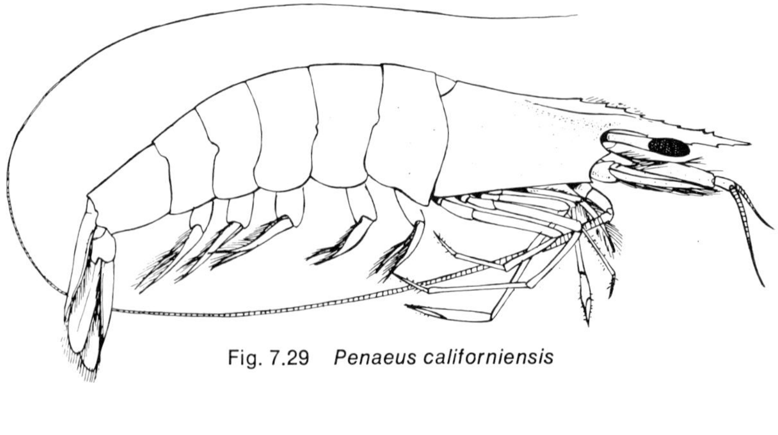

Fig. 7.29 *Penaeus californiensis*

Genus *Sicyonia*

These shrimp are usually brought up with *Penaeus* in shrimp trawls dragged across the sea floor in 10 to 40 fathoms. In the northern Gulf they also have been collected at very low tide in large, sandy bays. There are several species of this genus in the Gulf. Steinbeck and Ricketts reported *S. penicillata* (Lockington) from throughout the Gulf, and Schmitt (1924) reported it from Isla del Carmen. Castro (1966) stated his specimens from Puerto Peñasco did not match perfectly with the description of *S. penicillata,* but were near it or *S. disedwardsi.* Burkenroad (1938) reported seven species from the Gulf, all being very similar to one another. These shrimp are called the Japanese shrimp, or the "target shrimp" because of the large, dark, brownish-purple spot on their side, just behind the head. *Sicyonia* may be further distinguished by the keel-like extension of the rostrum along the dorsal margin of the carapace and abdomen, and the dorsal teeth on the carapace.

Suborder Reptantia

Section Macrura

Preservation: The macrurans should be preserved in alcohol, although the lobsters may be air dried subsequent to preservation in formalin. Storing specimens in the dried condition is meant to be done for display purposes only, and should not be done if the animals are to be maintained as reference collections or taxonomic data.

Taxonomy: see Crustacea section.

Macrura of the Gulf: Only two genera of macrurans are commonly seen in the Gulf, *Panulirus,* the spiny lobster, and *Evibacus,* the flat lobster. Both are subtidal forms occasionally picked up by shrimp trawlers. Spiny lobsters are best captured with SCUBA gear, or from a boat with trawling apparatus or lobster traps. Some authors include the Thalassinidea (ghost shrimp and their allies) in the section Macrura, while others prefer to place them in the section Anomura. I have chosen the latter course, and the ghost shrimp are therefore included in the following section of the handbook.

Family Palinuridae

Panulirus sp.

The spiny lobsters or "sea crayfish" of the west coast all belong to this genus. There are several species that occur on the west coast of America, but the systematics of the genus are not well-known due to

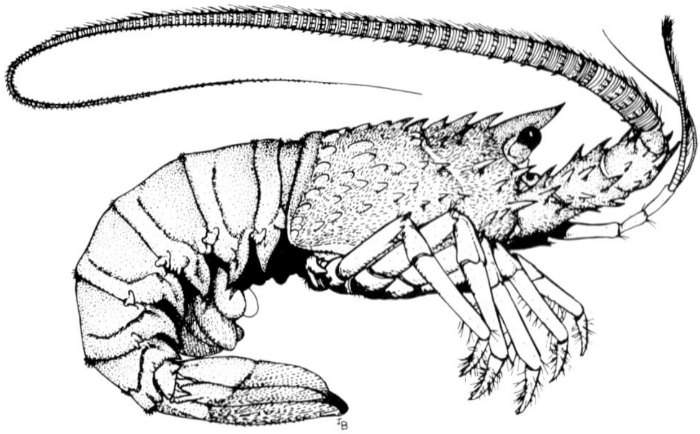

Fig. 7.30 *Panulirus* sp.

inadequate collections and other reasons. Steinbeck and Ricketts reported *Panulirus inflatus* Bouvier from the low intertidal at Punta Marcial and from the La Paz area. Dees (1968) reported *P. inflatus* as ranging from the Gulf of California to Ecuador. Boone (1931) reported its distribution as Baja California to Peru and the Galapagos Islands, and the distribution of *P. ornatus* as Peru and the Indo-Pacific. *P. interruptus* (Randall) is known from central California south to Isla de Cedros, on the west coast of Baja California (although Steinbeck and Ricketts reported it from Bahía San Francisquito in the northern half of the Gulf of California). *P. penicillatus* is reported from southern California to the Galapagos Islands and Costa Rica.

The spiny lobsters are not abundant in the Gulf, and are largely subtidal. The best lobster collecting method (other than commercial lobster traps and trawling) is by SCUBA diving. Migrations of these animals for several miles have been reported by a number of observers. The members of the migrating mass link themselves together, head to tail, forming a long chain of lobsters that slowly walk across the sea bottom, each following the member in front, and all ultimately dependent on the primary, leading lobster (Herrnkind, 1970).

Family Scyllaridae

Evibacus princeps Smith

The sand, flat, or boot lobster. This curious decapod has a very hard, flat carapace, that extends laterally beyond the margin of the body proper, like thin serrated wings. The abdomen can be tightly flexed under the thorax. There are five pairs of walking legs, no chelipeds. This strange lobster lives subtidally on sandy bottoms, and is occasionally picked up by the Gulf shrimp trawlers in their otter trawls, or washed ashore during a storm. It is prized for its delicious meat, but it is relatively scarce. There may be two species of flat lobster known from the Gulf, the other species being much larger and more laterally compressed. The Mexican shrimpers call the sand lobsters "langusta (or langosta) de arena" or "zapatera." The range is given by Rathbun (1910) as La Paz to Peru; however, Steinbeck and Ricketts reported it from Guaymas, in the lower Gulf, and I have collections from the Puerto Peñasco area, in the upper Gulf.

Section Anomura

Superfamily Thalassinoidea

Preservation: see general preservation, crustacea.

Taxonomy and Occurrence in the Gulf: The Gulf thalassinids resemble very small lobsters except that the first pair of legs is chelate, and the body is often soft and unpigmented as opposed to the rigid, colorful body of the true lobster. The rostrum is sometimes well-developed, and the tail fan is always well-developed. At least four genera of thalassinids occur in the Gulf intertidal: *Axius (Neaxius), Callianassa, Upogebia,* and *Callianidea.* The Gulf thalassinid fauna is only now being worked up. There are many undescribed and probably endemic species. Any specimens of this group would be welcomed by the University of Arizona, which has a specialist working on them.

Family Axiidae

Axius (*Neaxius*) *vivesi* (Bouvier) is a hard-bodied, well-pigmented, reddish species resembling a small crayfish. It was first described by Bouvier in 1895 from a unique specimen collected by M. Diguet in La Paz. It was later described more fully by J. G. De Man (1925a). The only published records since the original description are those of Steinbeck and Ricketts, who stated they found it in abundance east of La Paz in burrows in the gravelly mud flats, under and around the (mostly) dead heads of the coral *Porites,* and Westervelt (1967), who reported it from Puerto Peñasco, living in burrows in the rocky reef. I have found *Axius* to be present, but not abundant, at Puerto Peñasco and Bahía Cholla, where it lives in burrows and holes in the granite and Coquina reefs. I have also found it in great abundance at Bahía San Everisto (northeast of La Paz), and in the Cabo San Lucas region. The length and complexity of the burrows, coupled with the nature of the reef rock, make this animal especially difficult to capture, and this is no doubt one reason it has been so rarely reported in the literature. Martin Burkenroad has visited the Puerto Peñasco-Bahía Cholla area with considerable success in capturing *Axius* by enticing them out of their burrows with a bit of fresh meat (crushed snails, etc.), and then quickly spearing them with a small harpoon.

Family Callianassidae (ghost shrimp)

Ghost shrimp are found burrowing in muddy and sandy-mud bays and tidal flats, or occasionally along protected coastlines where rocks are imbedded in a soft substrate. They build elaborate burrows, often with one entrance under the edge of a large rock, and other entrances several feet away. Two genera of ghost shrimp occur in the Gulf: *Callianassa* and *Upogebia.*

Callianassa spp.

A genus of ghost shrimp similar to *Upogebia* (below), but with a less firm body and with the males possessing unequal chelae (the larger one being nearly as long as the body itself). Members of this genus, from the Gulf, are white to pinkish red, and are found in the same general habitat as *Upogebia.* The ghost shrimp pass quantities of the substrate through their guts to extract the organic detritus, much as earthworms do. They have been reported as turning over 30 inches of mud every 240 days. At least three species of this genus of ghost shrimp are known from the Gulf: *Callianassa seilacheri, C. rochei,* and *C. uncinata* (possibly a fourth species, the common Californian form, *C. californiensis*). Ghost shrimp often harbor a number of commensal and ectoparasitic animals in their burrow with them. The following animals have been found living with *C. californiensis* on the Pacific coast of the United States: *Clevelandia ios* (the arrow goby); *Typhlogobius californiensis* (the blind goby);

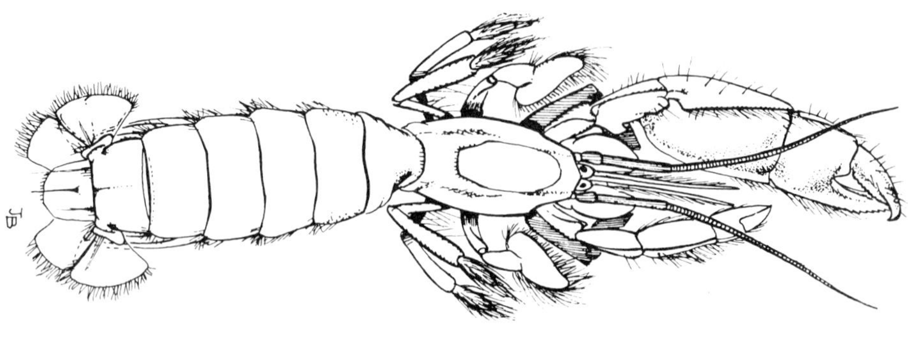

Fig. 7.31 *Callianassa californiensis* (from Allen, 1969)

Scleroplax granulata, Pinnixa franciscana, and *P. schmitti* (pea crabs); *Hesperone complanata* (a scale worm); *Betaeus complanta,* usually a male and a female (shrimp); *Cryptomya californica* (a small clam); and the ectoparasitic copepods *Hemicyclops callianassae, H. thysanotus* and *Clausidium vancouverense.* Members of the same or similar genera and species may be expected to be found living in the burrows of other species of ghost shrimp. I have collected only one member of this genus in the Gulf, from Bahía Adair. Mr. Burkenroad has identified it as *C. uncinata.*

Upogebia spp.

Members of this genus of ghost shrimp are found burrowing in soft mud, under rocks, or at the base of mangrove roots. They build an elaborate burrow, several feet long, with numerous side branches and turn-arounds. The chelae are approximately equal in size in both sexes, and the body is generally more rigid than in *Callianassa.* Careful collectors will often find one male and one female ghost shrimp inhabiting each burrow, plus a host of small commensals (see *Callianassa).* There are at least two species of *Upogebia* known from the Gulf of California, *Upogebia pugettensis* (the "blue mud shrimp"), and *U. rugosa.* The former ranges from Alaska to La Paz, on both sides of the Gulf. In California, *U. pugettensis* often harbors a parasitic isopod, *Phyllodurus abdominalis.* However, to the author's knowledge, this parasite has not yet been recorded from the Gulf of California. Many members of this genus tend to feed on plankton and suspended detritus, as opposed to filtering food out of the mud, as is done in the genus *Callianassa.*

Superfamily Galatheoidea
Family Porcellanidae (Porcelain Crabs)

Preservation: See Brachyuran section.

Taxonomy: The only group of galatheids commonly encountered in the Gulf intertidal are the porcelain crabs (Porcellanidae). The following section deals only with this family (see *Crustacea of the Gulf,* p. 182). The taxonomy of the porcellanids is based primarily on external features of the carapace, legs, antennae, and abdomen. The main features used to distinguish the species of the Gulf are carapace and cheliped structures. A brief discussion of some of these structures that are mentioned in the key follows:

Epibranchial Spines: The carapace of crabs may be divided into various areas, corresponding to the internal anatomy, and usually visible externally as elevated regions, demarcated by shallow grooves. For the purpose of this key, the only zone necessary to recognize is the epibranchial region, which is that area from the lateral margin of the carapace (about midway from anterior to posterior) to an area about even with a line extending directly back from the eye. This is roughly the two anterolateral quarters of the carapace. The importance of this region is whether or not it possesses small spines (the epibranchial spines), and if so, how many. The spines are usually entirely dorsal, but they may occasionally be formed as lateral projections of the carapace.

Piliferous Striations: The carapace of some porcelain crabs has numerous shallow, transverse depressions, each lined with a row of short, stiff setae. The setae are usually microscopic.

Cheliped Characters: The first segment behind the manus (the manus is the claw portion of the cheliped) is the carpus. The important characters of the carpus are its length-to-width ratio and the number of spines it has along the anterior or inside (leading) margin. Whenever the terms "manus" and "carpus" are used in the key, they refer to the first leg (chelipeds) only.

Telson: The telson is that portion of the abdomen that consists of articulating plates or pieces. The majority of the porcelain crabs have seven free plates in the telson, but a few have only five or six. When using the key or species descriptions, you may assume the crab's abdomen is seven-plated unless it is otherwise stated.

See fig. 7.48 for remaining crab morphology.

The systematics used in this handbook are from Haig (1960), and many of the key couplets are extracted or modified from her excellent monograph of the eastern Pacific Porcellanidae.

Porcelain Crabs of the Gulf: There are about ten genera of Porcellanidae common to the Gulf intertidal. The dominant genera are *Petrolisthes* (with about 18 species in the Gulf) and *Pachycheles* (with about five species in the Gulf). The most commonly encountered species throughout the Gulf are the little hairy-clawed *Pachycheles setimanus* (with unequal chelae), the smooth-clawed *Petrolisthes armatus, Petrolisthes crenulatus,* and *Petrolisthes gracilis* (with subequal chelae), and the hairy-clawed *Petrolisthes hirtipes* (with subequal chelae). In the northern Gulf *Petrolisthes sanfelipensis* is quite abundant, while in the southern Gulf *Petrolisthes nigrunguiculatus* is often encountered.

Many of the porcelain crabs of the Gulf are unique to this region, and are found nowhere outside the Gulf of California. The vast majority of the

Porcellanidae of this region, however, have Panamic affinities. In fact, not one of the crabs treated in this section is known to occur north of Mexico. Westervelt (1967) stated 42 percent of the anomurans (porcellanids and pagurids) he studied from the Puerto Peñasco region were endemics, and 46 percent were Panamic.

Most porcelain crabs are found clinging to the undersides of rocks in the mid- and low intertidal zones, or hiding in the crevices and chambers of coral and sponge. Some species are commensal on hermit crabs (*Porcellana paguriconviva* and *Porcellana cancrisocialis*) or gorgonian coral (*Orthochela pumila, Ulloaia,* etc.). Three species in the key superficially resemble the mole crabs (superfamily Hippidea). They are: *Euceramus transversilineatus, Minyocerus kirki,* and *Ulloaia perpusillia.* These three species may be readily distinguished from the true mole crabs by the presence of pincers on the first legs (chelipeds), whereas the mole crabs are without pincers.

Several of the species included in the key are rare or uncommon in the Gulf. These forms are listed below with a note of explanation as to their rarity, and are marked with an asterisk in the key.

Orthochela pumila: Commensal on gorgonian coral; largely subtidal; only Gulf record is from Mazatlán.

Petrolisthes glasselli: Usually associated with coral; largely subtidal; not reported north of La Paz.

Petrolisthes lewisi: Rarely encountered.

Petrolisthes nobilii: Not reported north of La Paz.

Petrolisthes tonsorius: Known only from the mouth of the Gulf (Cabo San Lucas).

Pachycheles spinidactylus: Only Gulf record is Cabo San Lucas.

Porcellana corbicola: Known only from Roca Consag, northern Gulf, at 10 to 25 fathoms.

Megalobrachium erosum: Subtidal (5 to 25 fathoms).

Minyocerus kirki: Commensal on sea stars, apparently restricted to the genus *Luidia,* and mainly subtidal.

Ulloaia perpusillia: Known only from three localities: Puerto Peñasco, Costa Rica, and Panamá.

Polyonyx nitidus: Uncommon and largely subtidal. Often commensal in tube of *Chaetopterus,* the polychaete worm.

KEY TO THE PORCELLANID CRABS OF THE GULF OF CALIFORNIA

1 Carapace longer than broad **2**
-Carapace length equal to, or less than, width **8**

2(1) Chelipeds normally held more or less straight out in front of carapace; lateral margins of carapace with a series of about 12 to 15 minute, close-set spinules; always commensal on gorgonian coral .... ***Orthochela pumila****
-Chelipeds normally bent sharply at articulation of carpus and manus, with manus directed inward; lateral margins of carapace without numerous close-set spinules; rarely commensal on gorgonian coral ... **3**

3(2) Antennae minute and rudimentary, the total length less than or barely exceeding the width of the eye; commensal on sea stars .... ***Minyocerus kirki****
-Antennae not minute, rather large and easily seen, length much greater than width of eye; not commensal on sea stars **4**

4(3) Chelae unequal in size; manus tuberculate; carapace and legs very rough, with numerous large ridges and furrows .... ***Ulloaia perpusillia****
-Chelae subequal; manus not tuberculate; carapace and legs relatively smooth, not as above **5**

5(4) Carapace 1.5 times as long as broad (or more); a large, orbitlike concavity on hepatic margin; usually found burrowing through sand .... ***Euceramus transversilineatus***
-Carapace not 1.5 times as long as broad; no orbitlike concavity on hepatic margin; usually found clinging to rocks or in coral or sponge ... **6**

6(5) Telson of abdomen with 5 plates; fingers of chelae opening vertically; carpus without spines or a large lobe on the anterior margin; probably not commensal with hermit crabs .... ***Porcellana corbicola****
-Telson of abdomen with 7 plates; fingers of chelae opening horizontally (or oblique); carpus with a single spine or lobe on the anterior margin; usually commensal on hermit crabs **7**

7(6) Carapace with 2 to 3 epibranchial spinules (fig. 7.43); carpus with a spine-tipped lobe on anterior margin (fig. 7.44); manus with long-marginal hairs (fig. 7.44); carapace yellow to orange, with red and blue spots .... ***Porcellana cancrisocialis***
-Carapace with no epibranchial spines; carpus without a spine on the anterior margin but with a large lobe; manus without long, marginal hairs; carapace lavender with orange stripes .... ***Porcellana paguriconviva***

8(1) Chelae subequal in size **9**
-Chelae unequal in size **33**

9(8) Basal segment of antennae large, produced forward, and broadly in contact with anterior margin of carapace, leaving movable segments far removed from orbit; chelipeds stout, length of carpus never 2 times width **10**
-Basal segment of antennae short, not strongly produced forward to meet anterior margin of carapace, leaving movable segments with free access to orbit; chelipeds usually (but not always) thin and flat, length of carpus usually 2 times width **15**

10(9) Telson of abdomen with 5 plates **11**
-Telson of abdomen with 7 plates **12**

12(10) Dactylus of walking legs ending in 2 or more large, strong, fixed spines; carapace markedly broader than long; front nearly transverse in dorsal view .... ***Polyonyx nitidus****

-Dactylus of walking legs ending in a single spine, often with accessory movable spinules on posterior margin; carapace not markedly broader than long; front variable **13**

11(10) Carapace and legs covered with small, shallow pits; legs without stiff, marginal setae; carpus with 1 spine on anterior margin .... ***Megalobrachium garthi***

-Carapace and legs not covered with small pits, but rather covered with numerous tubercles; legs with stiff, marginal setae; carpus with 3 to 4 spines on anterior margin ... ***Megalobrachium tuberculipes***

13(12) Carapace and chelipeds heavily eroded and pitted; not known to occur intertidally .... ***Megalobrachium erosum****

-Carapace and chelipeds smooth to rough, but not heavily eroded and pitted; intertidal or subtidal **14**

14(13) Carpus with a narrow, blunt spine on proximal third of anterior margin and often with several small blunt spines beyond this .... ***Megalobrachium sinuimanus***

-Carpus without spines, or at most with a slight projection at anteroproximal angle .... ***Megalobrachium smithi***

15(9) Anterior margin of walking legs spined; epibranchial spine(s) present **16**

-Anterior margin of walking legs without spines; epibranchial spines absent **24**

16(15) Carapace with distinct transverse piliferous striations or grooves, interrupted at grooves defining the various body regions **17**

-Carapace smooth to rough, but without transverse striations as above ... **21**

17(16) Length of carpus 2 times width; manus often with distinct tubercles; carpus with 3 to 5 spines on anterior margin **18**

-Length of carpus less than 2 times width; manus without distinct tubercles; carpus with 4 to 6 spines on anterior margin **20**

18(17) Manus and walking legs with a fringe of very long setae (figs. 7.37-7.38) .... ***Petrolisthes hirtipes***

-Manus and walking legs without a fringe of long setae **19**

19(18) Anterolateral area of carapace with a group of spines in addition to usual epibranchial spine; carpus with 3 to 5 *simple* spines on the anterior margin (fig. 7.41) .... ***Petrolisthes sanfelipensis***

-No spines on dorsal surface of carapace, other than a single epibranchial spine; carpus with 3 *serrated* spines on anterior margin .... ***Petrolisthes edwardsii***

20(17) Two epibranchial spines; carapace length 4 to 11 mm .... ***Petrolisthes glasselli****

-One epibranchial spine; carapace length 3 to 6 mm .... ***Petrolisthes polymitus***

21(16) Length of carpus less than 2 times width; anterior margin of carpus with 4 to 6 spines .... ***Petrolisthes haigae***

-Length of carpus 2 or more times width; anterior margin of carpus with 3 to 6 spines **22**

22(21) Anterior margin of carpus with 4 to 6 spines; spines of carpus broad and close-set (figs. 7.39, 7.40) .... ***Petrolisthes hirtispinosus***

-Anterior margin of carpus with 3 to 4 spines; spines of carpus low and wide-set **23**

23(22) Epibranchial spine distinct; no short pubescence on outer margin of ventral surface of manus; carapace nearly as wide across midbranchial region as posteriorly (fig. 7.32) .... ***Petrolisthes armatus***

-Epibranchial spine absent or vestigial; ventral surface of manus with a narrow band of short pubescence along outer margin; carapace widest posterior to midbranchial region .... ***Petrolisthes nobilii****

24(15) Epibranchial spines small and vestigial or entirely lacking; anterior margin of walking legs spined .... ***Petrolisthes nobilii****

-Epibranchial spines always absent; anterior margin of walking legs without spines **25**

25(24) Merus of walking legs nodulate on anterior margin; carpus of females with 5 to 8 very small spines on anterior margin; carpus of male unarmed; carpus over 2.5 times as long as broad ... ***Petrolisthes tiburonensis***
-Merus of walking legs without nodules on anterior margin; carpus of females with 0 to 4 spines on anterior margin; carpus of male armed or unarmed; length of carpus variable **26**
26(25) Carpus not armed with strong teeth or tubercles **27**
-Carpus with strong teeth or tubercles on anterior margin **29**
27(26) Carpus over twice as long as wide (figs. 7.35, 7.36) .... ***Petrolisthes gracilis***
-Carpus twice as long as wide or less than twice as long as wide **28**
28(27) Margin of carapace, between eyes, forms a narrow, rounded projection; length of carpus nearly 2 times width; occasionally with a single spine on anterior margin of carpus; length of carapace 3 to 12 mm .... ***Petrolisthes tonsorius****
-Margin of carapace, between eyes, forms a wide, V-shaped projection; length of carpus much less than 2 times width; often with a row of very small granules on anterior margin at carpus; length of carapace 3 to 6 mm .... ***Petrolisthes schmitti***
29(26) Manus with a thick fringe of hair on outer margin; fingers dissimilar in the two chelae; carpus with wide-set, conical tubercles on the anterior margin **30**
-Manus without a thick fringe of hair; fingers not distinctly dissimilar in the two chelae; carpus with strong teeth on anterior margin **31**
30(29) Tips of chelae red-orange; epibranchial area often with a tubercle or 1 or 2 minute spinules; manus with numerous denticles; posterior margin of merus of walking legs 1 to 2 with 2 to 3 distal spines; carapace covered with short, transverse plications (figs. 7.37, 7.38) .... ***Petrolisthes hirtipes***
-Tips of chelae not red-orange; epibranchial area never with a tubercle or spines; manus with few or no denticles; posterior margin of merus of walking legs 1 to 2 without spines; carapace nearly smooth, not as above .... ***Petrolisthes nigrunguiculatus***
31(29) Telson of abdomen with 5 plates; outer orbital angle produced into a distinct tooth .... ***Petrolisthes hians***
-Telson of abdomen with 7 plates; outer orbital angle not strongly produced **32**
32(31) Manus with 2 longitudinal crests on dorsal surface, and a heavy, rounded crest along the outer margin .... ***Petrolisthes lewisi****
-Manus without 2 dorsal crests, and without a heavy rounded crest along the outer margin **33**
33(32) Anterior margin of carapace forms a 3-lobed shelf between eyes; carapace covered with minute granules and/or short plications; carpus about 1.5 times as long as wide; carpus with 3 or 4 spines on anterior margin; length of carapace 3 to 14 mm (figs. 7.33, 7.34) .... ***Petrolisthes crenulatus***
-Anterior margin of carapace forms a straight shelf, not 3-lobed; carapace nearly smooth, not as above; carpus less than 1.5 times as long as wide; carpus with 3 spines on anterior margin; length of carapace 3 to 7 mm .... ***Petrolisthes ortmanni***
34(8) Telson of abdomen with 5 plates **35**
-Telson of abdomen with 7 plates **36**
35(34) Carpus with 0 to 3 spines on anterior margin; carapace devoid of visible hair; manus without setae .... ***Pachycheles biocellatus***
-Carpus with 4 to 5 spines on anterior margin; carapace with numerous, fine setae; manus lightly setaceous ...... ***Pachycheles marcortezensis***
36(34) Carapace with a distinct tuft of hairs on front, or surface covered with distinct hairs **37**

-Carapace devoid of hairs (or at most with traces scarcely visible under magnification) (fig. 7.42) .... ***Pachycheles setimanus***

37(36) Front with a distinct tuft of hair; carpus with 3 to 4 spines on anterior margin .... ***Pachycheles spinidactylus****

-Front without a distinct tuft of hair; carpus with 2 to 3 spines on anterior margin .... ***Pachycheles panamensis***

SYSTEMATICS AND SPECIES DESCRIPTIONS

Porcellanidae

Superfamily Galatheidea
Family Porcellanidae

Genus *Orthochela*
Orthochela pumila Glassell

A small and rather strange-looking porcelain crab with very large, heavy chelipeds, with the manus held straight out in front of the carapace instead of folded inward, as in most crabs. The chelipeds are unequal in size, the largest being about 2.5 times the carapace length. The chelipeds are without hair, the carpus lacks spines on the anterior margin, and the palm is longer than the fingers. The fingers open vertically, and bear a short pubescence in the gape. The carapace is nearly 1.5 times as long as broad, and has a large, projecting, median lobe. The lateral margins have 12 to 15 spinules, but lack epibranchial spines. The color of this crab is variable. It has been recorded as yellow, with red lines and blotches; solid purple; blotched red and yellow; brown with white spots; and white with rust-colored spots. However, it seems that the crab will always match perfectly the color of the gorgonian coral upon which it is living. It has seven abdominal plates, and the carapace is about 2 to 4 mm long. *Orthochela pumila* is relatively uncommon, and is an obligatory commensal on gorgonian coral, restricting its habitat to the subtidal and the lowest of low tides. Its range is reported as Bahía San Hipolito on the outer Baja California coast, and Mazatlán on the mainland, south to Ecuador.

Genus *Petrolisthes*
Petrolisthes armatus (Gibbes)

A very common crab of the Gulf. The carapace is pale, brown to olive-black, occasionally with green markings, occasionally pubescent (but usually smooth and naked), and bearing a single epibranchial spine. The carpus of the cheliped is two to 2.5 times as long as wide, usually naked, and armed with 3 to 4 teeth along the anterior margin. The manus

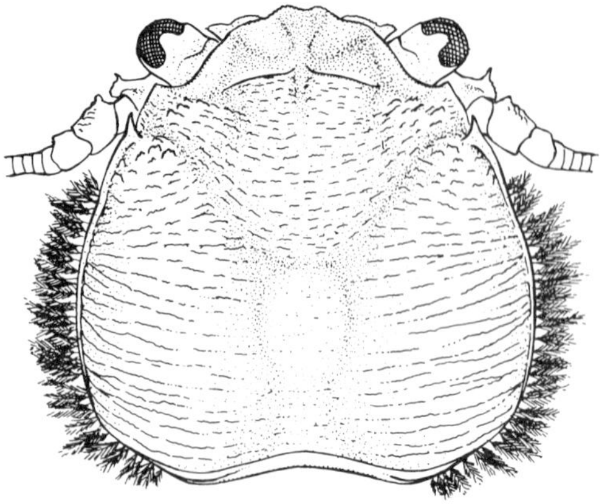

Fig. 7.32 *Petrolisthes armatus* (carapace)

is usually naked, although young individuals often have some setae. The gape of the fingers is pubescent. This species is about 4 to 15 mm long, and is common under stones in the mid- and high intertidal. It has been reported in mangrove beds, coral, sponge, oyster, and mussel beds, and under rocks from both the Atlantic and Pacific oceans. On our coast, it ranges from the upper Gulf to Peru and the Galapagos Islands. Feeding methods in the porcellanids have been observed and recorded by a number of authors, and is generally described as "cast-net" feeding, wherein the enlarged and highly setose third maxillipeds are waved through the water to collect plankton, and then brought in to be cleaned off by the second maxillipeds. Westervelt (1967) has a good discussion of porcellanid and pagurid feeding methods, and has recorded observations on *Petrolisthes armatus* and ten other species of porcelain crabs.

Petrolisthes crenulatus Lockington

A common species of crab throughout the Gulf. The carapace is covered with short plications, lacks epibranchial spines, and is about 3 to 14 mm long. The front is trilobate, and often bears long plumose hairs. The carpus of the cheliped is about 1.5 times as long as wide, with 3 to 4 teeth on the anterior margin. The manus has long plumose hairs, and the gape of the fingers bears a thick pubescence. A large species, easily recognized by its dull white carapace, covered with many rust-colored blotches and large teeth on the inner margin of the carpus. The fingers are entirely red, and the eyes are bright green. This species is common from Bahía Magdalena and the upper Gulf to Mazatlán and Islas las Tres Marías. It usually is found under stones in the low intertidal, but is also known to occur in coral heads.

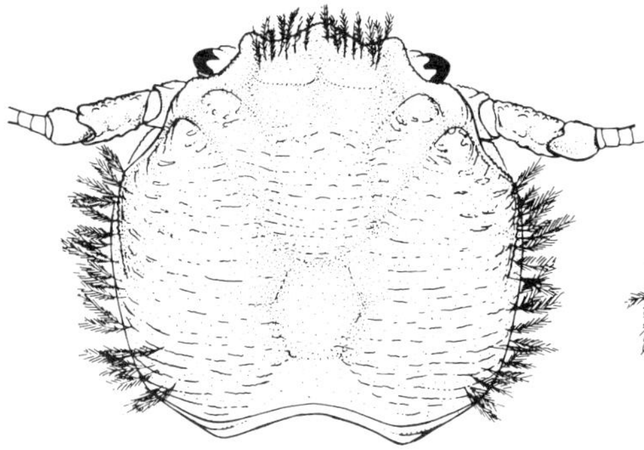

Fig. 7.33 *Petrolisthes crenulatus* (carapace)

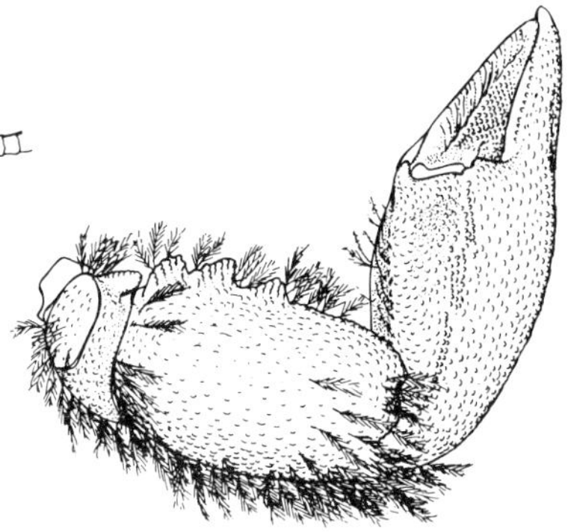

Fig. 7.34 *Petrolisthes crenulatus* (cheliped)

Petrolisthes edwardsii (Saussure)

This species is uncommon in the Gulf. The carapace is blotched with pink or red, and has strong, transverse, piliferous striations and a single epibranchial spine. The carpus of the cheliped is about two times as long as wide, and is armed with three denticulate spines on the anterior margin. The manus is broad and covered with tubercles (as is the carpus) and a few proximal spines. The fingers bear a short pubescence. This is a large species, reaching about 16 mm in length. *Petrolisthes edwardsii* occurs under rocks and in coral from Bahía Santa María on the outer Baja California coast, and the upper Gulf, to Ecuador and the Galapagos Islands. It has been recorded from the intertidal, as well as to depths of 20 fathoms.

Petrolisthes glasselli Haig

This species is uncommon in the Gulf. The carapace is about 4 to 11 mm long and covered with transverse, piliferous striations. It is the only member of this genus in the Gulf with two epibranchial spines. The carpus of the chelipeds is less than twice as long as wide, with four to five teeth along the anterior margin. The manus is broad, with a few flattened tubercles and occasionally a fringe of hairs. The fingers do not bear pubescence. Records indicate this species is usually associated with corals *(Pocillopora* and *Pavona),* and ranges from the mouth of the Gulf (Cabo San Lucas) south to Colombia and the Galapagos Islands.

Petrolisthes gracilis Stimpson

A common species in the Gulf. The carapace is about 3 to 11 mm long, bluish green with white spots, nearly smooth, and without an epibranchial spine. The carpus is 2.5 to three times as long as wide, and without spines along the anterior margin. The manus is long and slender, with no hairs, but the fingers have a thick pubescence in the gape. This species is found under stones from the upper Gulf as far south as La Bahía Tangola-Tangola, México.

Petrolisthes hians Nobili

The carapace of this species is about 2 to 6 mm long, rough anteriorly and smooth posteriorly, and lacks an epibranchial spine. The chelipeds are short and stout. The carpus is very wide, less than 1.5 times as long as broad, the anterior margin bearing three to four broad, rounded teeth. The manus is covered with rough plications, with a crest on the outer margin, and completely devoid of pubescence. This is the only species of *Petrolisthes* in the Gulf with a five-plated telson. It is usually found associated with coral and sponge, or occasionally under stones, in the low intertidal and subtidal. The range is reported as Guaymas to Ecuador, in addition to a record from Bahía Santa María on the outer Baja California coast.

Petrolisthes hirtipes Lockington

A common and easily recognized brown porcellan crab of the Gulf. The legs all bear a long fringe of hair, and the chelae have many large tubercles. The carapace is about 3 to 11 mm long, finely plicate and lightly tuberculate, and lacks an epibranchial spine (although a tubercle or one to two small spinules may be present). The carpus is about two times as long as wide, with a series of small tubercles in the proximal half. The manus is tuberculate and highly setaceous, with a short pubescence in the gape of the fingers. The species is fairly common under stones and in crevices of reefs, coral, and sponges in the low intertidal from Bahía Magdalena to Cabo San Lucas, and throughout the Gulf of California.

Petrolisthes hirtispinosus Lockington

A common form, found under rocks in the intertidal throughout the Gulf (and probably restricted to it). The carapace is nearly smooth, and bears a single epibranchial spine. The carpus is a little over two times as long as wide, the anterior margin with four to six serrate-edged spines. The manus has a longitudinal crest of flattened granules, a short fringe of hairs (easily rubbed off), and a short pubescence in the gape of the fingers. The carapace is light red, mottled with yellow and white, and the length is reported as 2 to 13 mm.

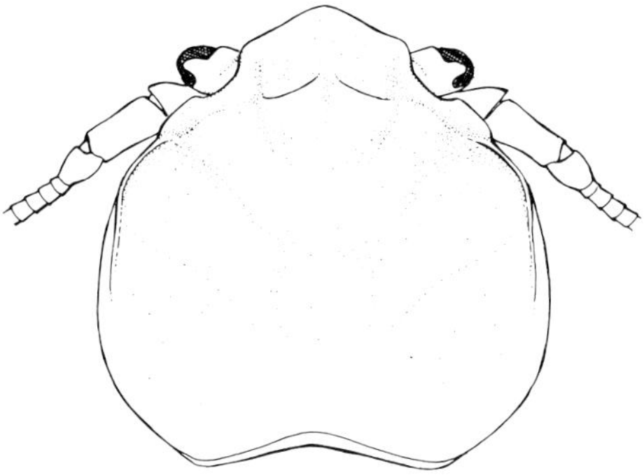

Fig. 7.35 *Petrolisthes gracilis* (carapace)

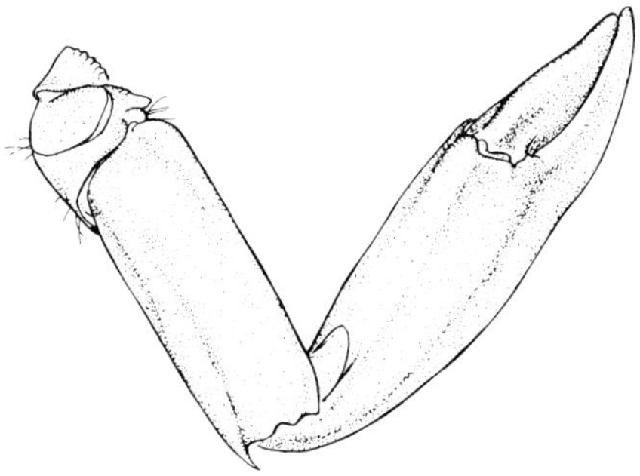

Fig. 7.36 *Petrolisthes gracilis* (cheliped)

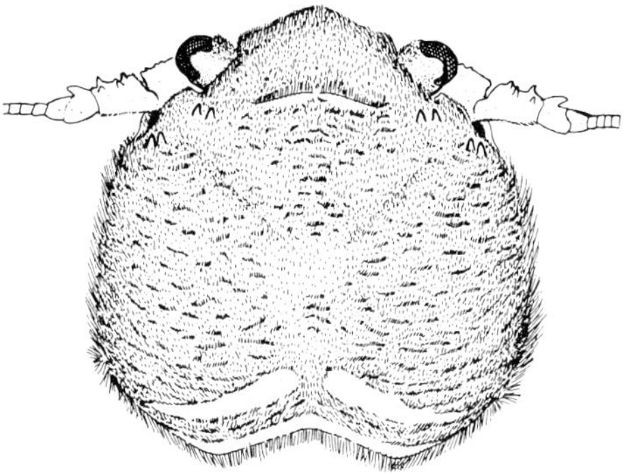

Fig. 7.37 *Petrolisthes hirtipes* (carapace)

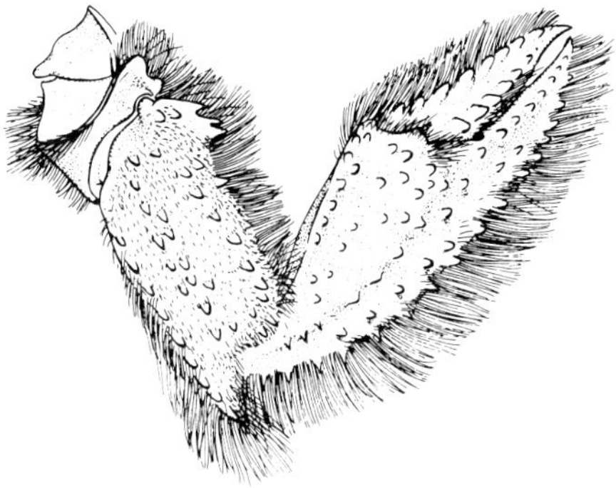

Fig. 7.38 *Petrolisthes hirtipes* (cheliped)

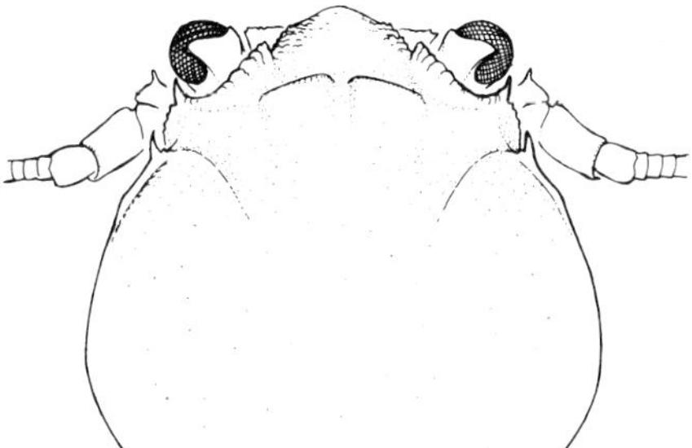

Fig. 7.39 *Petrolisthes hirtispinosus* (carapace)

Fig. 7.40 *Petrolisthes hirtispinosus* (cheliped)

Petrolisthes lewisi (Glassell)

A rare species in the Gulf. The carapace is nearly smooth, lacks epibranchial spines, and is about 3 to 6 mm long. The carpus is about 1.5 times as long as wide, and bears three to four spines on the anterior margin. The manus has three heavy crests, defined by deep grooves. The gape of the fingers has a short pubescence. Miss Janet Haig informs me the range of the Gulf subspecies *P. lewisi lewisi* is Isla del Carmen, Gulf of California to Bahía Tangola-Tangola, México. Another subspecies, *P. lewisi austrinus,* is reported from Nicaragua to Ecuador.

Petrolisthes haigae Chace

This species has a smooth carapace, a single epibranchial spine, and is reported as 3 to 13 mm long. The carpus is less than two times as long as wide, and bears four (occasionally five to six) spines on the anterior margin. The manus is granular, with a longitudinal crest of larger, flattened granules extending from the base of the fingers. This species is rarely found under stones, preferring to hide in the interstices of coral (*Pocillopora*). Its range is recorded as Guaymas to Ecuador and the Galapagos Islands. Until 1962, this species was known as *Petrolisthes marginatus,* an east coast form it closely resembles.

Petrolisthes nigrunguiculatus Glassell

This crab has a smooth carapace, about 3 to 10 mm long, without an epibranchial spine, and is mottled with brown and red. The carpus is about two times as long as wide, and bears four spines on the anterior margin. The manus has a thick fringe of hair on the outer margin and in the gape of the fingers. Usually found under stones in the intertidal.

Petrolisthes ortmanni Nobili

Petrolisthes ortmanni ranges from Bahía San Carlos in the Gulf, to Peru. The carapace is nearly smooth, without an epibranchial spine, and is about 3 to 7 mm long. The carpus is less than 1.5 times as long as wide, with three broad teeth on the anterior margin. The manus is smooth to granular, with a thick pubescence in the gape of the fingers. *P. ortmanni* has been taken from stones and coral in the low intertidal zone.

Petrolisthes polymitus Glassell

An uncommon species, with the carapace bearing distinct, transverse, piliferous striations and a single epibranchial spine. The length of the carapace is about 3 to 6 mm. The carpus is about 1.5 times as long as wide, with four to five broad, serrated spines on the anterior margin. The manus and fingers have flattened rugae and granules covering the surface

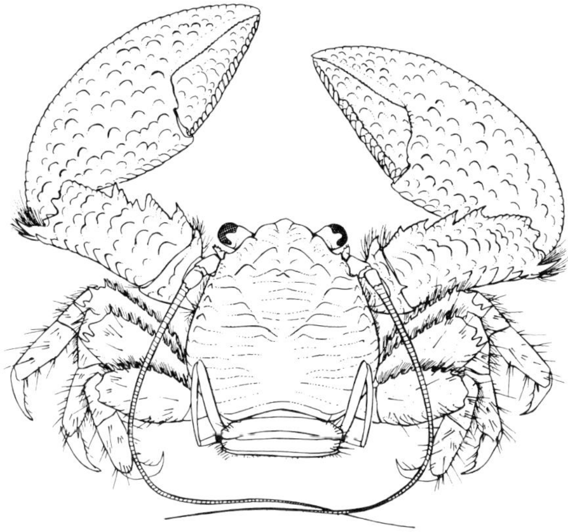

Fig. 7.41 *Petrolisthes sanfelipensis*

with a row of spines and scattered hairs on the outer margin. This species ranges from the southern Gulf to Ecuador and the Galapagos Islands, and occurs under rocks and in coral and sponges.

Petrolisthes sanfelipensis Glassell

A common species with a bright pink or red carapace, about 3 to 11 mm long, having distinct, transverse, piliferous striations on the dorsal surface, and a cluster of short spines in the anterolateral corner of the carapace (near the epibranchial spine). The carpus is about two times as long as wide, and bears three to four spines on the anterior margin. The manus is long and slender, covered with small granules, and bears a row of spines and a fringe of scattered plumose hairs along the outer margin. The fingers bear a short pubescence in their gape. The color in these crabs becomes greatly intensified when preserved in alcohol. This species is common in the low tide zone under rocks, and in sponge and gorgonian coral throughout the Gulf of California. It has been reported also from Bahía Magdalena on the west side of Baja California.

Petrolisthes schmitti Glassell

An uncommon species, reported from only a few locations in the upper Gulf of California. The carapace is finely granulate, with no epibranchial spine. The carpus is less than two times as long as wide, the anterior margin possessing numerous enlarged granules or spinules. The manus is somewhat swollen on the dorsal surface, with several large granules and a smooth crest, defined by a broad groove near the outer margin. The fingers bear a thick tuft of pubescence in the gape. The carapace is about 3 to 6 mm long.

Petrolisthes tiburonensis Glassell

A somewhat uncommon species, reported from only a few localities, all in the Gulf, from San Felipe and Puerto Peñasco south to Punta Trinidad. The carapace is chocolate-brown, about 6 to 12 mm long, and covered with rough granules and plications. There are no epibranchial spines. This species shows a great deal of sexual dimorphism. The carpus in males is over three times as long as broad, in females about 2.5 times as long as broad. The carpus in males is unarmed, but in females there are about five spinules on the anterior margin. The manus in males is unarmed, but in females bears numerous spines along the outer margin.

Petrolisthes tonsorius Haig

The only records of this species from the Gulf are from Cabo San Lucas (near the mouth). The carapace is smooth, about 3 to 12 mm long, and lacks an epibranchial spine. The carpus is about 1.5 times to two times as long as broad, and lacks spines on the anterior margin. The manus is without setae, except for a fine pubescence in the gape of the fingers. This species is found under rocks in the intertidal from Cabo San Lucas to Ecuador and the Galapagos Islands.

Genus *Pachycheles*

Pachycheles biocellatus (Lockington)

Uncommon in the Gulf. The carapace is smooth, about as broad as long in males, slightly broader in females. The chelipeds are unequal, and devoid of hair. The carpus is less than two times as long as broad and has none to three lobed spines on the anterior margin. The manus is smooth, without hair. The abdomen of this species has only five plates, and the males lack pleopods. The carapace is about 3 to 7 mm long, and red with white markings. This species is found under rocks from the lower Gulf, south to Ecuador and the Galapagos Islands.

Pachycheles marcortezensis Glassell

This species is rare intertidally, not uncommon subtidally, about 3 to 5 mm long, with the carapace covered with setae. The carpus is less than two times as long as wide, and has four to five spines on the anterior margin. The manus has four longitudinal rows of small granules, and scattered setae. The fingers are devoid of setae. The chelae are barely unequal in size. The abdomen has only five free plates, and males lack pleopods. The species is found under rocks and in coral, from Bahía de Santa María, on the ocean side of Baja California, and Isla Angel de la Guarda south to Arena Bank.

Pachycheles panamensis Faxon

A fairly common species, about 3 to 8.5 mm long, with the carapace covered with short setae. The carpus is covered with short rugae, or granules, and short setae, its anterior margin with two to three teeth. The manus and fingers are unequal in the two chelipeds, and covered with long setae. The abdomen has seven plates, and the males have a pair of pleopods. The range is from Bahía Santa María, and the middle of the Gulf south to Ecuador, occurring under stones and in coral in the low intertidal and subtidal to depth of about four fathoms.

Pachycheles setimanus (Lockington)

A very common crab in the low intertidal of the Gulf. The chelipeds are unequal in size, and both bear a very heavy tuft of long, brown setae. The carapace is smooth, without hair, and small (about 3 to 8 mm long). The carpus is less than twice as long as broad, and bears three teeth on the anterior margin. The gape of the fingers is without pubescence. This is a dull white to pale tan crab, common under large stones throughout the Gulf (and apparently restricted to it).

Fig. 7.42 *Pachycheles setimanus*

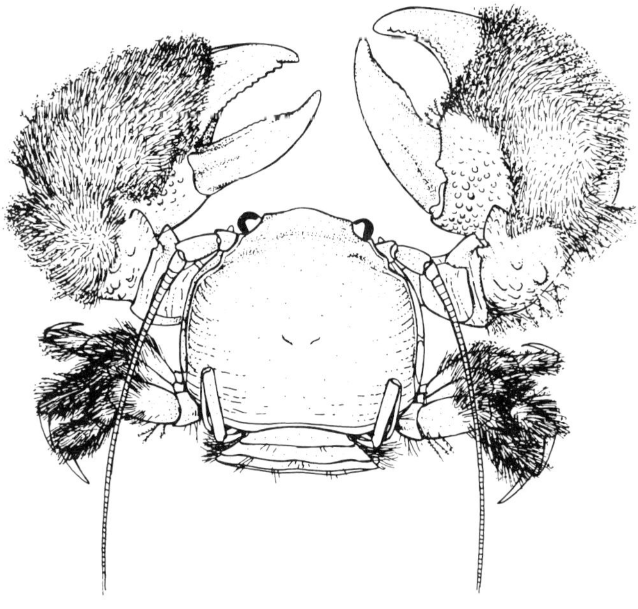

Pachycheles spinidactylus Haig

This species has a tuft of hair on the front of the carapace, and is about 3 to 8 mm long. The chelipeds are unequal in size, and the carpus is covered with small, conical tubercles with tufts of long, stiff, setae surrounded by clumps of long, plumose hairs. The manus is similar to the carpus. The abdomen is seven-plated in the males (which lack pleopods), but only five or six-plated in the females. This species ranges from the mouth of the Gulf (Cabo San Lucas) to Colombia (also reported from Bahía Santa María on the open coast of Baja California).

Genus *Porcellana*
Porcellana cancrisocialis Glassell

The carapace is about 2 to 7 mm long, naked or covered with long hair, with two to three epibranchial spinules. The dorsal surface is orange to yellow, with red and blue spots. The carpus is less than 1.5 times as long as broad, and bears a large lobe with a single spine on the inside margin. The manus has long hair on its margin. The anterior margin of the carapace has three large spines between the eyes. There are seven abdominal plates. This species is free-living, or commensal with hermit crabs. It has been found from the upper Gulf of California south to Ecuador. I have found it to be a common commensal on *Petrochirus californiensis* in the northern Gulf. A single hermit crab may often be host to six to ten of these beautiful little crabs.

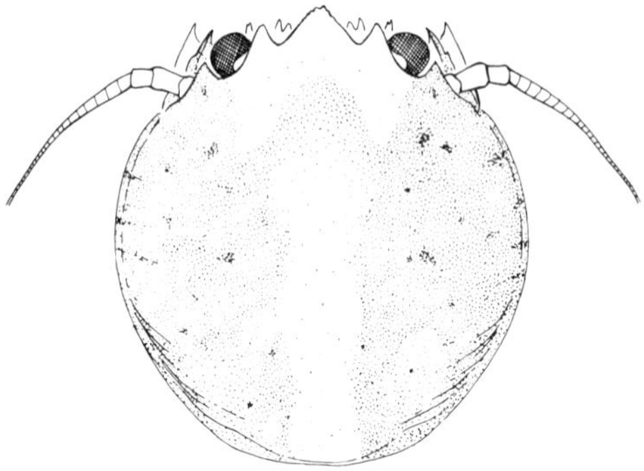

Fig. 7.43 *Porcellana cancrisocialis* (carapace)

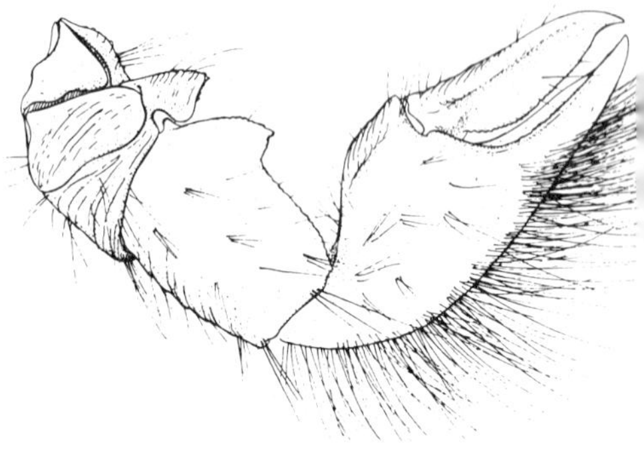

Fig. 7.44 *Porcellana cancrisocialis* (cheliped)

Porcellana corbicola Haig

A rare species, known only from the type locality, Roca Consag, at 10 to 25 fathoms. The carapace has a tuft of plumose hairs between the eyes, and a spine at the epibranchial angle. The carpus is about 1.5 times as long as wide, the anterior margin without spines. There are only five abdominal plates, and the length is reported as 6 mm.

Porcellana paguriconviva Glassell

This species has a nearly flat carapace, lavender with longitudinal orange stripes, and faint transverse plications. The carpus is slightly longer than broad, with no teeth, but a rounded lobe on the anterior margin. The chelipeds are smooth, with little or no setae, and the abdomen is seven-plated. The species is commensal with large hermit crabs (*Aniculus elegans, Petrochirus californiensis,* and *Paguristes digueti*). It is known to occur throughout the Gulf of California, south to Panamá. There is also a record from Bahía Magdalena on the west coast of Baja California.

Genus *Megalobrachium*
Megalobrachium erosum (Glassell)

The carapace of this species is about 2 to 6 mm long, heavily eroded, with strong, transverse ridges and shallow pits. The chelipeds are large, slightly unequal, and naked or covered with a light pubescence. The carpus is about 1.5 times as long as broad, with a single, low tooth on the anterior margin. The manus bears longitudinal rows of pits, and the abdomen is seven-plated. This species ranges from Isla Angel de la Guarda to La Paz (it has also been recorded from Punta Malarrimo to Bahía Magdalena on the west coast of Baja California), and is known from the subtidal only, from 5 to 25 fathoms.

Megalobrachium garthi Haig

This species resembles *Megalobrachium erosum,* but has a more rounded carapace. Carapace and chelipeds are covered with small, shallow pits (microscopic) and scattered hair. The carpus is about 1.5 times as long as broad, with a narrow blunt tooth, or lobe, on the anterior margin. The manus bears three longitudinal crests, defined by broad grooves. The gape of the fingers is without pubescence. The abdomen of this species has only five free plates. The carapace length is about 2 to 8 mm. This species ranges from the middle of the Gulf to Colombia. There is also a record from Bahía Santa María on the west coast of Baja California.

Megalobrachium sinuimanus (Lockington)

This crab is not uncommon in the Gulf, but is apparently restricted to it. The carapace is nearly smooth, rounded, covered with fine granules, about 2 to 6 mm long, and white to buff in color. The carpus is about 1.5

times as long as broad, with a narrow, blunt tooth on the anterior margin. The carapace and chelae are practically devoid of hair, and the abdomen is seven-plated. Found under stones and in coral in the low intertidal and subtidal.

Megalobrachium smithi (Glassell)

A fairly common species in the Gulf, and probably restricted to it. The carapace is roughened, granular near the margins, and about 3 to 7 mm long. The carpus is about 1.5 times as long as broad, and is without spines on the anterior margin. The manus is flattened with four crests, defined by broad grooves. The fingers are devoid of pubescence, and the abdomen is seven-plated. Individuals are found under rocks in the intertidal.

Megalobrachium tuberculipes (Lockington)

This species has a gray carapace, with a dark median blotch, and is covered with tubercles. It is an unusually small crab, rarely over 4 mm long. The carpus is less than 1.5 times as long as wide, has a single large spine and two to three smaller denticles along the anterior margin. The chelipeds are tuberculate, and the manus, carpus, and carapace all bear numerous bands of long hair. The overall appearance of this crab is that of a bumpy, rough, hairy creature. These crabs are found on sponge-incrusted sea fans, or on the sponges themselves, subtidally, and occasionally in the low tide zone. They range from Bahía Hipolito on the west coast of Baja California and the upper Gulf south to Ecuador.

Genus *Euceramus*

Euceramus transversilineatus (Lockington)

This curious creature resembles a mole crab, but has small chelipeds and weak pincers. The carapace is about 1.5 times as long as broad, with distinct transverse striations, and is convex, about 5 to 12 mm long. The dorsal surface is pink to buff, with red to brown striations. The chelipeds are covered with long, scattered setae, hooked at the tips. The carpus is short and stout, with one to four narrow spines on the anterior margin. The abdomen is seven-plated. These crabs are found buried in the sand in the low intertidal and subtidally. The range is from the upper Gulf south to Panamá. (There are also records from Bahías Santa María and Magdalena, on the west coast of Baja California.)

Genus *Minyocerus*
Minyocerus kirki Glassell

This species also resembles a mole crab, with the carapace about one and a third times as long as broad, and highly convex. It is easily recognized by its apparent lack of antennae, which are actually rudimentary, their total length being less than, or scarcely exceeding, the width of the eye. The chelipeds are subequal, small, and have a long narrow manus (much longer than the fingers). The carpus is short and stout, with a strong spine at about the center of the anterior margin, often followed by several spinules. The abdomen is seven-plated. The dorsal surface is usually greenish, with white or yellow blotches and is about 2 to 5.5 mm long. These small porcelain crabs are obligate-commensals on sea stars, possibly restricted to the genus *Luidia*, and nearly always subtidal. It has been reported from *Luidia columbia, L. phragma, L. foliolata,* and at least one species of serpent star. This species ranges from the upper Gulf of California to Nicaragua.

Genus *Ulloaia*
Ulloaia perpusillia Glassell

A rare species of crab known from only three localities; Puerto Peñasco; Puntarenas, Costa Rica; and Panamá. A strange species, it is found among gorgonian corals, sponges, and bryozoa from the low intertidal to eight fathoms. It is easily recognized by its elongate carapace (3 to 4 mm in length), and large protuberances and setae all over the body. The carapace and chelae are roughly granular. The manus is small and weak, with long plumose hairs. The carpus is slightly longer than wide, and armed on the anterior margin with a median lobe or broad tooth. The gape of the fingers has a long tuft of plumose hairs.

Genus *Polyonyx*
Polyonyx nitidus Lockington

The carapace is broader than long, smooth or lightly plicate, nearly oval, and lacks epibranchial spines. The chelae are smooth, slightly unequal in size, and have a fringe of long setae along the outer margin. The carpus is 1.4 to 1.8 times as long as broad, and without spines. The walking legs are all nearly smooth, fringed with fine plumose hair along the margins. The carapace is about 4 to 15 mm long. Its range is from the upper Gulf, south to Panamá. In Puerto Peñasco it has been collected from the tube of *Chaetopterus,* the polychaete worm, and Janet Haig has informed me the Panamá record was also as a commensal in *Chaetopterus.*

Superfamily Paguroidea (hermit crabs)

Preservation: The hermit crabs present a special problem, as they are prone to pulling themselves far into their shell prior to death. To keep them extended, they can be allowed to die slowly in putrid sea water. A dead hermit, fully extended, can usually be pulled gently from the shell, or the shell can be carefully chipped away from the animal. A number of methods will evict the hermits while they are still alive. Placing them in a mixture of fresh water and sea water and adding a few drops of formalin often causes them to crawl out of their shell. A hole can be drilled or chipped in the shell near the top, and a thin wire inserted to probe the hermit and make him crawl out. A less successful method is simply holding a match under the shell, hoping the heat will force the hermit out. Quite often, however, this latter method results in fried crab, as opposed to freed crab. The hermits should be preserved in alcohol along with their shell.

Taxonomy: The taxonomy of this group is based largely on the anatomy of the appendages, especially the head appendages and legs, and, to a lesser extent, on the anatomy and morphology of the reproductive system, both internal and external. Despite the amount of collecting of this group in the Gulf, some of the systematics remain yet to be worked out. Miss Janet Haig of the Allan Hancock Foundation has been working on the anomurans for some time, and is the leading authority on this group for the Pacific coast of North America. Regarding the Gulf fauna, Miss Haig states (personal correspondence), that a few species of *Paguristes* and *Pagurus* are undescribed, and have been badly confused with other species. The species grouped around *Pagurus lepidus* and *Paguristes anahuacus* present an incredibly complex situation. She also informs me that these are mostly very small species, none of which are included in this handbook. Miss Haig presently is working out the systematics of the Pacific coast pagurids, in preparation for a much-anticipated monograph. Because of the difficulty in working with these animals, the following key is based principally on characters of the head appendages, including color, hopefully rendering the specimens identifiable without extracting them from their shells.

Hermits of the Gulf: A good number of species of hermits are found in the Gulf, many very small, others rarely encountered. Only five are included in the key. I hope these represent five of the most commonly encountered types. The commonest pagurid of the Gulf intertidal is likely *Clibanarius digueti,* the little red-and-white hermit. In the southern Gulf, its abundance is perhaps rivaled by *Calcinus californiensis.* On the mud flats and estuaries, *Paguristes sanguinimanus, Clibanarius panamensis,* and *Petrochirus californiensis* are fairly common. The first of these three is more common in the northern Gulf, the second more common in the southern Gulf, and the third is found throughout the Gulf. Perhaps the most interesting hermit of the Gulf is *Petrochirus californiensis.* This abundant hermit is a real giant, the carapace alone reaching 50 mm in length. This big fellow has lumpy, heavy claws, and is strikingly colored. Despite its size and massive chelae, it is a very retiring and gentle species of hermit, quite fitting the vernacular of this group. *Petrochirus californiensis* is common during low tides in tidal flats, and extends its range subtidally at least as deep as 40 fathoms. I have found it common throughout the year on the tidal flats of Bahía Cholla, and while SCUBA diving over sandy bottoms off Punta Peñasco. About 75 percent of the individuals of this species I have collected in the Puerto Peñasco area were living in the shells of *Muricanthus,* and harbored several of the commensal porcelain crabs *Porcellana cancrisocialis,* as well as numerous amphipods and an occasional scaleworm. In 1966 and 1968, in the Guaymas area, I witnessed great hordes of the "land" hermit crabs *(Coenobita)* moving daily across the beaches and up to the coconut groves north of the city (near the Jergens plantation). These mass movements appeared to be most common in the hot summer months. I also collected two of these "land" hermits on Isla San Francisco in the winter of 1972.

A small, rocky intertidal hermit that occasionally turns up at Puerto Peñasco, but isn't included in the key, is *Pylopagurus roseus;* and a subtidal hermit, common in the northern Gulf, that builds a soft, flexible, artificial shell to live in is *Pagurus gladius* (Benedict).

KEY TO THE COMMON INTERTIDAL HERMIT CRABS OF THE GULF OF CALIFORNIA

1 Chelipeds equal 2
-Chelipeds unequal 3

2(1) Legs white to pink, with longitudinal purple stripes; eyestalks usually white; chelipeds with little or no hair ...... ***Clibanarius panamensis***
-Legs red with white spots; eyestalks usually red to brown; chelipeds quite hairy .... ***Clibanarius digueti***

3(1) Very large crabs, carapace length greater than 20 mm; usually inhabits large muricanthid shell; antennae with red-and-white bands (fig. 7.45) .... ***Petrochirus californiensis***
-Smaller crabs, carapace length less than 15 mm; usually not in muricanthid shell; antennae without red-and-white bands 4

4(3) Legs white; eyestalks white; maximum carapace length greater than 9 mm .... ***Pagurus albus***
-Legs not white; eyestalks not white; maximum carapace length less than 9 mm .... ***Calcinus californiensis***

SYSTEMATICS AND SPECIES DESCRIPTIONS

Hermit Crabs

Calcinus californiensis Bouvier

A common, small- to medium-sized hermit of the lower Gulf and the west coast of Mexico. The general color is a deep red. The chelipeds are unequal in size, broad and naked. The carapace reaches about 8 mm in length.

Clibanarius digueti Bouvier

This is probably the commonest intertidal hermit of the Gulf, especially the northern Gulf. *Clibanarius digueti* is a small red crab with brown eyes, and white or pale blue spots on its red legs and chelipeds. The chelipeds are equal in size, elongate, and hairy.

Clibanarius panamensis Stimpson

A medium-sized hermit with white eyestalks. The legs are pinkish white with long, purple stripes. The chelipeds are equal in size. This species apparently prefers estuaries, mangrove swamps, and tidal flats.

Petrochirus californiensis Bouvier

A giant, very heavy-clawed hermit, inhabiting large gastropod shells such as *Strombus, Muricanthus, Fasciolaria, Murex,* and *Hexaplex*. The chelipeds are slightly unequal in size. The general body color is a dull brick red, the antennae being red-and-white banded. The chelipeds are

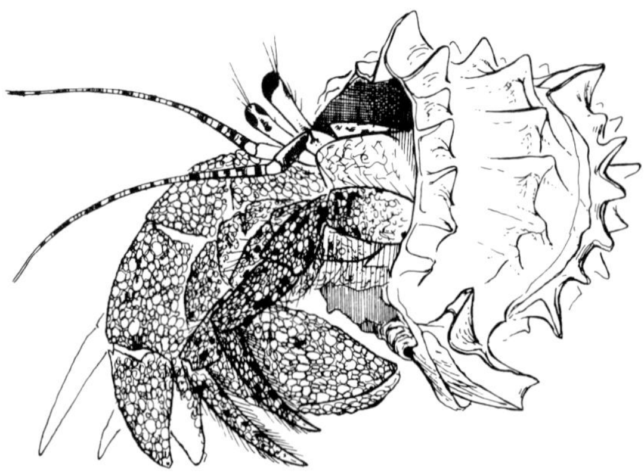

Fig. 7.45 *Petrochirus californiensis*

very rough, and covered with many small denticles. This is probably the largest hermit of the Gulf intertidal, the carapace reaching 40 to 50 mm in length. *Petrochirus californiensis* ranges from the upper Gulf to Ecuador, and is found most often in the low intertidal zone of mud flats, or collected off the shrimp trawlers from 20 to 40 fathoms. See introductory remarks of this section for further information.

Pagurus albus Benedict

A medium-sized hermit crab, with conspicuous white legs, chelipeds and eyes. The chelipeds are unequal in size. Pickens (1970) says *Pagurus albus* is a very aggressive species, and Glassell (1937) says it is an active crab and a rapid mover. It is usually found in the shells of *Polinices* and *Natica*. It may be restricted to the Gulf. I have found it to be common and abundant on tidal flats throughout the Gulf, generally in the low intertidal. The carapace is usually 8 to 11 mm long.

Superfamily Hippoidae (mole crabs)

The mole or sand crabs are curious little creatures that have a high, convex, oval carapace, the first pair of legs without pincers (simple), and the abdomen bent under the thorax. They are found only on sandy beaches, or on beaches of very small pebbles (see general Crustacea section). Mole crabs are surprisingly difficult to capture. They are rapid burrowers, despite the fact that they always burrow in reverse. In fact, they also run and swim backwards. On the California coast, fishermen use *Emerita analoga* (Stimpson) for bait. Schmitt (1921) gave the range

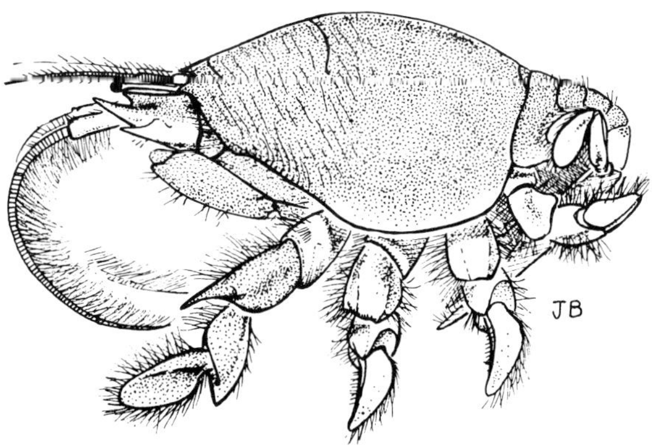

Fig. 7.46 *Emerita analoga*
(from Allen, 1969)

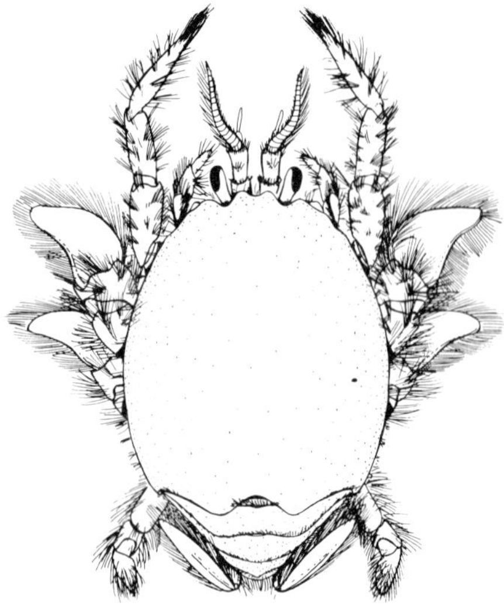

Fig. 7.47
Hippa pacifica

of this species as Oregon to Baja California, and from Peru to Chile, and stated the intermediate region (from Baja California to Panamá) is populated by *Emerita emerita* (Fabricius). This latter species, however, has since been synonomized with *Emerita rathbunae* Schmitt (1935). It is this species of *Emerita* that is probably found in the Gulf. *E. rathbunae* can be distinguished from *E. analoga* by the presence of two median spines on the anterior margin of the carapace in the former, while *E. analoga* has three. I have found only one species of mole crab from the Sonoran coast of the Gulf, *Hippa pacifica* (Dana).

Section Brachyura (true crabs)

Preservation: The true crabs should be preserved in 60 to 95 percent alcohol, as formalin will soften the calcareous carapace. Care should be taken not to break off the appendages when handling individuals; however, narcotizing is generally not necessary. The larger brachyurans may be left in the preservative, or removed (after several days) and air dried. Dried specimens do, however, become quite brittle and must be handled with care. Unfortunately, alcohol tends to bleach the color from the exoskeleton, so color notes always should be taken prior to preservation.

Taxonomy: The taxonomy of the brachyuran crabs is based on a large series of taxonomic characters. The majority of these characters are external and morphological features of the carapace, legs, and mouth parts. The use of the male's first pleopod as an important taxonomic character has resulted in the discovery of many significant taxonomic relationships among the higher taxa of this group. Fig. 7.48 will aid the user of the handbook in locating specific anatomical parts of his specimen mentioned in the key. Reference to the chelipeds of fiddler crabs (genus *Uca*) in the key or species descriptions refers specifically to the large cheliped of the male. In all genera, references to the manus, arm, or palm refer to the chelipeds, and not the walking legs, unless otherwise designated. When the number of anterolateral spines on the carapace is given, that number includes the outer orbital spine. The systematics used in this handbook are from Rathbun (1918–1937) and Garth (1958, 1966) (see Arthropod section for further taxonomy).

Brachyurans of the Gulf: Glassell (1934) listed 197 species of true crabs from the Gulf. Steinbeck and Ricketts collected 54 species (mainly intertidal). Rathbun (1918 to 1937) described over 200 brachyurans from this region. Garth (1960) reported 230 species from the Gulf. This large number of species makes the Gulf an extraordinarily rich collecting ground for students of the true crabs, and indicates this region possesses an ideal environment for their requirements. The author has found 56 species of intertidal brachyurans to be common enough in the Gulf to be included in the key that follows.

The Gulf of California possesses an especially rich fauna of swimming and fiddler crabs. Garth and Stephenson (1966) reported 17 species of Portunidae, and Rathbun (1930) reported more than eight species of fiddlers from the Gulf. By comparison, the Pacific coast of North America north of Mexico has but one fiddler (*Uca crenulata*), and probably only two swimming crabs (*Portunus xantusii* and *Callinectes arcuatus*). The Gulf pea crab fauna (*Pinnotheridae*) is not well-known, although Rathbun (1918) reported five genera, and Glassell (1934–1936) later reported 18 species he had found to be endemic to the Gulf. Because of the great variety and numbers of species known to occur in the Gulf, the key includes only the most common species of intertidal and shallow water crabs, mainly those collected by the author himself. The serious student of the Brachyura is referred to Rathbun's monumental series published by the United States National Museum (1918, 1925, 1930, and 1937), and the excellent monographs by Garth (1958) and Garth and Stephenson (1966). Dr. Garth is undoubtedly the foremost living authority on the Pacific Brachyura, and students of the true crabs are eagerly awaiting further volumes of his *Brachyura of the Pacific Coast of America.*

The most commonly observed large crab of the rocky shores of the lower Gulf is probably the famous Sally Lightfoot crab, *Grapsus grapsus,* while in the upper Gulf *Eriphia squamata* emerges as the most commonly encountered large crab. Of the small to medium sized crabs (those usually found under rocks, etc.), *Leptodius occidentalis, Tetragrapsus jouyi, Ala cornuta,* and the hairy crabs of the genus *Pilumnus* are most common to the Gulf intertidal. Sandy beaches, bays, and estuaries are inhabited by the swimming crabs, of which *Callinectes bellicosus* and *Portunus xantusii* are most often collected. Also, high on sandy beaches throughout the Gulf, are the fascinating ghost crabs, *Ocypode occidentalis.* In mud and sandy-mud bays, the fiddlers abound. Several species are common in the northern Gulf: *Uca brevifrons, Uca crenulata,* and *Uca monilifera,* while in the southern Gulf *Uca macrodactylus, Uca princeps,* and *Uca mordax* appear to be the most common fiddlers. The masking crabs are represented in the Gulf intertidal principally by *Ala cornuta* and *Dromidia larraburei.* Of the six species of spider crabs included in the key, only two of the genera, *Ala* and *Epialtus,* are commonly found intertidally, the others being uncommon in the intertidal zone and most easily collected by skin diving or by sorting through the catch of the shrimp trawlers. The common kelp crabs of the Pacific coast of the United States (*Pugettia*) are largely replaced in the Gulf by the smaller Sargassum crabs, *Epialtus minimus, Acanthonyx petiveri, Podochela* spp., and *Epialtoides paradigmus.* It should be kept in mind that almost any crab common to the low intertidal zone will also be found subtidally, at least in enough water to make SCUBA collecting profitable.

Although not included in the key, the giant land crab *Cardisoma crassum* Smith has been consistently found near the small fishing village of Boca del Rio, Sinaloa. These large, red-white-and-blue crabs dig burrows in the dry mud from a quarter mile to several miles inland, along the banks of the river. The burrows are deep (to 6 feet), but easily spotted by the large mound of dried mud that the crab deposits above the entrance. Anyone visiting the mouth of the Sinaloa River will be missing a beautiful sight if he doesn't travel up the river to observe these crabs.

The majority of the brachyuran fauna of the Gulf has its affinities with the Panamic Province. Of the 230 species of true crabs reported from the Gulf by Garth (1960), 131, or 52 percent, were considered by him to be Panamic, while 80 (32 percent) are indigenous to the Gulf, and 19 (8 percent) represent northern forms. The occurrence of species from the Californian Province is rare. The familiar *Pachygrapsus crassipes* of the north Pacific coast of America is reported as occurring in the Gulf, and I have found it only once, at Guaymas. It is likely that these, and other members of the colder water Californian fauna, occur in deeper water and must be collected subtidally, except in the colder winter months when they may move into the intertidal in the northern Gulf. The trend of colder water animals occurring deeper as one moves south (in the Northern Hemisphere) is often seen on the west coast of the United States and Baja California (see Introduction). The vast majority of Californian or temperate forms are restricted to the colder northern third of the Gulf, disappearing in the southern Gulf and on around Cabo San Lucas, to reappear midway up the west coast of Baja California. Other southern California-northern Baja California brachyura found in the Gulf include *Hepatus lineatus, Randallia ornata, Lophopanopeus frontalis, Uca crenulata,* and *Cancer amphioetus.*

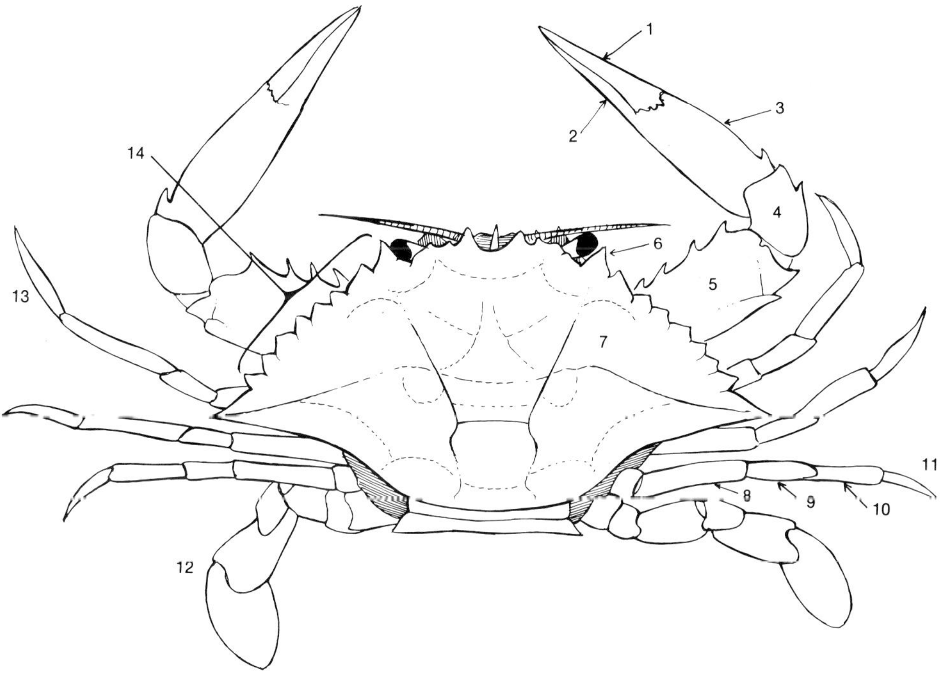

Fig. 7.48 Crab morphology

1 Movable finger
2 Fixed finger
3 Manus (palm, hand)
4 Carpus (wrist)
5 Merus (arm)
6 Outer orbital spine
7 Anterolateral or epibranchial region
8 Merus (of walking leg)
9 Carpus (of walking leg)
10 Propodite (of walking leg)
11 Dactylus (of walking leg)
12 Last walking leg
13 First walking leg
14 Anterolateral spines

KEY TO THE COMMON INTERTIDAL BRACHYURAN CRABS OF THE GULF OF CALIFORNIA

1 Small crabs, carapace width usually less than 2 cm; eyes reduced; carapace usually membranous and with reduced pigmentation; commensal in worm tubes, in holothurians, and on sand dollars (the pea crabs) **52**
-Larger crabs, carapace width usually greater than 2 cm; eyes not reduced; carapace not membranous, usually well-pigmented; free living, or occasionally commensal in sponges **2**

2(1) Last pair of legs flattened and paddlelike, with the tips rounded for swimming; carapace with a large lateral spine (Portunidae, the swimming crabs) **3**
-Last pair of legs not flattened and paddlelike, but pointed for walking; carapace without a large lateral spine (often with several small spines) **10**

3(2) Eyestalks very long, one-third or more of carapace width (fig. 7.68) .... ***Euphylax robustus***
-Eyestalks short, much less than one-third carapace width **4**

4(3) Entire carapace covered with pale red spots; supraorbital fissures broadly opened and V-shaped .... ***Arenaeus mexicanus***
-Carapace not as above; supraorbital fissures not broadly opened **5**

5(4) Anterolateral teeth alternately large and small; distal margin of merus of swimming leg with a large spine (fig. 7.69) .... ***Cronius ruber***
-Anterolateral teeth not distinctly alternating large and small; distal margin of merus of swimming leg with or without a large spine **6**

6(5) Abdomen of male triangular-shaped; anteroexternal angle of merus of third maxillipeds not strongly produced laterally; wrist of chelipeds with a large inner and outer spine **7**
-Abdomen of male ⊥-shaped; anteroexternal angle of merus of third maxillipeds strongly produced laterally; wrist of chelipeds with outer spine but without inner spine **9**

7(6) Posterodistal border of merus of swimming legs without spines or spinules; maximum width of carapace less than 35 mm .... ***Portunus tuberculatus***
-Posterodistal border of merus of swimming legs with spinules and 0 to 1 spines; maximum width of carapace greater than 40 mm **8**

8(7) Posterodistal border of merus of swimming legs with spinules but no spines; length of wrist and manus combined less than twice carapace length (fig. 7.67) .... ***Portunus xantusii***
-Posterodistal border of merus of swimming legs with spinules and one spine; length of wrist and manus combined 2 to 3 times carapace length .... ***Portunus asper***

9(6) Median pair of frontal teeth low and rudimentary, often entirely absent; length of largest lateral spine of carapace usually equalling 1/12 (.06 to .08 percent) carapace width; maximum carapace width 155 mm (fig. 7.66) .... ***Callinectes bellicosus***
-Median pair of frontal teeth well-developed, about half length of outer pair (fig. 7.64); length of largest lateral spine of carapace usually equalling 1/7 to 1/10 (.10 to .14 percent) carapace width; maximum carapace width 115 mm (figs. 7.63 to 7.65) ... ***Callinectes arcuatus***

10(2) Manus of one cheliped, in males, more than 3 times longer than opposing manus (*Uca*, the fiddler crabs) **11**
-Manus of opposing chelae subequal in both males and females **18**

11(10) Anterior margin of carapace between eyestalks wide, greater than 1/10 greatest width of carapace **12**
-Anterior margin of carapace between eyestalks narrow and spatuliform, less than 1/10 greatest width of carapace **17**

12(11) Fingers shorter than palm (fig. 7.59); inner surface of palm with a row of denticles near base of fingers, but without an oblique stridulating ridge; carapace usually less than 10 mm wide .... ***Uca latimanus***
-Fingers equal to or longer than palm; inner surface of palm with an oblique stridulating ridge; carapace usually greater than 10 mm wide ... **13**
13(12) Movable finger about twice as long as palm; known only from the vicinity of the mouth of the Colorado River .... ***Uca coloradensis***
-Movable finger much less than 2 times length of palm; range other than above **14**
14(13) Base of chelae red underneath; anterior part of sidemargins of carapace straight or nearly so, continuing backward with an angular turn (fig. 7.58) .... ***Uca crenulata***
-Base of chelae not red underneath; anterior part of sidemargins of carapace convex, curving gradually backward **15**
15(14) Eyestalks long, eyes reaching anterolateral corners of carapace; carapace less than 15 mm wide; length of fingers subequal to length of palm .... ***Uca macrodactylus***
-Eyestalks shorter, eyes not quite reaching anterolateral corners of carapace; carapace usually greater than 15 mm wide; length of fingers much greater than length of palm **16**
16(15) Anterior margin of carapace, between base of eyestalks, at least 1/3 width of entire anterior margin of carapace (fig. 7.60) .... ***Uca mordax***
-Anterior margin of carapace, between base of eyestalks, less than 1/3 width of entire anterior margin of carapace .... ***Uca brevifrons***
17(11) Movable finger of large claw widest at base, tapering gradually; outer base of large claw tuberculate (fig. 7.62) .... ***Uca princeps***
-Movable finger of large claw widest medially; outer base of large claw smooth (fig. 7.61) .... ***Uca monilifera***
18(10) Length of carapace, including rostrum, greater than width **19**
-Length of carapace less than width **24**
19(18) Carapace length subequal to rostral length; length of walking legs greater than 2 times carapace length; eyestalks long and conspicuous (fig. 7.81) .... ***Stenorhynchus debilis***
-Carapace length greater than rostral length; length of walking legs less than 2 times carapace length; eyestalks short, not conspicuous ... **20**
20(19) Rostrum present and dorsoventrally flattened; carapace not setaceous ... **21**
-Rostrum lacking, or, if present, not dorsoventrally flattened; carapace setaceous **22**
21(20) Arm long, extending beyond tip of rostrum; male with 5 free abdominal segments; carapace usually with a yellow median blotch .... ***Epialtoides paradigmus***
-Arm short, not extending to tip of rostrum; male with 6 free abdominal segments; carapace without a yellow median blotch (fig. 7.82) .... ***Epialtus minimus***
22(20) Dorsal surface of carapace nearly flat; first walking legs about the same length as second; carapace without a median row of prominent dorsal spines; carapace subtriangular (fig. 7.83) .... ***Herbstia camptacantha***
-Dorsal surface of carapace convex, strongly inflated; first walking legs much longer than second; carapace with a median row of prominent dorsal spines; carapace suboval **23**
23(22) Fingers pale or white; carapace with a median row of 6 large dorsal spines; rostrum prominent (fig. 7.84) .... ***Libinia mexicana***
-Fingers black; carapace with a median row of 2 to 3 small dorsal spines; rostrum reduced to 2 small rostral horns (fig. 7.86) .... ***Stenocionops angusta***

24(18) Dorsal surface of carapace and legs covered with bright yellow circles, bordered by reddish brown or orange (figs. 7.88, 7.89) .... ***Hepatus lineatus***

-Dorsal surface of carapace and legs not colored as above 25

25(24) Chelae covered with large, oval tubercles 26

-Chelae not covered with large, oval tubercles 27

26(25) Carapace highly setaceous; last pair of legs reduced and folded anteriorly on back; crab carries empty pelecypod valve on back; carapace unicolored tan (fig. 7.92, 7.93) .... ***Hypoconcha lowei***

-Carapace not setaceous; last pair of legs not as above; without pelecypod valve; carapace mottled (fig. 7.70) ... ***Eriphia squamata***

27(25) Dorsal surface of carapace setaceous (setae may be very short, requiring magnification) 28

-Dorsal surface of carapace not setaceous 36

28(27) Carapace entirely covered by living ascidian or sponge (fig. 7.91); fifth leg modified and folded anteriorly on back for holding sponge (fig. 7.90) .... ***Dromidia larraburei***

-Carapace not covered by a living sponge (although small bits of sponge and algae may be lodged on the back); fifth legs not as above 29

29(28) Carapace triangular 30

-Carapace rectangular, square, or oval 32

30(29) Posterior area of carapace with 2 or 4 large holes; carapace bare; posterior margin of carapace without numerous tufts of golden setae ... 31

-Posterior area of carapace without large holes; carapace covered with small bits of sponge and algae; posterior margin of carapace with numerous tufts of golden setae (fig. 7.85) .... ***Ala cornuta***

31(30) Posterior area of carapace with 2 large holes; carapace broader than long (fig. 7.87) .... ***Speleophorus schmitti***

-Posterior area of carapace with 4 large holes; carapace longer than broad .... ***Speleophorus digueti***

32(29) Fingers of walking legs with 6 distinct longitudinal rows of dark setae (fig. 7.55) .... ***Sesarma sulcatum***

-Fingers of walking legs not as above 33

33(32) Fingers of manus white; carapace and legs deeply eroded with patterned vermiculations, giving the crab a spongy appearance; dorsal setae small and inconspicuous (fig. 7.73) .... ***Glyptoxanthus meandricus***

-Fingers of manus black or dark; carapace and legs not as above; dorsal setae conspicuous 34

34(33) Carapace visible beneath dorsal setae; dorsal surface red to purple 35

-Carapace not visible beneath dorsal setae; dorsal surface tan (fig. 7.77) .... ***Pilumnus limosus***

35(34) Body setae long and stiff; anterolateral margins of carapace with 4 long sharp spines (including outer orbital spine); abdomen mottled with reddish purple (fig. 7.78) .... ***Pilumnus townsendi***

-Body setae short and soft; anterolateral margins of carapace with 5 short, blunt spines (including outer orbital spine); abdomen unicolored (fig. 7.76) .... ***Pilumnus gonzalensis***

36(27) Carapace entirely covered with many coarse granules; anterolateral margin of carapace forming a single large spine; in burrows on sandy beaches, well above water line (fig. 7.57) .... ***Ocypode occidentalis***

-Carapace not covered with many coarse granules; anterolateral margin of carapace forming no spines *or* 2 or more spines; not inhabiting burrows as above 37

37(36) Anterolateral corners of carapace rounded, not forming spines 38

-Anterolateral corners of carapace forming 2 or more blunt or pointed spines 40

38(37) Chelae with several longitudinal rows of denticles; 4th abdominal segment of male with 2 large tubercles .... ***Hepatus kossmanni***
-Chelae without rows of denticles; 4th abdominal segment of male without 2 large tubercles **39**
39(38) Carapace covered with small pits; small crabs, usually less than 30 mm wide; chelae black (fig. 7.80) .... ***Xanthodius hebes***
-Carapace smooth; large crabs, usually more than 30 mm wide; chelae white to pale (fig. 7.56) .... ***Gecarcinus quadratus***
40(37) Fingers of manus black or dark brown **41**
-Fingers of manus white or pale (at least distal quarter of fingers) **44**
41(40) Anterolateral margin of carapace forming 4 to 5 blunt or pointed spines ... **42**
-Anterolateral margin of carapace forming 1 to 3 blunt or pointed spines ... **43**
42(41) Anterior margin of carapace, between eyes, forming 2 humped, elevated flanges; outer orbital tooth of carapace forming single, pointed spine (fig. 7.74) .... ***Leptodius occidentalis***
-Anterior margin of carapace, between eyes, forming 2 flat, unelevated flanges; outer orbital tooth of carapace indented, forming two blunt spines (fig. 7.75) .... ***Panopeus purpureus***
43(41) Width of carapace, between eyes, less than ½ greatest width of carapace, carapace mottled pale red and blue-green; common throughout Gulf (fig. 7.71) .... ***Eurytium affine***
-Width of carapace, between eyes, more than ½ greatest width of carapace; carapace entirely red; not known to occur north of La Paz (fig. 7.79) .... ***Trapezia ferruginea***
44(40) Anterolateral margin of carapace forming a single spine, the outer orbital spine; chelae with a small patch of stiff black setae (fig. 7.54) .... ***Aratus pisonii***
-Anterolateral margin of carapace forming 2 or more spines (including outer orbital spine); chelae without a small patch of stiff black setae **45**
45(44) Walking legs with long, heavy black setae; upper surface of chelae highly denticulate (fig. 7.50) .... ***Goniopsis pulchra***
-Walking legs without long, black setae; upper surface of chelae generally not highly denticulate **46**
46(45) Anterolateral margins of carapace form 3 or more pointed or blunt spines (including outer orbital spine) **47**
-Anterolateral margins of carapace form 2 spines (including outer orbital spine) **49**
47(46) Width of carapace, between eyes, much less than ½ greatest width of carapace; carapace ovate, outer orbital angle not forming a sharp spine (fig. 7.72) .... ***Eurytium albidigitum***
-Width of carapace, between eyes, more than ½ greatest width of carapace; carapace square or rectangular; outer orbital angle forming a sharp spine **48**
48(47) Fourth (last) walking leg shorter than preceding legs; carapace rectangular, width greater than length; last (3rd) spine on lateral margin of carapace smaller than previous spines, but not minute (fig. 7.53) .... ***Tetragrapsus jouyi***
-Fourth (last) walking leg as long as, or longer than, preceding legs; carapace nearly square, length subequal to width; last (3rd) spine on lateral margin of carapace minute .... ***Geotice americana***
49(46) Width of carapace, between eyes, less than ½ greatest width of carapace; manus with numerous spines or tubercles **50**
-Width of carapace, between eyes, greater than ½ greatest width of carapace; manus without numerous spines or tubercles **51**
50(49) Third pair walking legs longest; manus with several longitudinal rows of small denticles; lateral margins of carapace highly convex; anterior margin of carapace folding down at a sharp angle (90°) between eyes; fingers with broad spooned tips (fig. 7.51) .... ***Grapsus grapsus***

-Second pair walking legs longest; manus with small denticles, but not arranged in rows as above; lateral margins of carapace straight, or nearly so; anterior margin of carapace folding down at slight angle (20° to 40°) between eyes; fingers acute, not spooned (fig. 7.49) .... ***Geograpsus lividus***

51(49) Distal end of merus of last 2 pairs of legs with numerous marginal spines; carapace dark red, occasionally with purple blotches; carapace width 10 to 17 mm .... ***Pachygrapsus transversus***

-Distal end of merus of last 2 pairs of legs without marginal spines; carapace dark red with green lines and spots; carapace width 25 to 50 mm (fig. 7.52) .... ***Pachygrapsus crassipes***

52(1) Ischium of external maxillipeds fused with merus to form a single piece, which lies obliquely or transversely **53**

-Ischium not fused with merus, lying longitudinally or slightly oblique **55**

53(52) Dactylus of first 3 pairs of walking legs bifurcate, with a long outer finger and a short inner finger; carapace subpentagonal; usually commensal on sand dollars .... ***Dissodactylus* spp.**

-Dactylus of first 3 pairs of walking legs not bifurcate; carapace not subpentagonal; never commensal on sand dollars **54**

54(53) Carapace much wider than long; legs diminish in length from first to last .... ***Parapinnixa* spp.**

-Carapace longer than wide; legs not diminishing in length from first to last .... ***Pinnotheres* spp.**

55(52) Third pair of legs longest .... ***Pinnixa* spp.**

-Second pair of legs longest .... ***Tetrias* spp.**

SYSTEMATICS AND SPECIES DESCRIPTIONS

Brachyura

Subtribe Brachygnatha
Superfamily Brachyrhyncha (grapsoid and cancroid crabs)
Family Grapsidae
Subfamily Grapsinae

Geograpsus lividus (H. Milne-Edwards)

A yellowish red crab, with lines or patches of darker red or purple on the carapace. The width is about 25 to 35 mm. This species is found on both coasts of temperate America, living among the loose stones and rocks of the high tide, and in splash zones in tropical and subtropical America. On the west coast, it ranges from the upper Gulf to Chile and the Galapagos Islands, and on the east coast from Florida to Brazil. I have collected it in the southern Gulf living among the roots of mangrove trees at Topolobampo, and on the rocky shores of Cabo San Lucas. (Topolobampo, incidentally, is a Yaqui Indian word meaning "watering place of the tiger.") Crane (1947) remarked that this crab was principally a

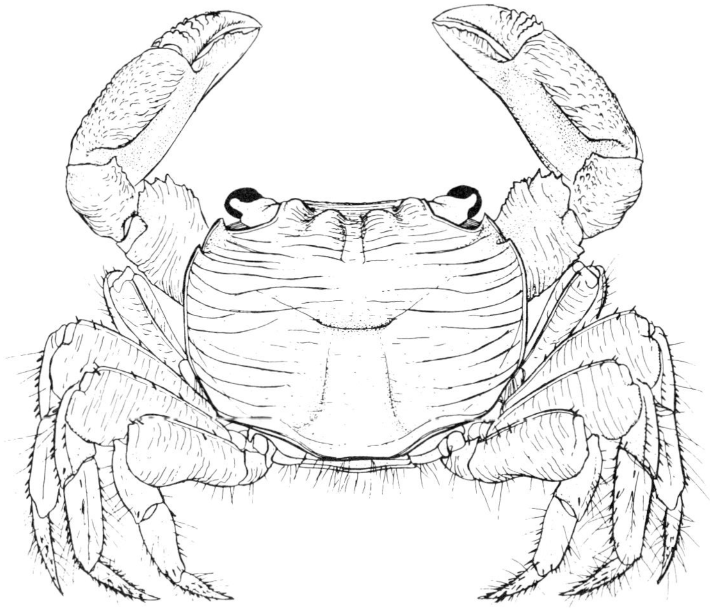

Fig. 7.49 *Geograpsus lividus*

carnivore, and Boone (1927) gave the following excerpt from William Beebe's book, *Galapagos — World's End,* "Everything about them was broad and flat and thin as tissue, and it needed all one's attention to catch one, for they could slip between closed fingers and into the merest crack of a crevice, like magic. When cornered, they gave up all attempt at escape and waved their claws valiantly and even rushed forward to attack a finger."

Goniopsis pulchra (Lockington)

This species is analogous to the only other member of the genus in the western hemisphere, *Goniopsis cruentata,* which lives in the same habitat on the east coast of America. *G. pulchra* is brown or reddish, with purplish variegations, similar to *Geograpsus,* but with slimmer legs and longer, thinner claws. The carapace of *G. pulchra* is about 35 to 45 mm wide. This crab is abundant in mangrove swamps, in estuaries (even where the water is quite fresh), and on roots and along tree trunks in damp, humid areas throughout the mid- and southern Gulf. It often builds horizontal burrows into the mud banks of estuaries or lagoons. It is also found among the rocks of the high tide zones, usually where the

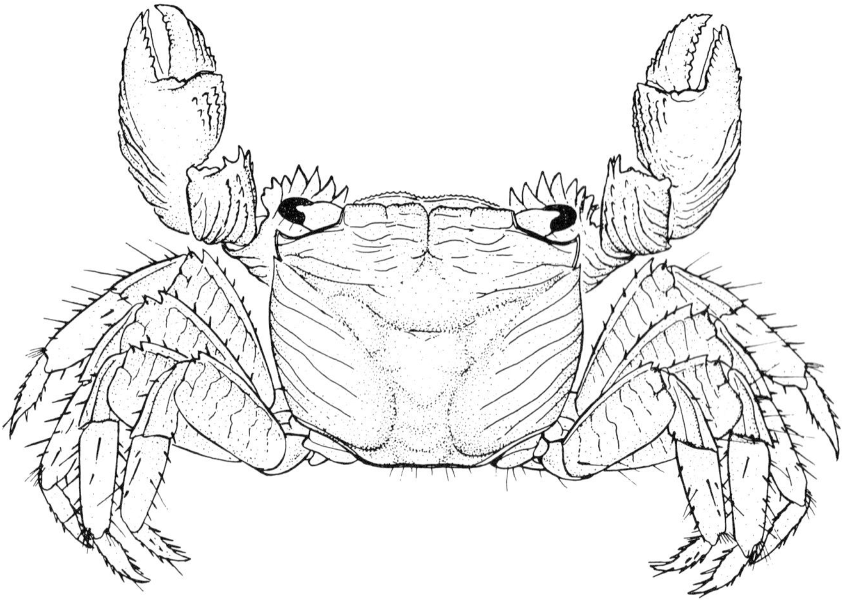

Fig. 7.50 *Goniopsis pulchra*

rocks are imbedded in a muddy substrate. *Goniopsis* prefers to remain very near the water, rarely wandering about the rocks as does *Grapsus grapsus*. It ranges from Bahía Magdalena and the mid-Gulf to Peru. I have collected it at Bahía Kino, Bahía Lobos, Guaymas, and on most rocky shorelines or mangrove swamps between Santa Rosalía and Cabo San Lucas.

Grapsus grapsus (Linnaeus)

The famous Sally Lightfoot crab is one of the most common rocky shore inhabitants of the lower Gulf intertidal, whose abundance is reflected in its technical name (literally, Crab crab). *Grapsus grapsus* is variegated deep red and pale green, or occasionally solid red. The carapace is usually 40 to 90 mm in width. The fingers are spoon-shaped, and the front of the carapace is strongly vertical. Sally Lightfoot is a very agile crab, preferring the continual wave shock of the low splash zone of exposed outer coasts, although it often follows the tide line down as the tide recedes. It scampers about the rocks, being knocked into the sea by the waves and climbing quickly back onto the nearest exposed rock. Sally Lightfoot seems to prefer a diet of algae although, like so many animals, it will alter its eating habits in response to changes in the environment. It has been seen feeding at various times on a number of small invertebrate types. Boone (1927) mentioned members of the Galapagos Islands population of *Grapsus grapsus* catching and feeding on the ticks of the iguanid lizards. Crane (1947) stated these crabs occasionally catch and eat martins, and are preyed upon in turn by nocturnal raccons, birds, and other predators.

In spite of their numbers, these attractive crabs are very difficult to capture. John Steinbeck said of them in *Sea of Cortez,* "They seem to be able to run in any of four directions; but more than this, perhaps because of their rapid reaction time, they appear to read the mind of their hunter. They escape the long-handled net, anticipating from what direction it is coming. If you walk slowly, they move slowly ahead of you in droves. If you hurry, they hurry. When you plunge at them, they seem to disappear in little puffs of blue smoke . . . It is impossible to creep up on them. Man reacts peculiarly but consistently in his relationship with Sally Lightfoot. His tendency eventually is to scream curses, to hurl himself at them, and to come up foaming with rage and bruised all over his chest."

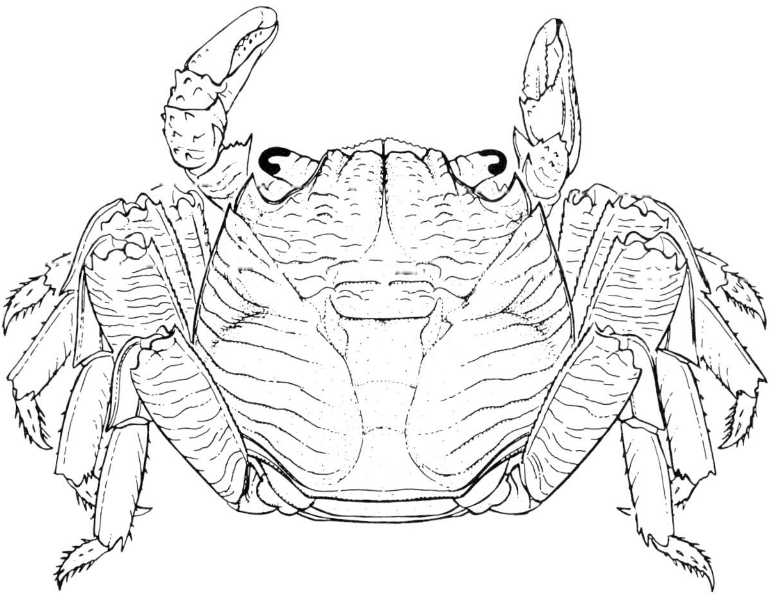

Fig. 7.51 *Grapsus grapsus*

Had Mr. Steinbeck done his collecting during the dark of the night, he may have had better luck with Miss Lightfoot, for these crabs occasionally "sleep" on the rocks well above the water line, often becoming so inactive they can be picked up by hand. Yet their antics prompted Hans Sloane, as long ago as 1725, to christen the Jamaican members of this species "Mardi Gras" (*Natural History of Jamaica*). One of the most colorful accounts of these acrobats I have read can be found in Boone (1927). A passage from William Beebe's *Galapagos — World's End* brings tears of joy to many a marine biologist: "Lying flat on the gentle slope of a huge cube, twenty feet each way, I watched the waving tentacles of anemones far below me. Suddenly a scarlet curtain swept across the whole face of the rock as an army of crabs skittered into view . . . These hosts of Sally Lightfoots were the most brilliant spots of color above water in these islands, putting to shame the dull drab hues of the terrestrial organisms and hinting of the glories of colourful animal life beneath the surface of the sea."

Grapsus grapsus is cosmopolitan in tropical and subtropical America and Africa. On the west coast, it ranges from Bahía Kino and the middle of the Gulf to the Galapagos Islands and Chile. I have found it occasionally at Guaymas, and in abundance at Mazatlán and on all rocky points south of Santa Rosalía in Baja California, including all the southern islands of the Gulf.

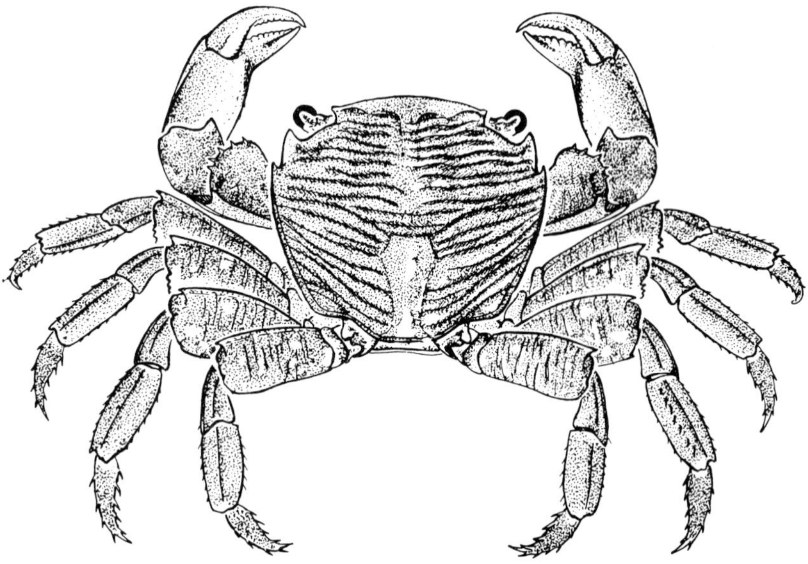

Fig. 7.52 *Pachygrapsus crassipes*
(from Allen, 1969)

Pachygrapsus crassipes Randall

The striped shore crab is one of the most common rocky shore inhabitants of the California, Oregon, and Washington coasts, but is rarely encountered in the Gulf. Dr. John S. Garth considers the records of this crab from Chile and the Galapagos Islands as doubtful. It also occurs in Japan and Korea. This species has a dark red carapace with transverse green lines. The carapace is about 40 to 60 mm wide. Students of the California coast will recognize *Pachygrapsus* as an inhabitant of their high and midintertidal zones. However, in the Gulf, it is probably restricted to the low intertidal and subtidal zones. I have collected only one individual of this species in the Gulf, from Guaymas. Dr. T. H. Bullock informs me he has seen crabs of this species on the rocks of Bahía Cholla in past years.

Pachygrapsus transversus (Gibbes)

A small to medium sized crab, resembling a darker and smaller version of its big brother, *Pachygrapsus crassipes*. There are two anterolateral teeth (including the outer orbital tooth) as in *P. crassipes,* and the carapace is transversed by furrows or depressions. The sides are more strongly convergent posteriorly than in *P. crassipes. P. transversus* is found among stones and the roots of mangrove trees on both coasts of North America, as well as on Oriental shores. It has also been found on coral outcroppings, sandy beaches, and even far out at sea, clinging to floating clumps of *Sargassum* in the North Atlantic. Its range on the west coast of America is from California and the lower Gulf to Peru and the Galapagos Islands. I have collected it in the Gulf at Topolobampo, where it lives in great numbers in the mud and stones and throughout the mangrove swamps. Crane (1937) stated these little crabs are preyed upon by the tidepool blennies and gobies.

Subfamily Varuninae

Tetragrapsus jouyi (Rathbun)

A small species of crab, abundant throughout the Gulf, although often overlooked, possibly because of its small size and nondescript appearance. The carapace is dark and nearly square, slightly broader than long, with four sharp anterolateral spines (including the outer orbital spine). The legs are slender and setaceous. The chelae are equal, but larger in males than in females. Rathbun (1930) stated this species is rare and known from only the type locality of Guaymas. However, I have found *Tetragrapsus jouyi* to be common under stones in bays and

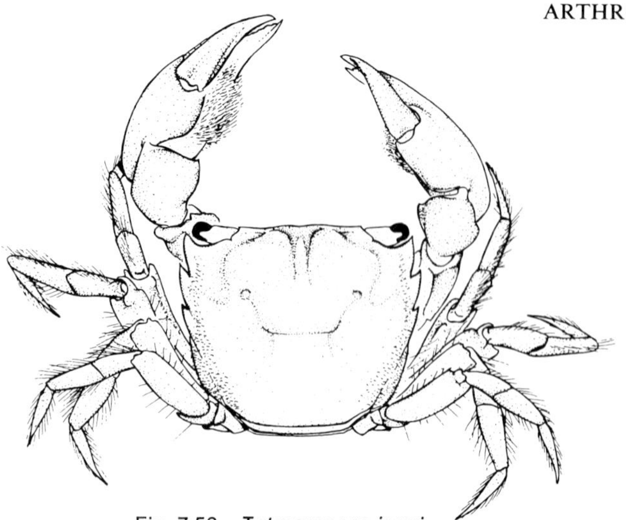

Fig. 7.53 *Tetragrapsus jouyi*

other quiet waters from Puerto Peñasco to Mazatlán, Islas Espíritu Santo and San Francisco, and La Paz. At Kino Bay in 1971, nearly every rock turned over had a handful of these little crabs half-buried in the sand under it.

Subfamily Sesarminae

Aratus pisonii (Milne-Edwards)

This is a curious little crab, with the sides of the carapace converging strongly posteriorly. Its most distinctive feature is perhaps the long tuft of stiff, black bristles on each cheliped. The carapace is mostly smooth and dark, mottled red-brown-green. The legs are reddish. The carapace is about 20 to 25 mm wide. *Aratus* is reported from mangrove swamps and among rocks in estuaries from the lower Gulf to Panamá. I have found this crab only at Bahía San Carlos (Guaymas).

Sesarma sulcatum Smith

This is the only member of a large genus of marsh crabs (at least ten species on the west coast) I have turned up in the Gulf. It is a large crab, 40 to 50 mm wide, with the carapace highly grooved and pitted. The carapace and broad, flat legs are very hairy. The anterolateral margins of the carapace form two sharp teeth (including the outer orbital tooth). The chelae are oval and heavy in the male, thin and small in the females. The range is Guaymas to the Panama Canal Zone. I have collected this species only at Guaymas.

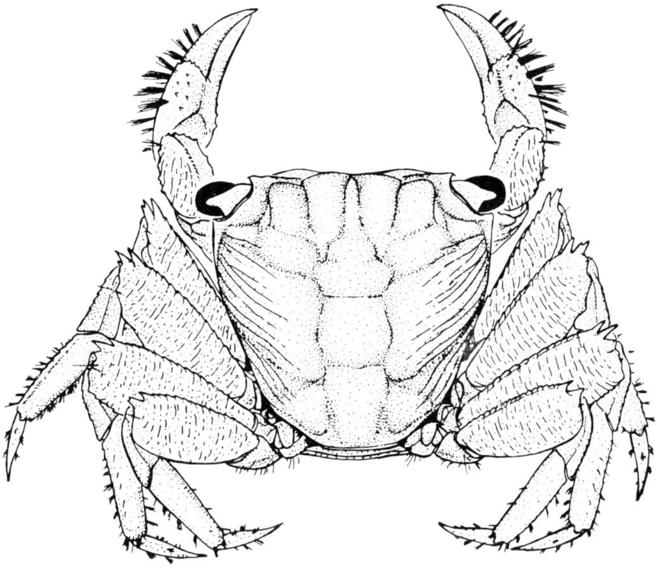

Fig. 7.54 *Aratus pisonii*

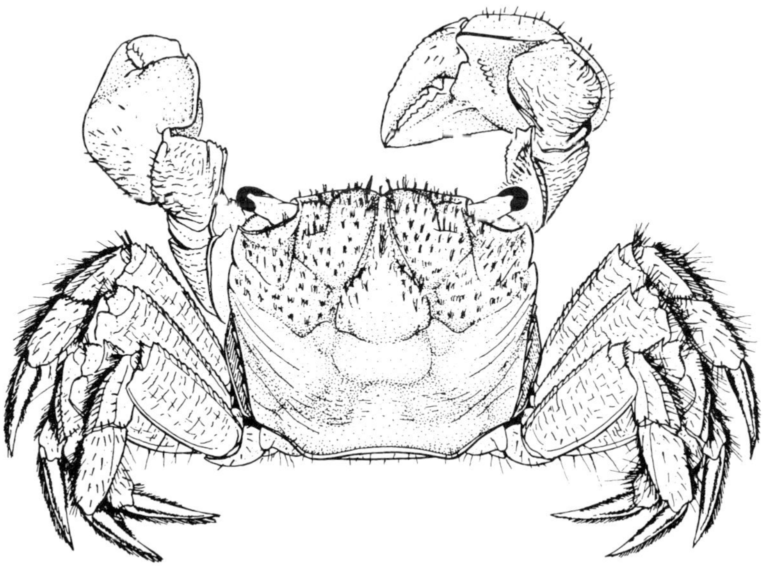

Fig. 7.55 *Sesarma sulcatum*

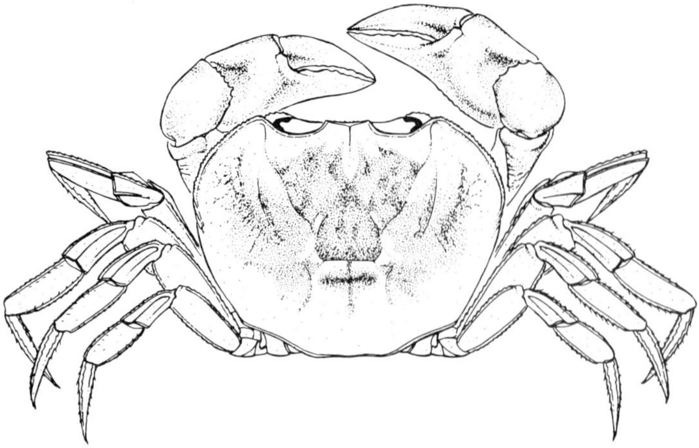

Fig. 7.56 *Gecarcinus quadratus*

Family Gecarcinidae

Gecarcinus quadratus **Saussure**

One of the smallest members of its genus, this crab inhabits estuaries, mud flats, and mangrove swamps. The carapace is about 45 to 60 mm wide, smooth, with a central depression. The dorsal surface is brown, with a reddish purple, median blotch. The eyes are on short stalks, near the center of the anterior margin of the carapace. The species ranges from the upper Gulf to Colombia, where it has been reported from both the Pacific and Atlantic oceans. I have collected it at Puerto Peñasco, Guaymas, and Mazatlán.

Family Ocypodidae (ghost crabs and fiddler crabs)

Ocypode occidentalis **Stimpson**

One of the darkest members of a genus that is usually a light sandy color to blend with its background, *Ocypode occidentalis,* the Gulf ghost crab, is still difficult to see on a sandy beach. The species has unequal chelae, long heavy eyestalks, and little coarse tubercles all over the carapace. When pursued, *Ocypode* runs rapidly across the beach to the soft, dry sand, where it will dive into its burrow to escape capture. Contrary to popular belief, the ghost crabs cannot burrow in the sand faster than a person can dig. They are easy to catch by simply following their burrow down as you dig them up. The size is quite variable, because the same beach may harbor several generations. Usually there are two to three size groups, or generations, on a beach. The larger crabs may reach 150 mm

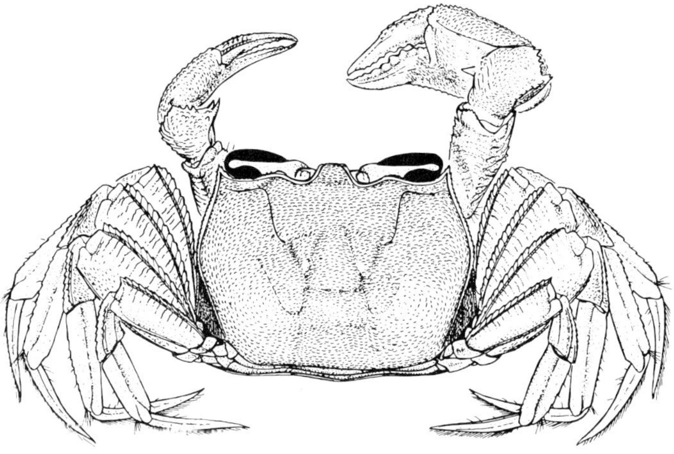

Fig. 7.57 *Ocypode occidentalis*

in width. Smaller individuals seem to fall into 35 to 45 mm and 15 to 20 mm categories. Ghost crabs range from the upper Gulf and the west coast of Baja California to Peru. I have found them to be common inhabitants of nearly every stretch of sandy beach in the Gulf.

Uca brevifrons (Stimpson)

A medium-sized fiddler crab, with short eyestalks and very thin fingers on the large cheliped. This species often wanders inland several miles in woody areas, and has even been found in fresh water. The carapace is about 30 mm wide. *Uca brevifrons* occurs from the northern Gulf to Panama. I have found it to be one of the more common fiddlers of the upper Gulf.

Uca coloradensis (Rathbun)

A common species in the Colorado River estuary, where it may be found burrowing in great colonies into the muddy shores above the high tide mark. This is a small fiddler, with long eyestalks and a fairly smooth cheliped. It is a drab olive color. *Uca coloradensis* has also been reported living in fresh water wells near the coast. The carapace averages 15 to 25 mm in width. The species is known only from the upper Gulf of California, and may be restricted to the vicinity of the mouth of the Colorado River.

Uca crenulata (Lockington)

This is the most common fiddler of southern California. It is a small species, with fairly long eyestalks, and a smooth cheliped that has two rows of small denticles at the base of the fixed finger. The carapace is about 15 to 20 mm wide and 10 to 12 mm long. This fiddler, and almost all others, prefers estuaries and sand-mud flats where it burrows well into the high intertidal zone, and, when possible, amongst vegetation. *Uca crenulata* occurs on both coasts of America. On the west coast, it ranges from southern California to the lower tip of Baja California and throughout the Gulf. It is one of the most common fiddlers of the Gulf. *U. crenulata* will often be a different color when collected at different times of the day. The reason for this is: *Uca* possesses three types of pigment-bearing cells, or chromatophores, in its body (red, black, and white), and can vary the presence of each to alter its appearance. This species, and probably almost all species of fiddler crabs, has both a tidal and a diurnal rhythm in its color phases (Bates, 1966). The red and black chromatophores will become more or less concentrated according to the tide level and time of day. Brown (1953) has shown that the daily dispersion of melanin in *U. pugnax* occurs on a time cycle that allows it to occur about 50 minutes later each day, in other words, following a semidiurnal tide cycle.

Wallace (1967) found that excessive crowding in *U. crenulata* inhibits the normal burrowing behavior at high tide. Straw (1967) reported from a behavioral study of several unidentified species of *Uca* from the northern Gulf that females and young males apparently have no permanent residence. They burrow at high tide, but abandon the burrow at low tide, to wander about in search of food or mating areas, not returning to their old burrows when the tide waters again return. Male fiddlers, however, seem to be quite territorial, and readily chase other males away from the vicinity of their burrows. Johnson and Snook (1955)

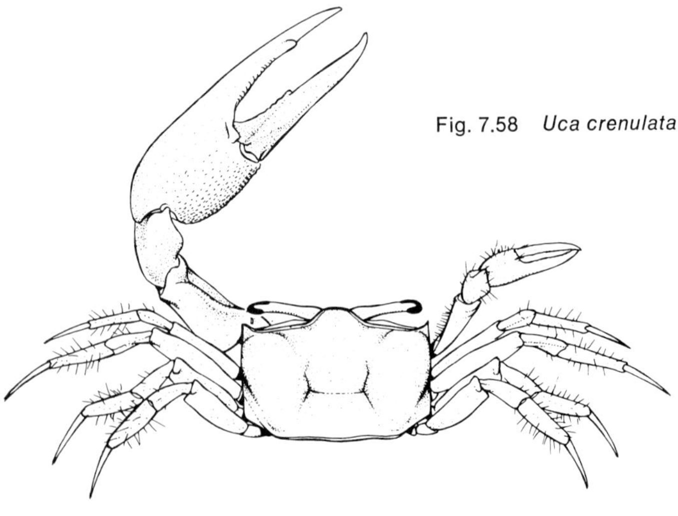

Fig. 7.58 *Uca crenulata*

reported witnessing battles between males that ended in one crab throwing the other over his back a distance of a foot or more. An excellent summary of the fascinating field of claw-waving and sound production behavior is given by Kilgore (1970) and Crane (1941).

Uca latimanus (Rathbun)

A small species, with a finely granulate carapace. The large cheliped is short and stout, the fingers being broad and flat and shorter than the palm. The third to sixth segments of the male abdomen are fused, though vestiges of the sutures remain. This species ranges from the upper Gulf to Ecuador. I have collected it only at Puerto Peñasco.

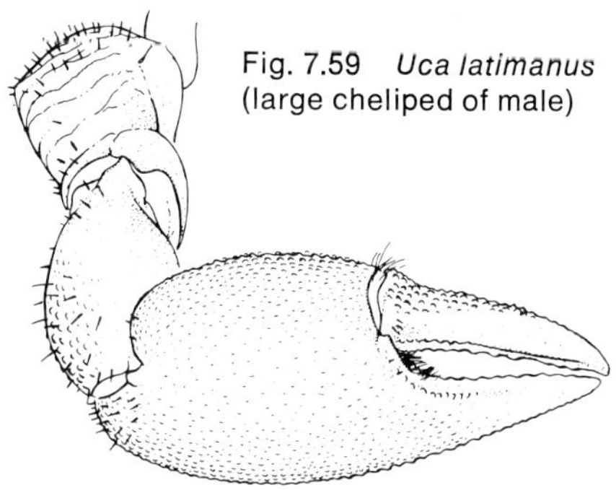

Fig. 7.59 *Uca latimanus* (large cheliped of male)

Uca macrodactylus (Milne-Edwards and Lucas)

A small fiddler crab, with fairly long eyestalks and a tuberculate cheliped. The carapace tapers strongly posteriorly and is about 12 mm wide. This species is abundant in the mud of mangrove swamps in the lower Gulf, where it is often found in great hordes; it occurs in the upper Gulf as well, and I have collected it often at Bahía Cholla.

Uca mordax (Smith)

A medium-sized fiddler, 30 to 35 mm wide, with long eyestalks and slightly roughened chelae. This species prefers estuaries, often digging holes in the mud, or living under stones. It is a common inhabitant of mangrove swamps on both sides of America. On the west coast it is reported from the lower Gulf of California only.

Fig. 7.60 *Uca mordax* (large cheliped of male)

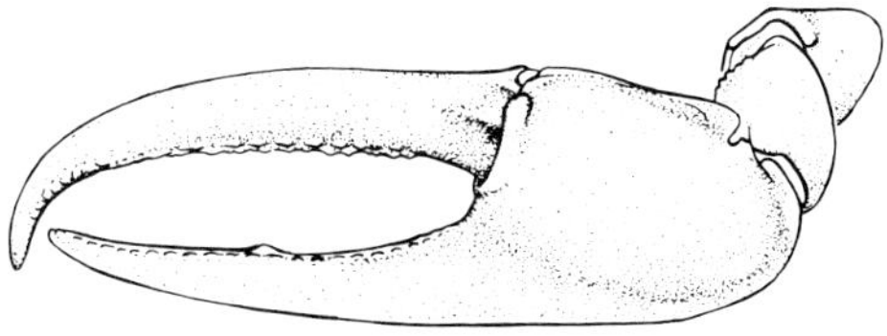

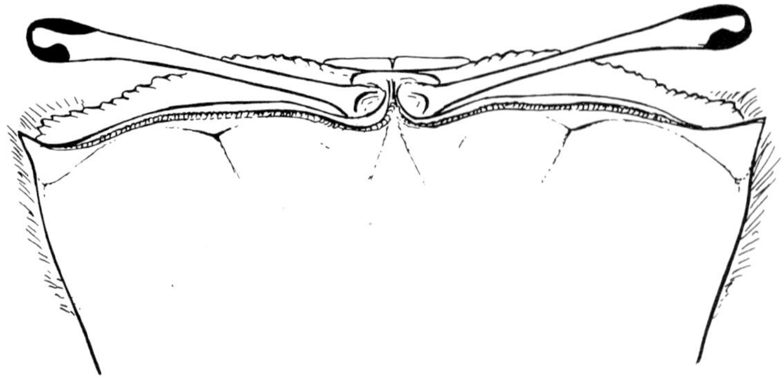

Fig. 7.61 *Uca monilifera* (anterior of carapace)

Uca monilifera Rathbun

A large pale blue fiddler crab, with long eyestalks and smooth chelae. The manus usually has a few small tubercles near the base, but is smooth and bare near the fingers. The carapace is about 40 to 50 mm wide. This species and the following are the only two fiddlers of the Gulf with these extremely long eyestalks, leaving the anterior margin of the carapace, between the eyes, as only a very thin downward extension. *Uca monilifera* is probably restricted to the northern Gulf of California, and is especially abundant in the regions near El Golfo de Santa Clara and Puerto Peñasco.

Uca princeps (Smith)

A large, dark blue fiddler with long eyestalks, resembling *Uca monilifera,* but with the small tubercles on the chelae extending all the way to the fingers. It inhabits salt marshes and estuaries, but is occasionally found under rocks in the middle and low tide zones. The carapace is usually 35 to 45 mm wide. Its range is from the upper Gulf of California and the west coast of Baja California to Peru. It has been found to be especially common in the southern Gulf, where I have seen mass emergences of thousands of these crabs from their burrows on mud flats near the small fishing villages of Boca del Rio and Yavaros.

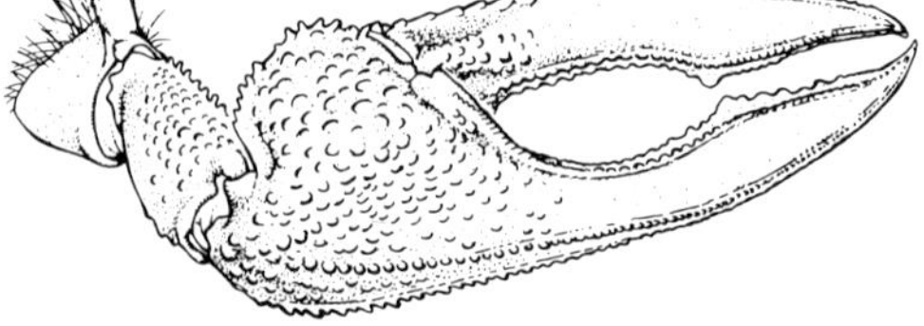

Fig. 7.62 *Uca princeps* (large cheliped of male)

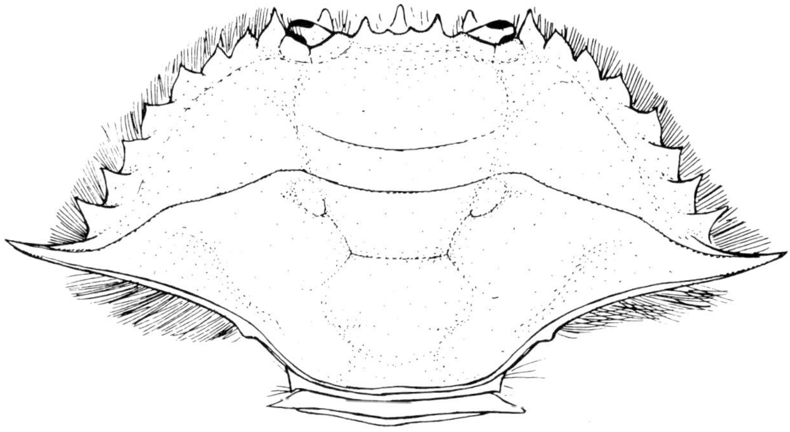

Fig. 7.63 *Callinectes arcuatus* (carapace)

Family Portunidae (swimming crabs)

Callinectes arcuatus Ordway

The lateral spines on this species are very long and pointed. The arms of the chelae usually have 3 to 4 long spines on the inside margin. The carapace is very convex and finely granulate, especially in the median region. The width of the carapace on adults is 100 to 170 mm, but young individuals are commonly found whose carapaces are 40 to 80 mm wide. This species is common from southern California and the upper Gulf to Chile. It is often confused with *Callinectes bellicosus* (below). Two morphological features that may be used to tell these two swimmers apart are: the rows of denticles on the outside of the manus, *C. bellicosus* having several, *C. arcuatus* only one or two; and the length of the carapace spines between the eyes (see key). This, and all the following species of swimming crabs, are found intertidally, but are most easily collected from shrimp trawlers, by skin diving in shallow sandy waters, or with beach seines.

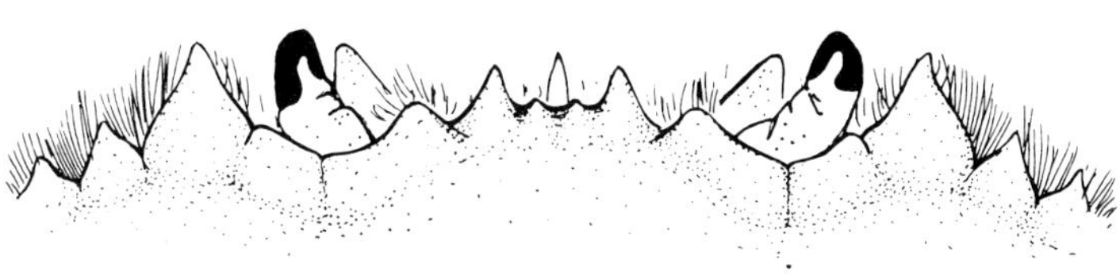

Fig. 7.64 *Callinectes arcuatus* (anterior margin of carapace)

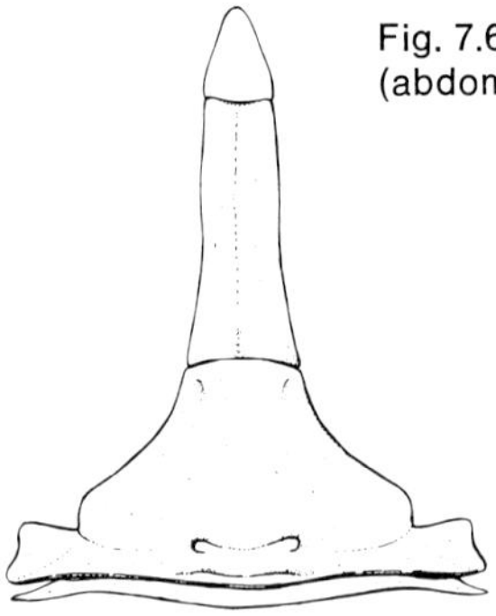

Fig. 7.65 *Callinectes bellicosus* (abdomen of male)

Fig. 7.66 *Callinectes bellicosus* (anterior margin of carapace)

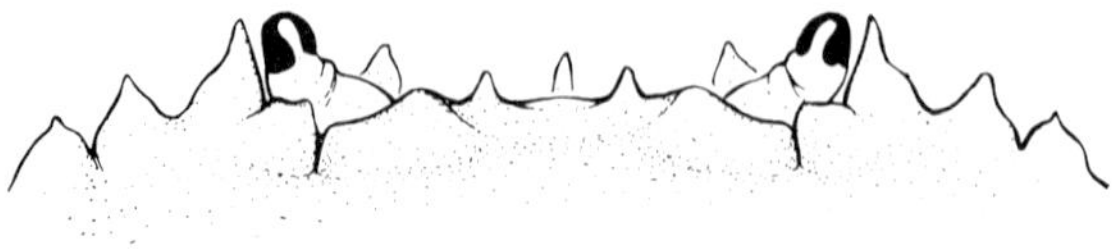

Callinectes bellicosus Stimpson

The largest lateral spine of this swimming crab is shorter than in *Callinectes arcuatus,* in respect to the body width. The carapace is evenly convex, and the arms of the chelipeds usually have 5 to 6 long spines on the inside margin. This and *C. arcuatus* are very common crabs in the Gulf, frequently found in shallow water over sandy bottoms of bays and lagoons. *C. bellicosus* ranges from San Diego, California, and the upper Gulf as far south as Mazatlán, and possibly farther, and is the most commonly encountered swimming crab of the Gulf intertidal. *C. bellicosus* commonly buries itself in the sand until only the eyes and antennae are visible, forming incurrent and excurrent water channels to supply the gills. I have found immature individuals, 1 to 3 cm wide, buried as deep as 20 inches on an exposed tidal flat. This is a very aggressive species, and has been seen feeding on fiddler crabs and small fish. Jenkins (1966) reported finding a rather old and sickly individual at Puerto Peñasco with a heavy infestation of small goose barnacles (6 to 18 mm tall) in the gill region. I have since found numerous healthy, mature individuals thus infected at Puerto Peñasco.

Portunus asper (A. Milne-Edwards)

The chelipeds of this swimming crab are robust, and the postero-distal border of the merus of the fifth leg is serrated and has a single spine. This is a large species, the males reaching nearly 100 mm across the carapace. The anterior border of the arm has 4 to 5 spines, the distal spine being well-separated from the others. The inner spine of the wrist is well-developed. The manus and wrist are elongate, making the cheliped of the males over three times as long as the carapace. The dorsal surface is covered with a very short pubescence. This species is synonomous with *Portunus panamensis,* and ranges from the mid-Gulf of California (Canjamé, Sonora) to Chile.

Portunus tuberculatus (Stimpson)

A small swimming crab, with the posterodistal border of the merus of the swimming legs without spines or spinules The carapace is usually 10 to 30 mm wide, light brown, and marbled with black. The base of the lateral spines is often tinged with red. The postlateral corner of the carapace is spinous, and the entire carapace is covered with rows of small, elevated tubercles. The anterior border of the arm has four spines, the distal spine being well-separated from the others. The inner and outer spines of the wrist are both well-developed. This swimmer ranges from the mid-Gulf to Ecuador and the Galapagos Islands, and is largely subtidal.

Portunus xantusii (Stimpson)

This species represents a complex assemblage of three subspecies. Through the examination of over 3,500 specimens, and the statistical methods applied by Stephenson (Stephenson, 1965; Garth and Stephenson, 1966), these cited workers were able to resolve the complex relationship of what now consists of one species with three subspecies (and over eight synonyms). It is comforting to the practical naturalist to see a group of nearly identical animals competently placed together in what appears to be a natural phylogenetic scheme, as opposed to being split up and given three or four or even five different species names that tend to disguise their natural affinities and confuse students. The three subspecies of *Portunus xantusii* are: *P. x. xantusii* (Stimpson), *P. x. minimus* Rathbun (=*P. pichilinquei*), and *P. x. affinis* (Faxon). The posterodistal border of the merus of the swimming legs of this species has spinules, but no spines. The inner spine of the wrist is short, and the postlateral corner of the carapace is rounded. The width of the carapace and length of the last anterolateral tooth are variable, and serve as a partial basis for distinguishing the subspecies. The arm has four to six spines on the anterior border, occasionally more in very large specimens. The three subspecies

Fig. 7.67 *Portunus xantusii* (abdomen of male)

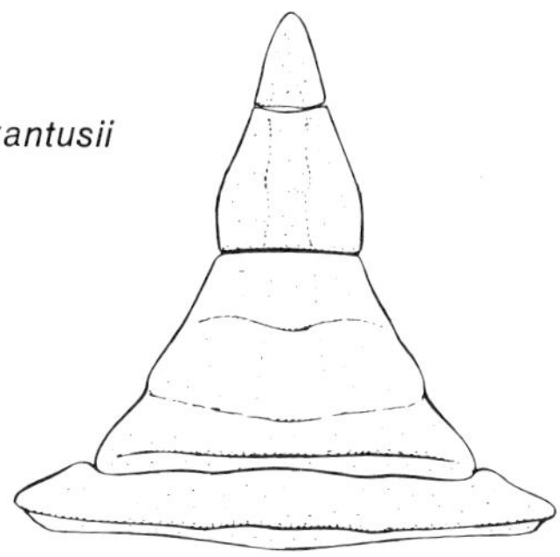

are nearly allopatric, with only a slight overlap in distribution. *P. x. xantusii* is the only one occurring north of Bahía Magdalena on the west coast of Baja California, extending from Santa Barbara, California, to Cabo San Lucas (also reported from Bahías Concepción and Agua Verde in the Gulf); *P. x. minimus* is the common form in the Gulf, ranging from El Golfo de Santa Clara to Cabo San Lucas; and *P. x. affinis* is the southern form, ranging from Cabo San Lucas to Ecuador (reported also from Isla Isabel in the Gulf [Garth, 1960]). The carapace of *P. x. xantusii* is approximately 8 to 73 mm wide, that of *P. x. minimus* about 13 to 43 mm wide, and that of *P. x. affinis* about 11 to 54 mm wide. Swimming crabs of this species captured north of Cabo San Lucas in the Gulf of California will probably be *P. x. minimus*.

Euphylax robustus A. Milne-Edwards

This interesting and distinctive swimming crab has extremely long eyestalks, the eyes placed at or near the anterolateral corners of the carapace. The chelipeds are robust and the carapace is dull, not polished as in most of the swimming crabs. The arms have three well-developed spines on the anterior margin, and the wrist has a well-developed spine on the inner and outer sides. The carapace is usually 30 to 60 mm wide, gray with tinges of blue-green, and with only four anterolateral spines, the second from the front being much shorter than the others. This species ranges from El Golfo de Santa Clara, at the head of the Gulf, to Peru. I have collected it in abundance from shrimp trawlers at Puerto Peñasco.

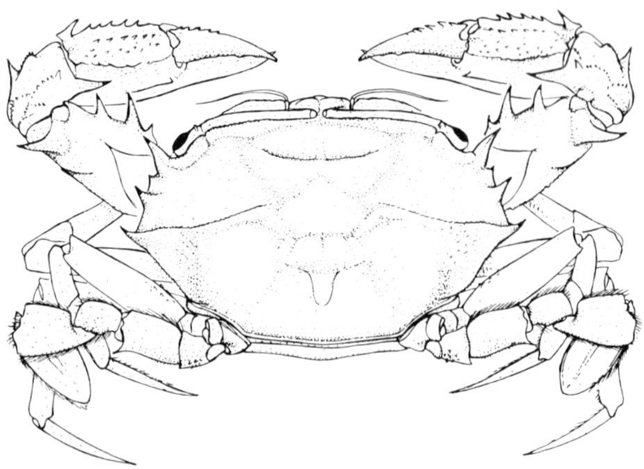

Fig. 7.68 *Euphylax robustus*

Arenaeus mexicanus (Gerstaecker)

The most beautiful swimming crab of the Gulf, and one of the most attractive crabs I have ever seen, this swimmer has the carapace covered with pale red or pink circles having white to pale green centers. The ground color is a pale reddish brown. The coloring is nearly identical to that of the box crab, *Hepatus lineatus*. There are six to eight anterolateral teeth. The chelae are stout, and the wrist has a short outer spine or tubercle. This species ranges from Bahía Magdalena and the mid-Gulf to Peru. I have collected it only at Huatabampito (near Yavaros, Sonora). *Arenaeus mexicanus* seems to prefer open, sandy beaches.

Cronius ruber (Lamarck)

An attractive swimming crab, with a dark carapace and purplish red to purplish blue shading on the back and legs. The swimmerets are rust-colored, and the entire carapace is covered with transverse rows of thick, short setae. The outer surface of the wrist has three strong spines, and the posterior border of the merus of the swimming leg has a very strong spine. The chelae are heavily spined, darkest near the tips, but white on the extreme tip. The carapace is about 15 to 80 mm wide. The anterolateral spines are alternately large and small, the second, fourth, and sixth being the shorter. This species ranges from the upper Gulf to Chile and the Galapagos Islands. It also occurs on the Atlantic seacoast. I have collected it in several feet of water over sand at low tide in Bahía Cholla and at Bahía Kino.

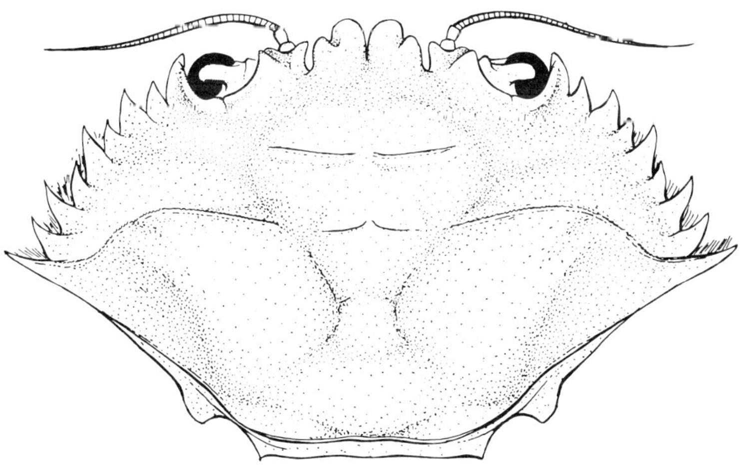

Fig. 7.69 *Cronius ruber* (carapace)

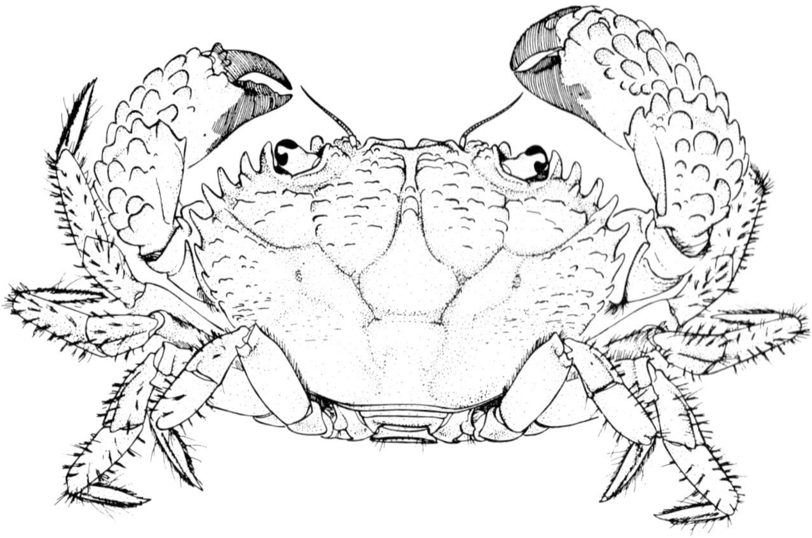

Fig. 7.70 *Eriphia squamata*

Family Xanthidae

Eriphia squamata Stimpson

This is one of the most common crabs, and a dominant animal, of the rocky intertidal of the upper Gulf. The chelae are highly tuberculate, the tubercles being large and round to oval-shaped, ciliated on the margin. The carapace is dark green or gray, mottled, and the legs are usually banded with black. The carapace is usually 30 to 55 mm wide. Students familiar with the Californian fauna might at first mistake this crab for *Paraxanthus taylori,* common in similar habitats in southern California. *Eriphia squamata* occurs in great numbers in the mid- and low intertidal of rocky shores in the northern Gulf, and less frequently in the southern Gulf. It commonly hides in holes or crevices during the day, but emerges at night to forage for food. It has been seen grazing on algae and detritus, as well as attacking smaller crabs and worms. Crane (1947) examined the stomachs of numerous individuals, and found they had been feeding primarily on small worms, crustacea, and algae. This hardy crab is also said to be a fairly common inhabitant of mangrove swamps, where its carnivorous tendencies should be most easily seen.

Eriphia squamata ranges from the upper Gulf to Ecuador and the Galapagos Islands, but I have rarely found it south of Topolobampo and Bahía San Luis Gonzaga. Cox (1965) reported from his observations on the behavior of this species that individuals are apparently quite territorial, defending their territory by claw-waving gestures reminiscent of *Uca,* or by actual attacks. He also reported tidepool gobies commonly occurring around *Eriphia* when the latter is feeding on algae. Apparently the crab dislodges a good deal of detritus and other foodstuffs for the small fish to feed on as they dart in and out between the crab's chelae to snatch food floating in the water near the crab's mouth. The crab appears to be oblivious to the fish. The same study revealed that the algae *Valoniopsis pachynema* is the most common food item of *Eriphia* in the region of Puerto Peñasco. I have also observed *Eriphia* feeding on small fiddler crabs. A study by Beumer (1966) revealed that the upper lethal temperature for this species is about 40 to 44°C, while the lower lethal temperature is about 0°C (population from Puerto Peñasco).

Eurytium affine (Streets and Kingsley)

A medium-sized crab, 20 to 30 mm wide, closely resembling its relative, *Eurytium albidigitum,* but having much darker claws and a flatter carapace. The teeth of the anterolateral margin are flattened anteriorly, but become pointed toward the rear. The carapace is a mottled red-blue-green-tan. The chelae are slightly unequal in size. This species is quite common throughout the Gulf intertidal zone, in quite muddy bays, usually under stones. Its range is from the upper Gulf and west coast of Baja California as far south as Manzanillo (possibly to Ecuador), reappearing again in the Galapagos Islands.

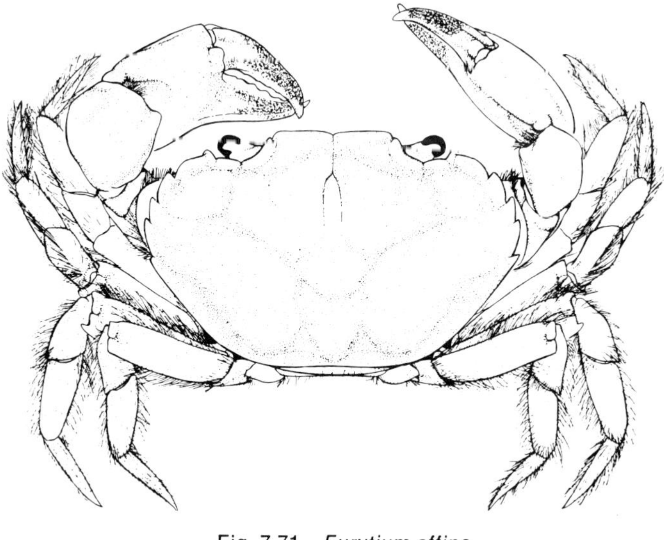

Fig. 7.71 *Eurytium affine*

Eurytium albidigitum Rathbun

A medium to large crab with a mottled blue-gray carapace, usually with white spots. The carapace is nearly 1.5 times as broad as long. This species is common in mud flats and estuaries where it burrows into the mud, or keeps well under the rocks in the high tide zone. *Eurytium albidigitum* is carnivorous on small fiddler crabs, and is especially common where there is a lot of vegetation. It has been found consistently throughout the upper Gulf, but is not known to occur below San Felipe and Desemboque.

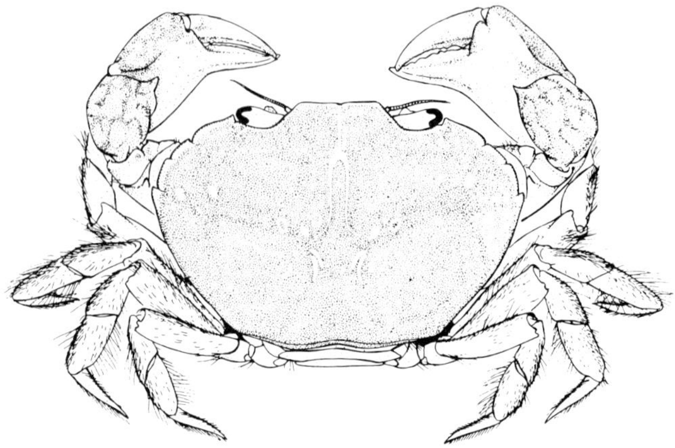

Fig. 7.72 *Eurytium albidigitum*

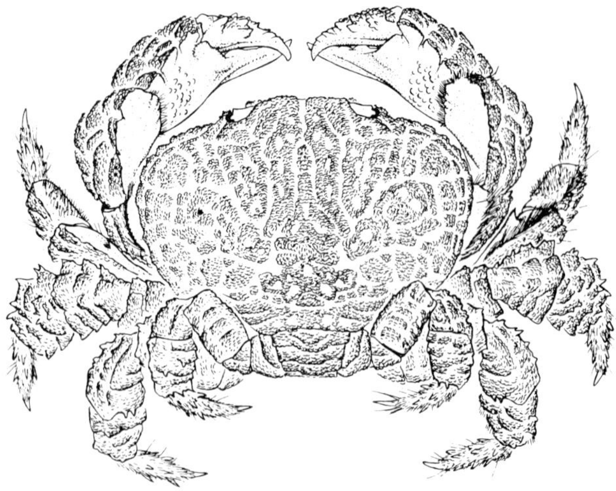

Fig. 7.73 *Glyptoxanthus meandricus*

Glyptoxanthus meandricus (Lockington)

This easily distinguished crab has the surface of its carapace and legs highly vermiculated with deep furrows, ridges, and pits. It is a small- to medium-sized crab, 25 to 35 mm wide, pale, usually variegated with yellow and carmine. It lives under large stones in the low intertidal, and is common throughout the Gulf, at least as far south as Guaymas. *Glyptoxanthus labyrinthicus* was synonymized with *G. meandricus* in 1930 by Rathbun, but removed from synonomy in 1939 by Garth.

Leptodius occidentalis (Stimpson)

One of the commonest crabs of the mid- and low intertidal throughout the Gulf. This species has a varicolored carapace, mottled with gray, brown, white, violet, or greenish red, often with one to three darker patches (usually red) medially. The carapace often turns brilliant red-orange after preservation in alcohol. The palm of the chelae is light, but the fingers are dark brown to black. There is a distinguishing bilobed flange between the eyes, each lobe being dorsally convex. The sides of the body and the base of the legs are very hairy. The carapace is 25 to 45 mm wide, but juveniles are often found to be no more than 1 or 2 cm in width. Juveniles of this and several other species of crabs are commonly encountered living in the algal turf of large stones, and in the pores of thick sponges like *Geodia, Leucetta,* and *Pseudosuberites*. This crab is common among rocks and in mangrove swamps throughout the Gulf, south as far as Manzanillo, reappearing again in the Galapagos Islands.

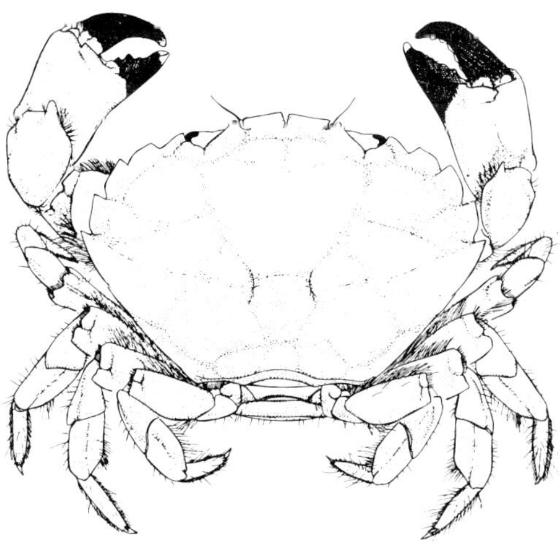

Fig. 7.74 *Leptodius occidentalis*

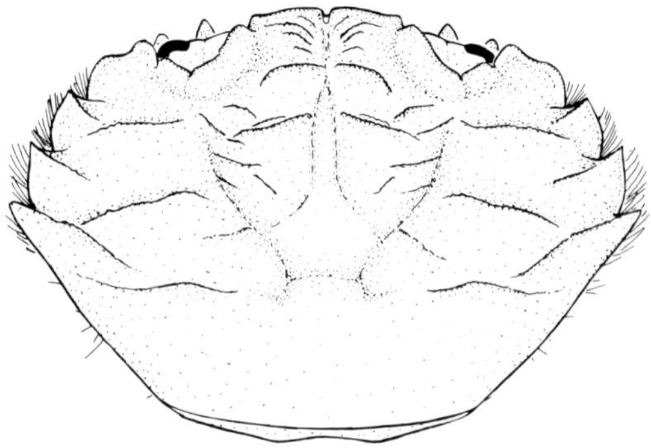

Fig. 7.75 *Panopeus purpureus* (carapace)

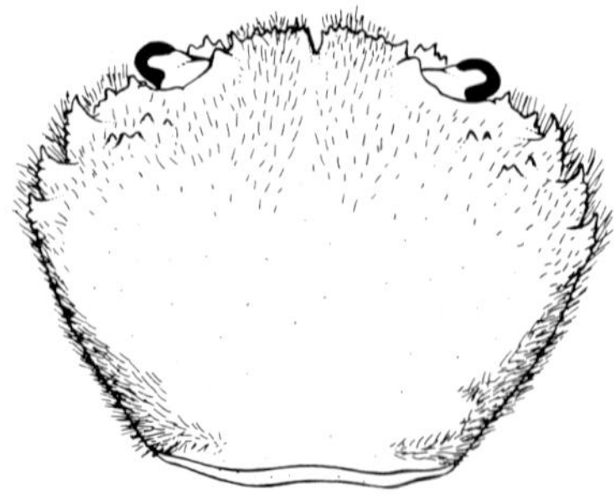

Fig. 7.76 *Pilumnus gonzalensis* (carapace)

Panopeus purpureus Lockington

A medium-sized crab, 20 to 50 mm wide, closely resembling its cousin *Leptodius*. Both of these genera, as well as *Eurytium affine* and *Xanthodius hebes,* have the anterior margin of the carapace, between the eyes, more or less extended to form an indented flange or protrusion. There are four fairly strong anterolateral teeth on the mottled red or brown carapace of *Panopeus*. The fingers of the chelae are brown, with white on the tips. This species ranges from the southern Gulf to Peru, preferring the same habitat as *Eurytium affine*. I have often found the two species living together in the lower Gulf.

Pilumnus gonzalensis Rathbun

The most common hairy crab of the Gulf intertidal. It has short setae on the dorsal surface, spread thinly so as to leave the reddish purple carapace visible below. The carapace is flat, 15 to 20 mm wide, with five anterolateral spines, the first two being quite small. This species ranges throughout the Gulf of California to Tenacatita, and may be found at nearly every rocky point in the Gulf.

Pilumnus limosus Smith

The smallest and least common of the hairy crabs of the Gulf intertidal. This species has the carapace completely obscured by the short, thick, setae that form a soft, velvety tan carpet across the dorsal surface. There are four anterolateral spines. The upper and outer surfaces of both palms are rough and hairy. The carapace is 10 to 15 mm wide. This species ranges from the upper Gulf to Peru, but I have not seen it below Bahía Kino.

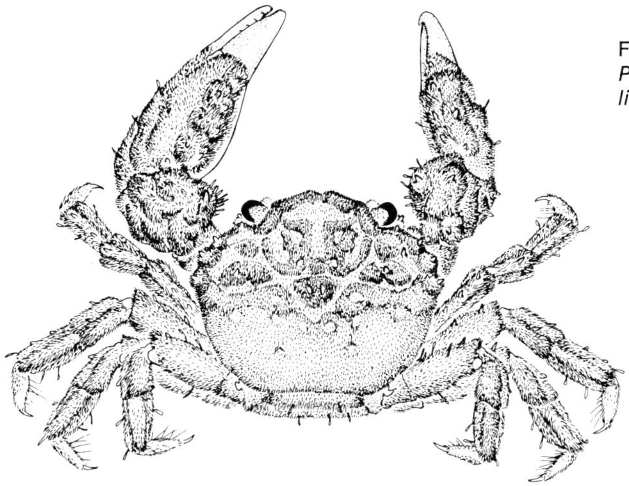

Fig. 7.77 *Pilumnus limosus*

Pilumnus townsendi Rathbun

The second most common of the three hairy crabs of the Gulf intertidal. This species has a carapace width of 15 to 35 mm. There are four anterolateral spines on the carapace. Less than half of the outer surface of the palm is smooth. The chelae have numerous, stiff, black-and-brown setae, and are covered with sharp spines. The carapace setation is of two types, a long stiff type and a shorter, softer type. The dorsal setae are not so thick as to obscure the purple carapace below. This species lives under the rocks in the low intertidal from the upper Gulf to Bahía Manzanillo. Students of the California seashore might at first mistake this species for the similar appearing *P. spinohirsutus,* whose range is from southern California to lower Baja California, but it has one less anterolateral spine.

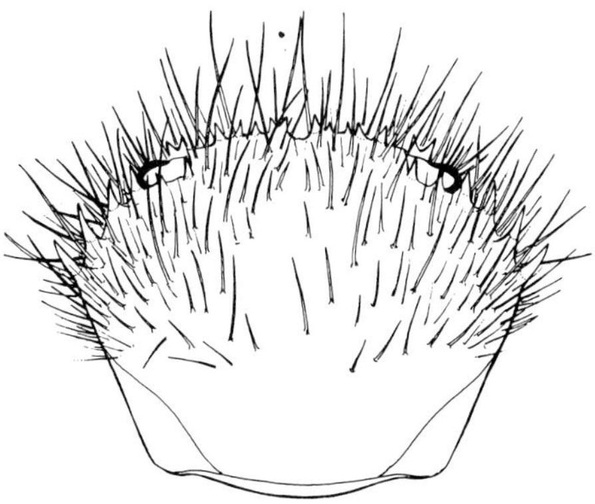

Fig. 7.78 *Pilumnus townsendi* (carapace)

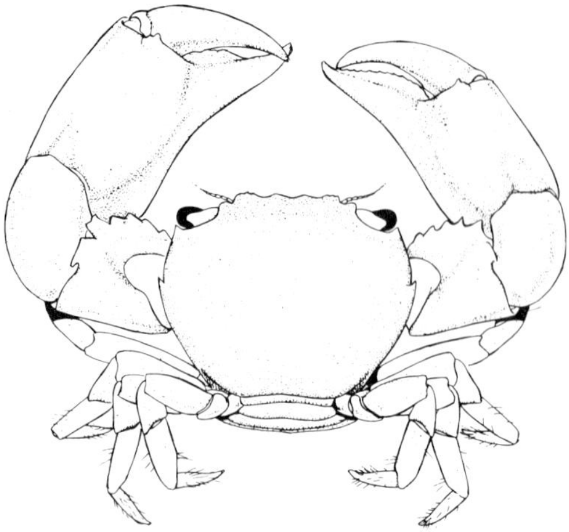

Fig. 7.79 *Trapezia ferruginea*

Trapezia ferruginea Latreille

A shiny little red to orange crab, either unicolored or with a few dark patches on the carapace, and with brown to gray fingers on the chelae. As the crab matures, the chelae turn progressively darker, although never becoming black. The chelae are very large for the size of the body. The carapace is only 10 to 20 mm wide. This species is barely included in the Gulf fauna. It ranges from Banco Arena and Cabo San Lucas to Colombia and the Galapagos Islands, and also occurs in the Indo-Pacific region and the Red Sea. I have collected it only at Cabo San Lucas, where it is probably strictly a coral commensal. Crane (1947) mentioned this species as found commonly in coral heads of *Porcillopora,* often one male and one female in a chamber. Crane also noted that the shrimp *Crangon ventrosus* is often found living in the coral with *Trapezia,* probably mimicking the crab's color. Two other crabs occur as members of coral colonies in the Isla Espíritu Santo-Cabo San Lucas area. These are *Trapezia digitalis* and *Domecia hispida;* I have not collected either.

Xanthodes hebes Stimpson

This attractive little crab has an almost perfectly oval carapace, 20 to 35 mm wide. The color is usually red, occasionally tinged with green. The carapace has no anterolateral teeth and the fingers of the chelae are black. Close examination of the carapace and chelipeds will reveal them to be highly punctate. This species ranges from the lower Gulf (and Bahía Magdalena on the outer coast of Baja California) to Isla de Maria Madre (off the Nayarit coast). I have collected this species only at Guaymas (1971), on the rocky point north of the Miramar Resort area, where it was fairly common.

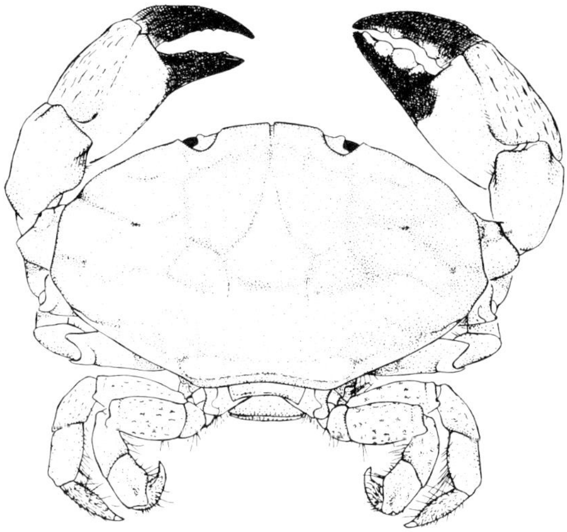

Fig. 7.80 *Xanthodes hebes*

Family Pinnotheridae (the pea crabs)

Pinnotheres Latreille

The members of this genus of pea crabs are commensal in pelecypods and occasionally gastropods, tunicates, annelids, and sea urchins. The majority of the taxonomy of this genus is based on characters of mature females, the males being much smaller and usually free-living. At least three species are known from the Gulf of California. Dr. W. L. Schmitt of the United States National Museum is presently revising the Pinnotheridae of the world, and his work, when published, will greatly add to our knowledge of these odd little crabs.

Parapinnixa Holmes

This genus is known from only North America, Mexico, and the Galapagos Islands. The single species reported from the Gulf (Rathbun, 1930) is known from only the original description, the type of which has since been destroyed.

Pinnixa White

A very large genus of pea crabs, living commensally in mollusc and annelid tubes and burrows, in the cloaca and rectum of holothurians, and freely in mud burrows. Only one or two species have been reported from the Gulf.

Tetrias Rathbun

A small genus of pea crabs known mainly from the Indo-Pacific region, with a few species occurring in the Eastern Pacific. Only one or two species have been reported from the Gulf.

Dissodactylus Smith

A small genus of pea crabs with a subpentagonal carapace, minute eyes, and a bifurcate dactylus on the first three pairs of walking legs. This genus is restricted to America. Its members are commonly found clinging to the outside of sand dollars and sea urchins with their bifurcate fingers. The sand dollars *Encope grandis* and *Encope micropora* in the northern Gulf usually have these little commensal crabs clinging to the spines in the lunules.

Superfamily Oxyrhyncha (spider crabs)
Family Majidae
Subfamily Inachinae

Stenorhynchus debilis (Smith)

This majid crab has a smooth, triangular carapace, long, thin legs, and a long slender rostrum that is cylindrical and as long, or longer than, the remainder of the carapace. The rostrum is armed with short pointed spines. This beautiful fragile crab is bright green in life, with the lateral portions of the back and legs covered with yellow streaks and a soft, gray

Fig. 7.81
Stenorhynchus debilis

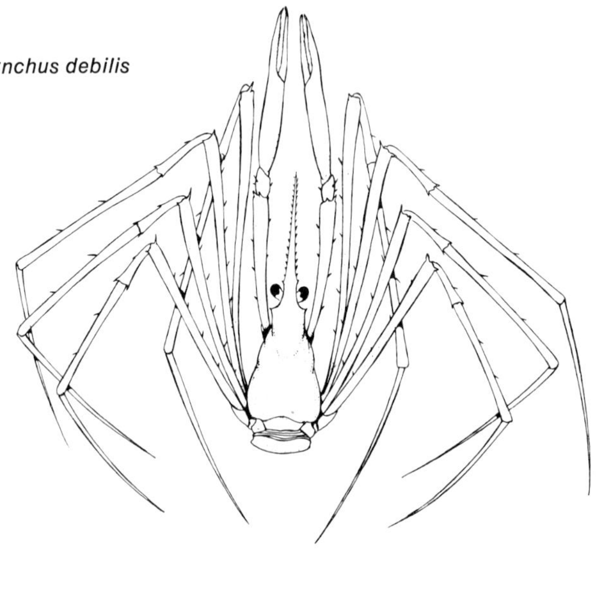

pubescence. The fingers of the chelae are reddish. The width of the carapace is usually 10 to 15 mm. The body, with its long rostrum and rounded margins, greatly resembles a giant teardrop. This species is mainly subtidal usually being collected with SCUBA gear or from shrimp trawls, but may occasionally be found intertidally during the lowest minus tides. *Stenorhynchus debilis* ranges from the upper Gulf to Chile and the Galapagos Islands. I have collected it at Isla Espíritu Santo in 25 fathoms, at Guaymas in a few feet of water at low tide, and at Puerto Peñasco clinging to *Sargassum* at low water.

Subfamily Acanthonychinae

Epialtus minimus Lockington

A small majid crab with an elongate, flattened, and bifurcate rostrum in the males, but a shorter, rounded, and more truncate rostrum in females. The carapace is 10 to 15 mm wide, and forms two large anteriolateral lobes. The dorsal surface is reddish yellow, occasionally spotted with brown, with three backward-pointing tubercles in the gastric region. The chelae of the males are massive, the meri grossly enlarged and tuberculate. This species is commonly found clinging to bunches of *Sargassum*, especially where this seaweed forms large beds or mats. It is reported from Laguna Ojo de Liebre (Scammons Lagoon) and throughout the Gulf, south at least as far as Acapulco. I have collected it only at Puerto Peñasco.

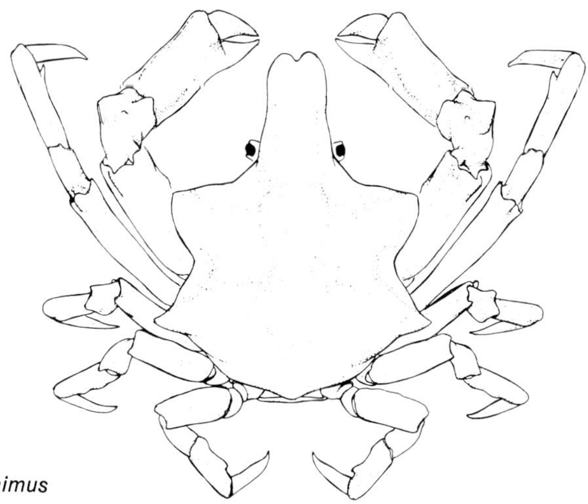

Fig. 7.82
Epialtus minimus

Epialtoides paradigmus Garth

A small majid crab, resembling *Epialtus* but with longer chelae, larger processes on the carapace, and five instead of six free abdominal segments in the males. The rostrum is emarginate, but not highly bifurcate. The dorsal surface is brown to red as in *Epialtus,* but usually differs from the latter in the presence of a large, drab yellow blotch in the midregion. The chelae are massive in the males, but partly obscured by the large anterolateral projections of the carapace. This species ranges throughout the Gulf to somewhere south of Mazatlan.

Subfamily Pisinae

Herbstia camptacantha (Stimpson)

This is primarily a subtidal form, occasionally found on rocks or coral heads in the low intertidal, or by diving. The carapace is about 15 to 25 mm wide, triangular-shaped, and covered with short spines, each on an elevated region of the back. The legs bear many heavy marginal spines, and the chelipeds are long and thin. This spider crab ranges throughout the Gulf, and down the west coast of Mexico at least as far as the state of Oaxaca. I have collected it at Puerto Peñasco, where it is often brought up in the shrimp trawls.

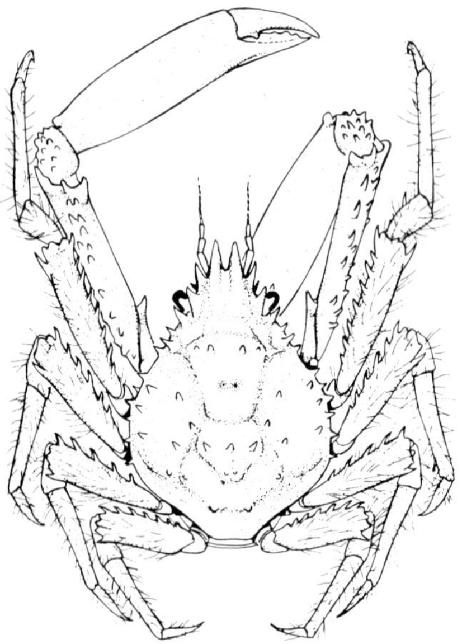

Fig. 7.83 *Herbstia camptacantha*

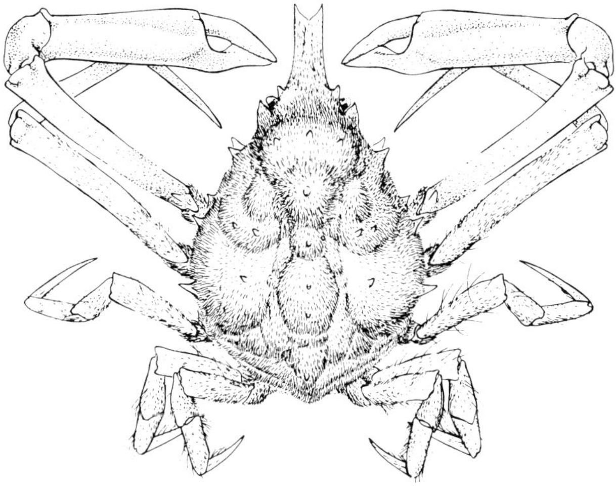

Fig. 7.84 *Libinia mexicana*

Libinia mexicana Rathbun

An odd-looking crab with an inflated, oval carapace, and a long tubular rostrum. The carapace is 50 to 65 mm long, bears a short brown pubescence, and has six large dorsal spines in a row down the middle, plus numerous other spines on both sides of this median row. The chelae are white, and much longer than the first walking legs, which are in turn longer than the second. This crab is endemic to the Gulf. It is common subtidally from El Golfo de Santa Clara to Bahía de San Ignacio in northern Sinaloa and possibly further south. I have collected it at El Golfo and Puerto Peñasco, from shrimp trawlers and with SCUBA gear. *Libinia mexicana* could easily be confused with the similar appearing *L. setosa*, a cognate species, but the former is restricted to the Gulf, while the latter occurs on the west coast of Baja California.

Subfamily Mithracinae

Ala cornuta (Stimpson)

The Gulf masking crab. One of the commonest small- to medium-sized crabs of the low intertidal, on rocky beaches throughout the Gulf. The carapace is 20 to 55 mm wide, moderately triangular, and has seven to nine tufts of stiff setae or bristles on the dorsal surface. There are also numerous tufts along the margins of the legs and carapace. The carapace is pale olive, or dull yellow-brown to pale red, often with small dark patches. The dorsal surface is usually covered with small bits of algae and sponge, and occasionally a solitary anemone or some compound ascidian. This material is placed there by the crab, enmeshed in the stiff bristles of the carapace. When *Ala* is kept in an aquarium and deprived of natural material for camouflage, it will substitute with most anything you give it, including paper, bread, and hamburger. The rostrum is one-fourth to one-fifth the length of the remainder of the carapace. The chelipeds are tuberculate and slender, a little longer than the first pair of walking legs. When the carapace is scraped clean, you can see four to five very large spines along the anterio-lateral margins. When threatened, *Ala* shows no defense behavior other than increasing the strength of its hold on the substrate. When picked up, the crab remains perfectly still, perhaps a death-feigning reaction. The main food appears to be algae. *Ala* is common throughout the Gulf, and ranges as far south as Peru. I have collected it at nearly every rocky point in the Gulf. It is the only species of masking crab I would call really common to the Gulf intertidal. *Dromidia,* which usually masks itself with a large sponge, is principally a subtidal form.

Stenocionops angusta (Lockington)

This is another species of spider crab only rarely encountered in the intertidal, but occurring subtidally to at least 30 fathoms. The body is strongly inflated, triangular, and about 20 to 40 mm wide. Garth (1958) reported all the collections of this species he has seen from sand, shell, or coralline bottoms. The chelae are long and slender, with black fingers. The carapace is similar to *Herbstia* in being covered with numerous short spines set on elevated regions. This species occurs throughout the Gulf and on the west coast of Baja California. I have collected it subtidally and among rocks, in a few feet of water during minus tides, at Puerto Peñasco.

Subtribe Oxystomata (oxystomatous crabs)
Family Leucosiidae
Subfamily Ebaliinae

Speloeophorus schmitti Glassell

Members of this unusual genus have large orifices in the posterior region of the carapace. Two species are common in the Gulf. *Speloeophorus schmitti* has two large holes, invisible from the dorsal view, and a triangular carapace that is much broader than long, the width being about 30 to 40 mm. This crab is found throughout the Gulf of California subtidally and in the low intertidal. I have collected it at San Felipe and Puerto Peñasco.

Fig. 7.85 *Ala cornuta*

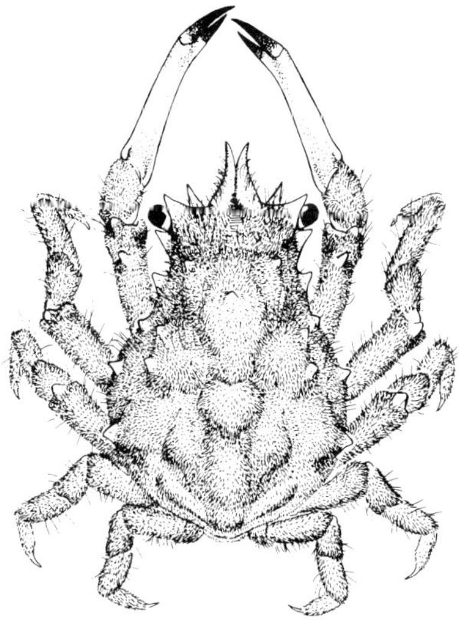

Fig. 7.86
Stenocionops angusta

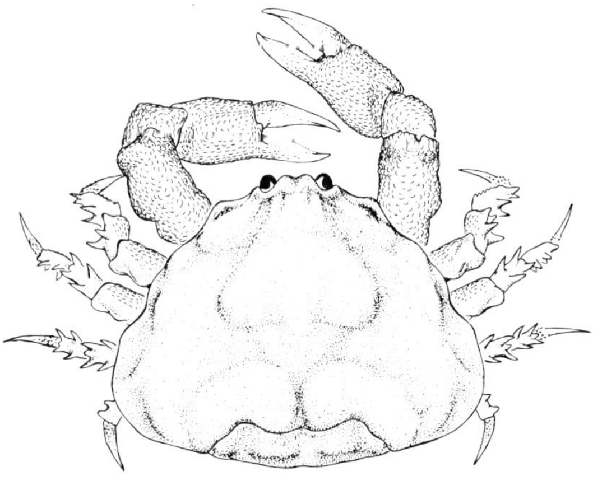

Fig. 7.87
Speloeophorus schmitti

Speloeophorus digueti (Bouvier)

This species has four large posterior orifices, two that are visible from the dorsal view, and two that are hidden. The carapace is roughly pentagonal, 5 to 20 mm wide, and longer than broad. This crab probably does not occur intertidally. Its range is given as the Gulf of California and the west coast of Baja California to Panamá. I have not collected this species.

Family Calappidae
Subfamily Matutinae

Hepatus lineatus Rathbun

This is certainly one of the most beautiful crabs of the Gulf. It has a narrow, high carapace, strongly arched, with the height equal to about one-third the width. The dorsal surface is 20 to 80 mm wide, covered with yellow to orange spots, and encircled by reddish purple lines. It is reported that these crabs occasionally decorate their already attractive carapace with sea anemones of the family Sagartiidae. The species is found in the low intertidal and subtidal, often buried under a few inches of sand. At certain times of the year, it is possibly the most commonly encountered brachyuran in the shrimp trawls of the northern Gulf. Its range is from

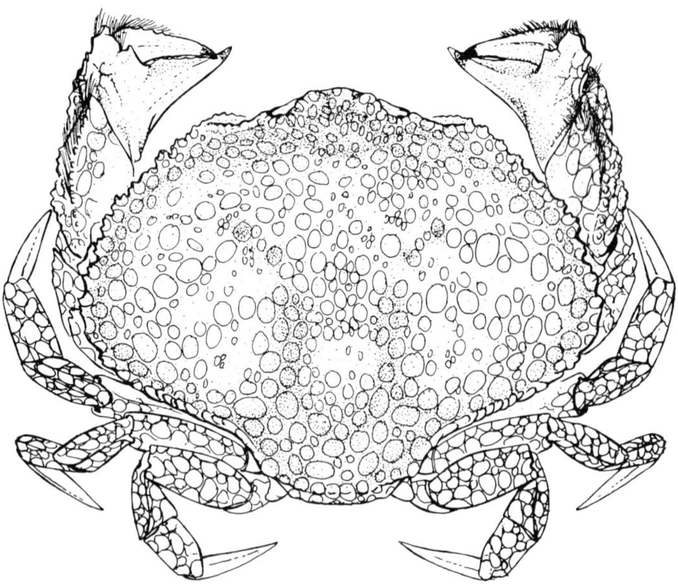

Fig. 7.88 *Hepatus lineatus* (dorsal view)

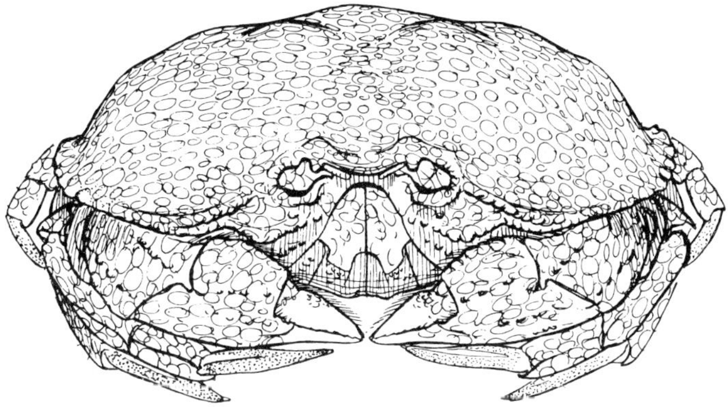

Fig. 7.89 *Hepatus lineatus* (anterior view)

the upper Gulf and the west coast of Baja California to La Paz and possibly farther south. I have collected it throughout the northern Gulf, in sandy bays at low tide and from shrimp trawls.

Hepatus kossmanni Neumann

This species has the carapace covered with narrow stripes of dull red making it nearly as attractive as *Hepatus lineatus*. However, this species loses most of its color in preservation, whereas *H. lineatus,* for some reason, retains most of its color. There are two large tubercles on the fourth abdominal segments of the male. The carapace is usually 60 to 70 mm wide. The species occurs in the low intertidal and subtidal, from the upper Gulf to Ecuador. I have collected it only at El Golfo de Santa Clara.

Subtribe Dromiacea
Superfamily Dromioidea
Family Dromiidae

Dromidia larraburei Rathbun

Members of this genus usually carry a sponge, a compound ascidian, or a sea anemone on their backs for concealment. The fifth legs are reduced and highly modified for this purpose, being folded anteriorly on the back. This is a medium-sized crab, 20 to 45 mm wide, with numerous small, ball-shaped tubercles on the upper surface of the palm. The carapace is tan and highly pilose. The species ranges from Monterey, California, to Peru and the Galapagos Islands. I have collected specimens subtidally and in the lowest intertidal at Puertecitos (with a large, living sponge on the back), and Puerto Peñasco (carrying various species of compound ascidian). These are fascinating crabs to keep in the aquarium. When they are deprived of their natural camouflage they will utilize nearly anything of the correct size, from pieces of algal fronds they cut up, to pieces of newspaper.

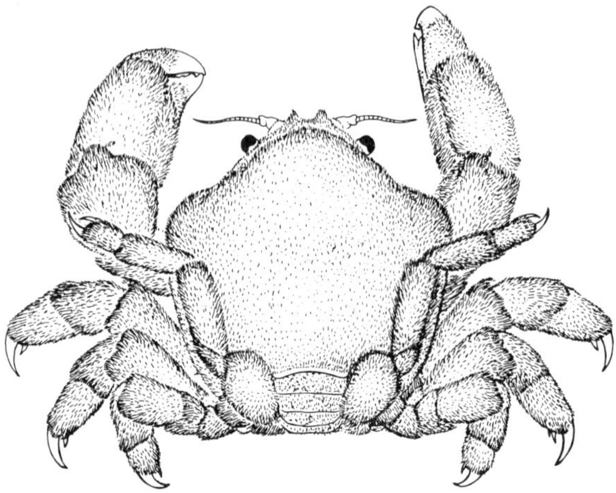

Fig. 7.90 *Dromidia larraburei*

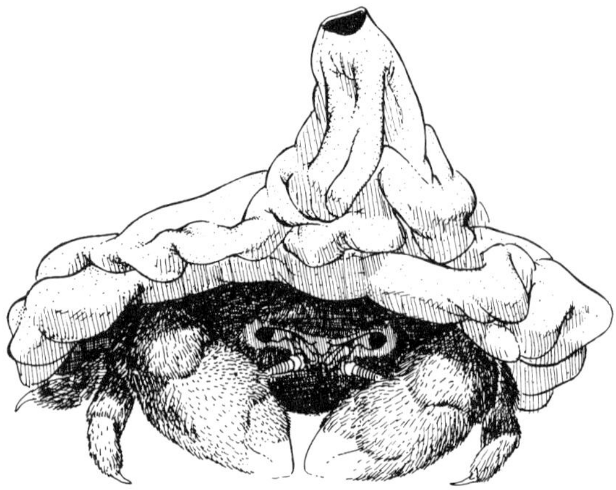

Fig. 7.91 *Dromidia larraburei* (with sponge on back)

Hypoconcha lowei Rathbun

A very hairy crab, 30 to 60 mm wide, that carries an empty pelecypod shell on its back by the use of its modified last pair of legs. The chelae are covered with numerous large, white, oval denticles. It is common in the Gulf, often brought up in the otter trawls of the shrimp fishermen, but also encountered in the low intertidal, especially when SCUBA diving. Its range is from the upper Gulf to Ecuador. I have collected it in abundance throughout the upper Gulf, from shore to 40 fathoms.

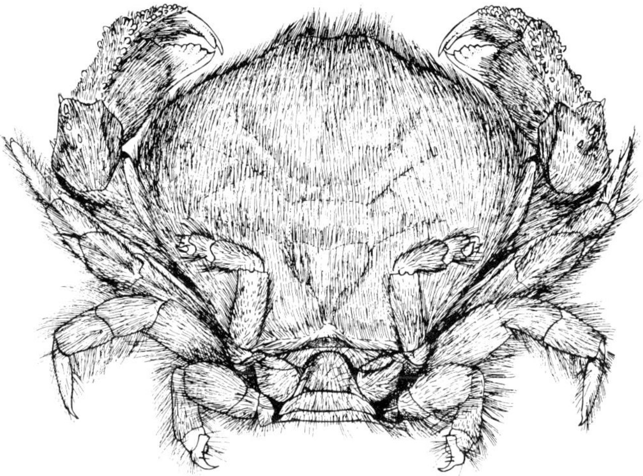

Fig. 7.92 *Hypoconcha lowei*

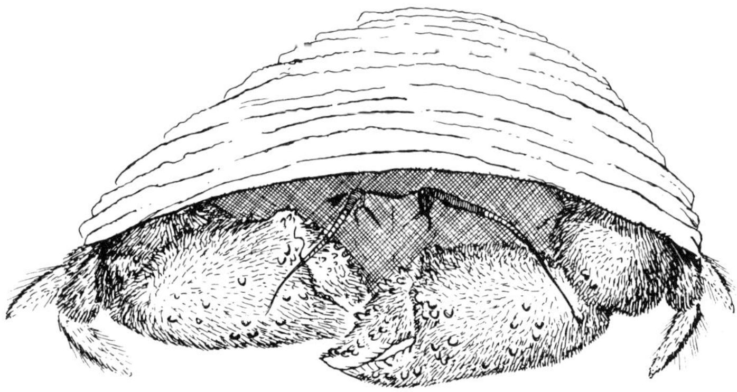

Fig. 7.93 *Hypoconcha lowei*
(with pelecypod shell on back)

Sipunculida (Peanut worms)

Preservation: Sipunculids will invariably retract their introvert when removed from the substrate, and must be relaxed prior to preservation. Narcotizing may be with magnesium sulfate, or by the putrid water method described for the holothurians. Preservation should be in alcohol or 10 percent buffered formalin, the formalin preserving the color better than alcohol. Steinbeck and Ricketts stated that these worms are especially sensitive to metals and their salts, and will retract the introvert when in contact with even small amounts of these compounds during relaxing procedures.

Taxonomy: The taxonomy of the Sipunculida, or peanut worms, is based primarily on internal anatomy, tentacle arrangement, and the position and morphology of the body papillae and introvert hooks. There has been no generally accepted arrangement of the genera into higher taxa. Fisher (1952) and Edmonds (1955, 1956) recognize 13 genera. Fisher uses the name *Golfingia* Lankester in place of *Phascolosoma,* and replaced *Physcosoma* Selenka with *Phascolosoma* Leuckart, in order to conform to the rules of zoological nomenclature (also *Dendrostomum* Grube instead of *Dendrostoma* Keferstein to retain the original spelling). The following classification follows this arrangement.

Sipunculids of the Gulf: Compared with most regions of the world, the sipunculid fauna of North America is relatively sparse. *Phascolosoma agassizii* is the dominant intertidal form from Alaska to central California, and extends into the Gulf of California with less frequency. *Phascolosoma puntarenae* is the dominant species from the lower Gulf of California to Panamá. *Sipunculus nudus* is an offshore form, cosmopolitan in the bays and estuaries of the world, and fairly common in the Gulf. Steinbeck and Ricketts reported six species of sipunculids from the Gulf, and found *Golfingia hespera* (reported as *Phascolosoma hesperum*) and *Physcosoma* sp. to be the most common peanut worms of this region. I have found *Themiste lissum* and *Phascolosoma perlucens* to be the most common species in the upper Gulf.

Sipunculids may be found buried just below the surface on sandy beaches, mainly under large rocks in the mid- and low intertidal. Some species are common inhabitants of sponges, and at least one (*T. lissum*) may be found burrowed into porous rocks and conglomerate reefs.

KEY TO THE INTERTIDAL SIPUNCULIDS OF THE GULF OF CALIFORNIA

1 Oral tentacles bushy and conspicuous when extended **2**
-Oral tentacles small and fingerlike, not bushy and conspicuous when extended **3**

2(1) Oral tentacles arise from 4 or 8 roots; body smooth and devoid of papillae; small (length of trunk and introvert 25 to 35 mm); tentacles surround mouth (fig. 8.2) .... ***Themiste lissum***
-Oral tentacles arise from more than 10 roots (the tentacular crown is thrown into numerous folds to accommodate the large number of tentacles, usually 50 or more); body with numerous papillae; medium to large (length of trunk and introvert 45 to 60 mm); tentacles dorsal to mouth .... ***Phascolosoma antillarum***

3(1) Body small and threadlike; introvert extremely filamentous and wrinkled; muscles of body wall divided into separate longitudinal bands; usually commensal in burrow of annelid or burrowing sea anemone .... ***Golfingia hespera***
-Body not small and threadlike; introvert not as above; muscles of body wall not divided into separate longitudinal bands; free-living **4**

4(3) Skin with rectangular cross-hatchings; tentacles surround mouth; nuchal organ dorsal to tentacles; length of trunk and introvert about 120 to 300 mm (fig. 8.1) .... ***Sipunculus nudus***
-Skin without rectangular cross-hatchings; tentacles arranged in a circle enclosing the nuchal organ and dorsal to mouth; length of trunk and introvert less than 150 mm (usually 25 to 70 mm) **5**

5(4) Introvert usually without brown bands; subepidermal, longitudinal muscle bands conspicuous; introvert with numerous, prominent, pointed papillae on dorsal surface (fig. 8.3) .... ***Phascolosoma perlucens***
-Introvert with numerous brown bands; longitudinal muscle bands not conspicuous; introvert without enlarged, conspicuous papillae as above **6**

6(5) Introvert with 15 to 25 rings of hooks; length of body and introvert 30 to 70 mm .... ***Phascolosoma puntarenae***
-Introvert with 25 to over 100 rings of hooks; length of body and introvert 70 to 150 mm .... ***Phascolosoma agassizii***

SYSTEMATICS AND SPECIES DESCRIPTIONS

Sipunculida

Sipunculus nudus Linnaeus

This is one of the best-known and most widely distributed peanut worms in the world. *Sipunculus nudus* is eurythermal, nearly cosmopolitan, in all but the coldest seas. The skin appears longitudinally ribbed, usually

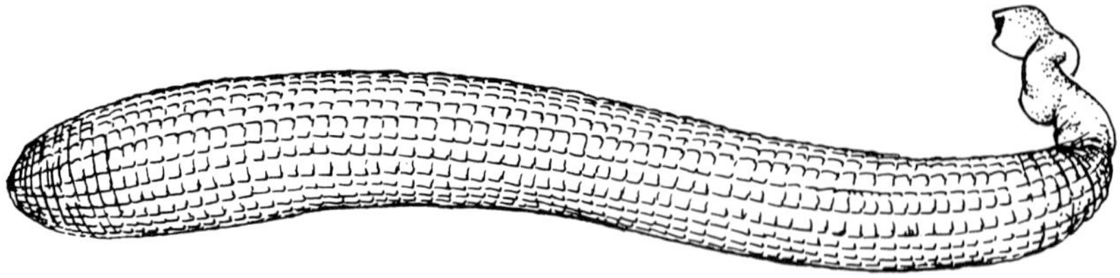

Fig. 8.1 *Sipunculus nudus*

cross-hatched to form numerous small rectangles, due to the underlying muscle bands. The introvert is short, and bears numerous scalelike papillae, that are somewhat triangular and have the tips pointing away from the mouth, closing together to overlap like the shingles of a roof. The tentacles of the introvert are fingerlike processes, borne on a horseshoe-shaped tentacular fold, and usually difficult to observe. The skin is white or flesh colored, and shines with a pearly iridescence when alive. Most individuals are about 200 mm long when extended (a subtidal form reported from Pacific Grove, California, was 300 mm long). This species is commonly found in the sandy mud of sloughs, estuaries, or bays, the largest colonies being subtidal. They feed by passing quantities of bottom sediment through their gut, much as the earthworms do, to extract organic matter. This species is recorded from Brazil north to the Carolinas, Europe, the Indian Ocean, and the western shores of the Pacific. On the west coast of America it is known from central California to Panamá, and throughout the Gulf of California.

Golfingia hespera (Chamberlin)

This unique sipunculid is a very thin, wrinkled, wormlike animal that is found commensally in the tubes of burrowing sea anemones *(Cerianthus, Edwardsiella),* polychaetes *(Mesochaetopterus),* and brachiopods *(Glottidia).* The introvert is exceedingly long and slender, up to eight times the trunk length, and usually randomly coiled or twisted when extended. The introvert bears numerous rings of minute hooks. When the long introvert is retracted, it entirely fills the body cavity with its irregular coils, displacing the viscera. The body wall is translucent, a bleached brown or yellow color. The body tapers to a point posteriorly, and bears numerous short, ovoid, or cylindrical papillae. There is a very small tentacular crown. The length of the trunk and introvert combined is 100 to 140 mm. This species has been previously reported as *Phascolosoma hespera.* Its range is from southern California to the lower Gulf of California.

Fig. 8.2 *Themiste lissum*

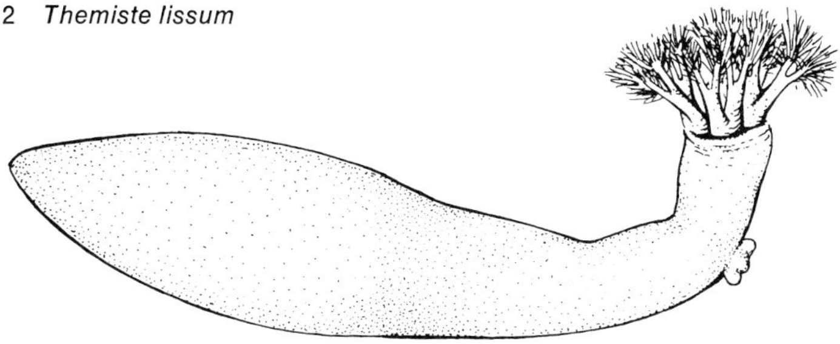

Themiste lissum (Fisher)

A very small sipunculid with a short, tan body, and a short, thick introvert. It is devoid of visible spines and papillae on both trunk and introvert. The tentacles are large and profusely branched, making it easy to identify among the other short-tentacled species of the Gulf. The tentacles arise from four (or eight) thick basal tentacles. The introvert is about one-fourth the body length, and the anterior third of it forms a smooth, translucent collar, while the rest is whitish and crowded with minute convex glands, not easily seen. This species is known only from the intertidal zone of the Gulf of California, where it is usually found burrowed into clay or soft stone. *Themiste lissum* is rarely seen unless one hacks away at the hard substrate, whereupon a broken chunk of rock will often reveal great masses of these worms in their narrow burrows. This species has recently been placed in the genus *Themiste,* and was known for many years as *Dendrostomum lissum.*

Phascolosoma agassizii Keferstein

A good-sized peanut worm with 24 or less short, digitiform tentacles. The introvert has 15 to 25 rings of tiny hooks anteriorly, is long (more than half the length of the trunk), and marked by irregular transverse bands of dark brown. The trunk is bluntly pointed posteriorly, and rough to the touch, due to the numerous convex or conical skin papillae that are a darker brown than the general skin tone. The body wall varies from slightly translucent to opaque, and the skin color from pinkish gray or yellowish gray to brown. Quite often, the trunk and introvert may be spotted with black, brown, or purple. The length of an extended animal is 70 to 150 mm. This is a dominant sipunculid of the intertidal zone in many regions, having adapted itself to a variety of habitats from the mid-tide zone to over 100 fathoms. In the intertidal zone, it may be found under rocks in sand or mud, in kelp holdfasts, or in rocky crevices filled with mud. In California, it is almost invariably found in the large mussel beds of *Mytilus californianus,* and in southern California it is quite common in the massive colonies of the tube-building polychaete worm *Phragmatopoma,* and also in association with terebellid polychaetes. This species has been previously reported as *Phymosoma agassizii, Golfingia agassizii,* and *Physcosoma japonicum.* Its range includes Alaska to San Quintín, Baja California, and the Gulf of California. It is recorded from temperate and tropical waters of both hemispheres.

Phascolosoma puntarena Grube

A small- to medium-sized sipunculid (to about 70 mm), similar to *Phascolosoma agassizii,* but with slightly longer papillae, especially on the dorsal surface of the introvert. The introvert has transverse bands of brown and bears 24 tentacles. The hooks of the introvert are borne in 25 to 100 or more rings. Some workers consider this species to be a variant of *P. agassizii,* but Fisher (1952) considers it more closely related to the tropical form, *P. nigrescens (P. agassizii* is a temperate, cold water species). Fisher also notes that the name *P. puntarena* has 7 years' priority over *P. agassizii,* should synonomy be considered. This species has been previously reported as *Sipunculus puntarenae* and *Physcosoma agassizii.* It ranges from the Gulf of California to Panamá.

Phascolosoma perlucens Baird

A medium-sized, slender species with a dorsal, preanal area of conspicuously enlarged, dark brown, sharp tubercles, and a similar area of tubercles at the posterior end. The introvert bears about 16 to 21 rings of small hooks, sharply bent apically, and usually has some dark brown patches on the dorsal surface. There are about 12 to 15 tentacles, usually olive colored. This species resembles *Sipunculus nudus* in having longitudinal muscle bands visible through the body wall. The body is highly papillated, and when extended is 40 to 70 mm long. The introvert is a

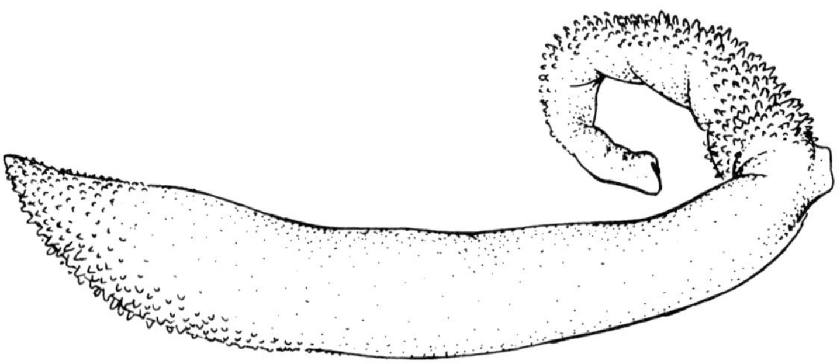

Fig. 8.3 *Phascolosoma perlucens*

dark brown to purple. The species is recorded from the Philippines and the general Indo-Pacific region and the Gulf of California, where it occurs under stones in the midintertidal zone. It was formerly known as *Phascolosoma dentigerum.*

Phascolosoma antillarum Grube & Oersted

A medium-sized, robust *Phascolosoma* with a brown trunk bearing numerous low, convex papillae that become large and crowded at the posterior end, and again in the anal region. The introvert is set off by its pale cream color, being about one-third the trunk length. There are no hooks on the introvert. This is one of two sipunculids of the Gulf that have long, conspicuous, and highly branched tentacles, the other being *Themiste lissum.* These two species are easily distinguished, because the latter is smooth and without papillae, while *Phascolosoma antillarum* is highly papillated. There are 50 to 200 oral tentacles, and the tentacular disk is thrown into a number of folds to accommodate this great number of feeding appendages. Extended length is about 50 to 60 mm. This species is reported from the lower Gulf of California to Chile and Hawaii, and on the east coast from the West Indies to Brazil.

Echiuroidea (Spoon worms)

Preservation: The spoon worms should be narcotized prior to preservation to prevent contraction of the proboscis. Knudsen gave the following technique, recommended by the *British Museum Collector's Guide* (1954): Place the specimens in finger bowls containing sea water, and sprinkle liberally with menthol crystals after they are fully expanded. After an hour or two, begin to introduce, drop by drop, enough alcohol to make a 10 percent solution. Preservation may be in 5 percent formalin or 80 percent alcohol.

Taxonomy: The echiuroids resemble the sipunculoids (peanut worms), but lack proboscis tentacles and possess a terminal anus, as opposed to the midbody or anteriorly placed anus of the sipunculoids. The body is usually sausage-shaped, with a solid, extensible proboscis just anterior to the mouth. The shape of this contracted proboscis has given the group the common name of "spoon worms." A ring of posterior setae usually is found near the anus, and a pair of small hooked setae is usually located behind the mouth. The systematics used in this handbook are from Fisher (1946). Fisher reported only three species of echiuroids from the Gulf of California, all three of which are known to occur intertidally. Fisher stated, "Echiuroids are burrowers in mud or sand, where

KEY TO THE ECHIUROIDEA OF THE GULF OF CALIFORNIA

1 Width of extended proboscis, at base, equal to length; body length usually less than 20 mm; longitudinal muscle bands not present (fig. 9.1) .... ***Thalassema steinbecki***

-Extended proboscis much longer than width; body usually greater than 25 mm in length; longitudinal muscle bands present and visible through epidermis **2**

2(1) Body quite transparent throughout length; longitudinal muscle bands weakly defined; 8 longitudinal muscle bands .... ***Listriolobus pelodes***

-Body translucent in midregion only; longitudinal muscle bands distinct, at least in anterior half of body; 16 to 17 longitudinal muscle bands (fig. 9.2) .... ***Ochetostoma edax***

they fashion more or less permanent tunnels. Sometimes they live under rocks; sometimes in mud-filled mollusc shells or sand-dollar tests, which afford some protection; or they inhabit the rock galleries excavated by boring clams. Their food consists of organic material contained in the mud, which they swallow in large quantities, or of lighter organic detritus selected by the usually long proboscis. In the same species from different localities the intestinal pellets vary with the character of the bottom. One genus (*Urechis*) has very specialized feeding habits and uses only finely divided material, including bacteria. It is probable that any small organism living in the surface film of mud will be eaten by echiuroids."

SYSTEMATICS AND SPECIES DESCRIPTIONS

Echiuroidea

CLASS ECHIURIDA

Order Echiuroinea

Family Echiuridae

Thalassema steinbecki Fisher

A small spoon worm, with a body length of about 12 mm and a proboscis of about the same length. The body is slightly translucent, but not transparent as in *Listriolobus pelodes*. The proboscis is very broad at the base, as wide as the entire length of the extended structure, which is ribbonlike distally. *Thalassema* does not possess differentiated muscle bands, as do the other two species of spoon worms reported from the Gulf. Dr. W. K. Fisher named this species after John Steinbeck, who sent him a single specimen found burrowed in the sandy mud, collected by himself and E. F. Ricketts, near La Paz. Other records for this species are from 10 fathoms off Isla Cerralvo (just south of La Paz), and farther south to Ecuador. It has been collected both intertidally and at depths of up to 165 fathoms. I have not collected this species.

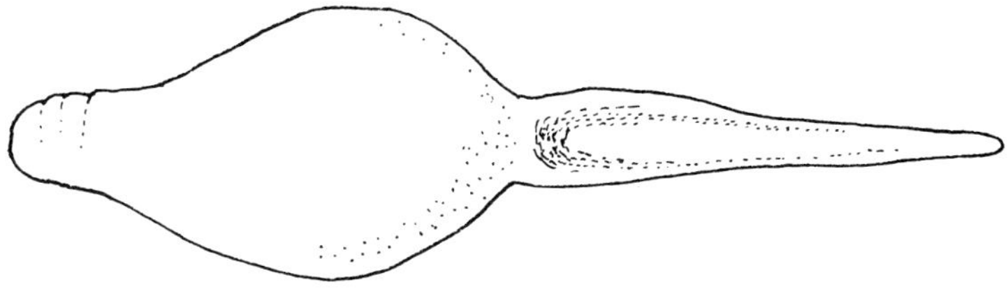

Fig. 9.1 *Thalassema steinbecki*

Listriolobus pelodes Fisher

A highly transparent species, with the proboscis much longer than wide, quite ribbonlike when extended. The body reaches at least 60 mm in length. The body wall is marked by eight weakly defined muscle bands. The proboscis in life is reported as yellow-orange, the body gray-violet with greenish raised specks. When fully extended, the proboscis length slightly exceeds the body length. Dr. Fisher reported this species from numerous localities in central and southern California, Los Frailes near the southern end of Baja California, and in the upper end of the Gulf of California in 5 to 120 fathoms. I have collected what may be this species at Bahía San Everisto on the east coast of Baja California, at approximately lat. 24°50′N, in December, 1971.

Ochetostoma edax Fisher

Probably the most common species of intertidal spoon worm in the Gulf, *Ochetostoma edax* is opaque only in the midregion of the body. The proboscis is longer than it is wide, and the body is 25 to 50 mm long. This species has 16 to 17 distinct muscle bands, merging into a continuous sheet posteriorly. The body is often covered with whitish papillae. The proboscis is about one-fourth the body length. Fisher reported that *O. edax* feeds upon the sand or coarser material in which it lives, and does not mold its excrement into definite fecal pellets. He reported the type locality as Bahía Pichalingue (near La Paz), and listed records from Isla Coronado, Isla Espíritu Santo, and Isla Angel de la Guarda, all in the Gulf. I have collected this species in large numbers at Bahía San Everisto on the east coast of Baja California, at approximately lat. 24°50′N. This particular little bay, which is opposite Isla San José, proved to be one of the richest collecting sites I found in the Gulf. Unfortunately it is only accessible by boat, or perhaps I should say fortunately, as this is probably one of the reasons it possessed such a grand fauna.

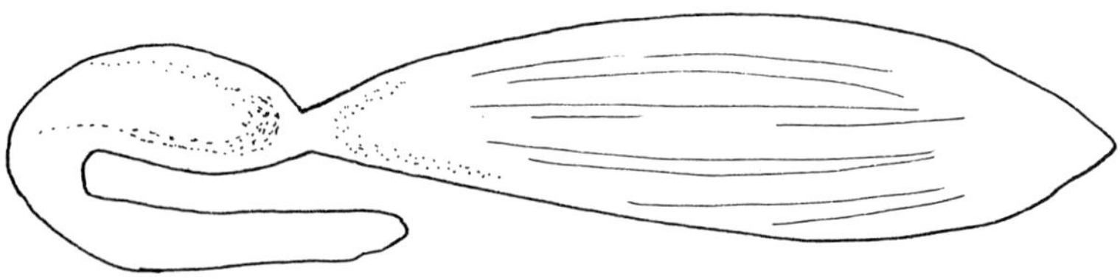

Fig. 9.2 *Ochetostoma edax*

Bryozoa

Preservation: To preserve fully expanded specimens of Bryozoa, proper narcotizing methods are mandatory. Relaxed colonies of Bryozoa may be narcotized with small amounts of one of the following chemicals: menthol, chloretone, Epsom salts, or clove oil. One or two drops of these relaxants should be added every couple of hours, until mechanical stimulation no longer induces retraction of the feeding tentacles. Preservation should be in alcohol, to prevent decomposition of the calcareous parts of the animal.

Taxonomy: The group Bryozoa actually consists of two distinct phyla, the Ectoprocta and the Entoprocta. However, due to the lack of agreement among taxonomists of this group as to the naming of the higher taxa, the two groups are often considered simultaneously. Also, due to their similarity in form and habitat, most workers study both phyla. The taxonomy of the bryozoans is based primarily on the morphology of the zooecial cells and other external structures. The taxonomic arrangement used in this handbook follows that of Osburn (1950, 1952, 1953). A comparison of the two phyla of Bryozoa is given:

Ectoprocta

Tentacular crown a lophophore.
Possess a true coelom.
Tentacles do not enclose anus.
Retractor muscles present to pull tentacles into zooecium.

Entoprocta

Tentacular crown not a lophophore.
Pseudocoelomate.
Tentacles enclose anus.
No retractor muscles (tentacles not retractile).

Taxonomic identification of bryozoan species is nearly impossible unless one is a specialist. Over 4,000 species of Bryozoa have been described, many of which are known only from the original unfigured description. Descriptions of Bryozoa are nearly always short and inadequate, and often so broad as to allow numerous species to fall under the same description. There are no published keys to the higher taxa of Bryozoa, although Osburn (1950, 1952, 1953) did give keys to the species of several larger families on the Pacific Coast of America.

Entoprocts are few in number and rarely encountered, the bulk of the bryozoans being members of the Ectoprocta. Although the entoprocts may be separated from the ectoprocts in the following key, none of the species have been discussed in the handbook.

KEY TO THE COMMON GENERA OF INTERTIDAL BRYOZOA IN THE GULF OF CALIFORNIA

1 Polyp naked, not enclosed in a chitinous or calcareous zooecium; tentacles not retractile; anus within the ring of oral tentacles .... **Entoprocta**

-Polyp enclosed in a chitinous or calcareous zooecium; tentacles retractile; anus not within the ring of oral tentacles **(Ectoprocta)** 2

2(1) Operculum wanting; zooecia simple calcareous tubes, generally with a circular aperture and lacking any accessory markings or appendages (Cyclostomata) 3

-Operculum present; zooecia often with more or less complex markings and appendages (Cheilostomata) 4

3(2) Colony discoidal, with tubular zooecia in rows radiating from the center of the disk .... ***Lichenopora***

-Colony dendritic, not discoidal, and composed of segments of tubular zooecia united by corneous joints .... ***Crisia***

4(2) Colony erect and branching .... ***Bugula***

-Colony encrusting 5

5(4) Margins of zooecia not raised .... ***Lagenipora***

-Margins of zooecia raised 6

6(5) Front wall wholly or partly membranous .... ***Membranipora***

-Front wall entirely calcareous .... ***Thalamoporella***

SYSTEMATICS AND SPECIES DESCRIPTIONS

Bryozoa — Ectoprocta

CLASS GYMNOLAEMATA

Order Chilostomata

Family Membraniporidae

Membranipora Blainville

Members of this genus form an encrusting colony, sometimes adhering closely to the substrate, other times raised in folds or curving sheets free from the substrate. The zooecia alternate, arranged in a linear series,

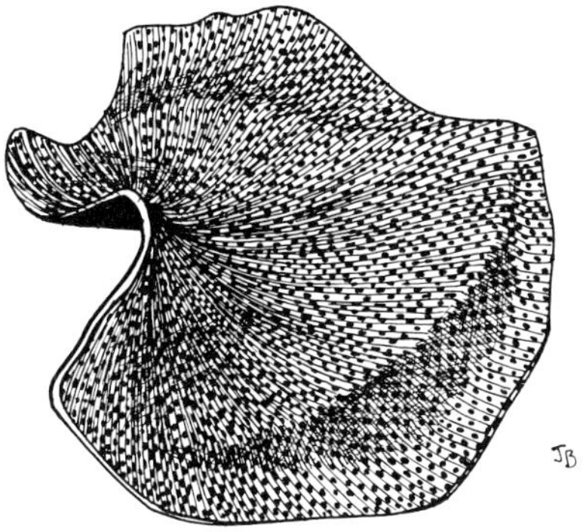

Fig. 10.1
Membranipora sp.
(from Allen, 1969)

usually radiating from a central primary zooecium. The margins of each zooecium are raised and are more or less calcareous. The aperture is membranous and depressed, and usually occupies the whole of the front of the zooecium. Species of this genus are often found encrusting the blades or fronds of kelp and algae. Osburn (1950) listed three species of *Membranipora* from the Gulf intertidal: *M. tenuis* Desor, *M. savarti* (Audouin), and *M. tuberculata* Bosc, the latter also being reported by Steinbeck and Ricketts. *Membranipora tuberculata* is the well-known "Gulf-weed" bryozoa. It forms a white lacework on floating *Sargassum* and shore algae along warm coastlines throughout much of the world.

Family Thalamoporellidae

Thalamoporella Hincks

Thalamoporella often closely resembles *Membranipora,* and growth forms for the two genera are very similar. *Thalamoporella gothica* (Busk) is very common in the Gulf, where it usually forms a white or pale encrustation on rocks in the low intertidal region. The avicularia of members of this genus are especially large and conspicuous. *T. californica* (Levinsen) is also common in the Gulf. These two species resemble one another greatly; however, the latter is a smaller form and is taxonomically differentiated on the basis of a number of very small and obscure details. I have found *T. gothica* to be common from Bahía Cholla to Guaymas, but have found *T. californica* only at the former locality.

Family Phylactellidae

Lagenipora Hincks

Members of this genus have "flask-shaped" zooecia, the zooecial body being entirely decumbent. The colonies may be entirely encrusting, or they may become erect and branching from the encrusting base. Steinbeck and Ricketts reported *Lagenipora erecta* O'Donoghue from Bajio de Punta Marcial, and Osburn (1952) reported *L. punctulata* (Gabb and Horn), *L. mexicana* Osburn, *L. spinulosa* Hincks, *L. marginata* Canu and Bassler, and *L. lacunosa* Bassler from the Gulf.

Family Bicellariellidae

Bugula Oken

This is a large and common genus of ectoprocts. The zooecia all face the same direction in members of this genus. At the point of origin on the dorsal side, they are usually prominently forked, the distal end of each extending above the base of the succeeding one. The stalked avicularia strongly resemble birds' heads or birds' beaks. The zooecia are typically globular, and are attached by a short stalk at the distal end, in the median line. In older colonies, the lower branches are often lost, and the radicles pass downward for attachment, giving the appearance of a stalk. Steinbeck and Ricketts reported *Bugula neritina* (Linnaeus) from various places in the southern Gulf, and Osburn (1950) reported this species as well as *B. longirostrata* Robertson and *B. californica* Robertson from the Gulf.

Order Cyclostomata

Family Crisiidae

Crisia Lamouroux

There are five to as many as 30 or more zooecia between the internodes of members of this genus. They are arranged very symmetrically, in two alternating series, the short projecting peristomes giving the edges a serrated appearance. Steinbeck and Ricketts reported an unnamed member of this genus from Cabo San Lucas, and Osburn (1953) reported *C. serrulata* Osburn, *C. operculata* Robertson, *C. maxima* Robertson, *C. elongata* Milne-Edwards, and *C. punctata* d'Orbigny from the Gulf.

Family Lichenoporidae

Lichenopora DeFrance

Members of this genus form minute disklike encrustations on rocks, shells, etc. Osburn (1953) reported *Lichenopora buskiana* Canu and Bassler from the Gulf, and *L. novae-zelandiae* (Busk) and *L. intricata* (Busk) from Mazatlán.

Brachiopoda (Lamp shells)

Preservation: The brachiopods, or lamp shells, should be preserved in 70 to 90 percent alcohol. Relaxation may be accomplished with magnesium sulfate.

Taxonomy: The taxonomy of this group is based largely on valve and hinge anatomy and the anatomy of the soft parts. Most authors divide the Brachiopoda into two classes, the Inarticulata (ecardines) and the Articulata (testicardines). These bivalve creatures greatly resemble the pelecypod molluscs, but differ by possessing a *lophophore,* a fleshy *peduncle,* and in having the valves fixed dorso-ventrally as opposed to right and left, as in the clams.

Brachiopods of the Gulf: There are probably at least three species of lamp shells in the Gulf: *Glottidia audebarti* (Hinds), *Glottidia palmeri* Dall, and *Terebratulina ungulcula* Carpenter. Members of the first genus have long thin shells, while members of the latter genus have wider, more typically clam-shaped shells. *G. audebarti* is a large species, with the distal half of the valves a brilliant blue-green. It is reported from the intertidal of Topolobampo. *G. palmeri* lacks this blue-green coloring. *T. ungulcula* is reported from the Cabo San Lucas region. Jones and Barnard (1963) reported the common United States Pacific species *G. albida* (Hinds) from as far south as Mazatlán, but they stated it is replaced in the Gulf by *G. palmeri.* I have collected only one brachiopod in the Gulf, subtidally, from 25 feet of water at El Golfo de Santa Clara. My brachiopod is probably *G. palmeri* (order Atremata, family Lingulidae).

Fig. 11.1 *Glottidia* sp. (from Allen, 1969)

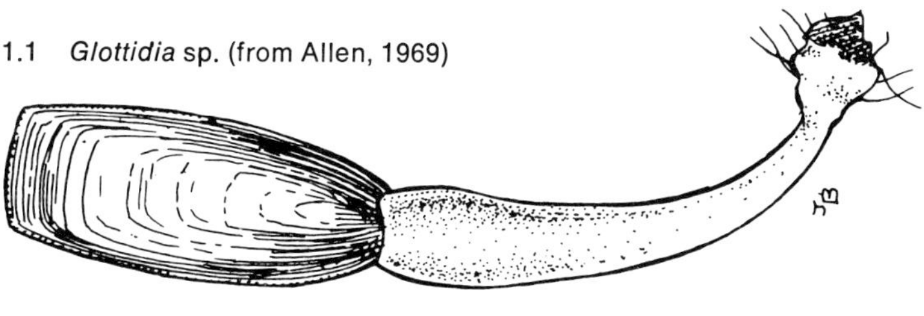

Echinodermata (Starfish, etc.)

Richard C. Brusca and Charles R. Smith*

Preservation: (see individual class.)

Taxonomy: (see individual class.)

Echinoderms of the Gulf: The vast majority of echinoderms of the Gulf of California are affiliated with the Panamic fauna. Fifty-two of the 58 species of echinoderms collected bv the East Pacific Expedition of the New York Zoological Society were considered by Clark (1940) to be Panamic in origin, 53 of the 65 species collected by the Templeton Crocker Expedition can be considered Panamic, and Clark (1948) reported 19 of the 22 species of echinoids he treated from the Gulf as having Panamic origins. Steinbeck and Ricketts reported all the asteroids, ophiuroids, and echinoids they collected as being affiliated with the Panamic fauna, while 13 of the 20 holothurians from their expedition were considered Panamic. Deichmann (1941, 1958) stated that the holothurians of the Gulf show close affinities to both the Panamic and West Indies faunas, probably the majority having their origin in the Caribbean.

The echinoderms of the Gulf are extremely numerous, both in individuals and species. The group constitutes one of the dominant taxa of the region, and its members are conspicuous inhabitants of all intertidal environs from El Golfo de Santa Clara to Cabo San Lucas and Mazatlán. The sun star *Heliaster kubinijii* is possibly the most common echinoderm of the Gulf, as well as one of the most conspicuous animals of the intertidal zone throughout this region. The holothurian *Selenkothuria lubrica* and several of the brittle stars (*Ophioderma teres, Ophioderma panamense, Ophiothrix spiculata, Ophiocoma aethiops,* and *Ophiocoma alexandri*) are also extremely common. In the northern end of the Gulf the asteroid *Othilia tenuispina* may be as common as *Heliaster,* particularly in the winter, while in the mid- and lower Gulf the asteroids *Pharia pyramidata* and *Phataria unifascialis* are almost as abundant as *Heliaster.* The blunt-spined sea urchin *Eucidaris thouarsii* and the sharp-spined

*California State Department of Public Health, Redding, California.

purple urchin *Echinometra vanbrunti* are the most common and widespread intertidal echinoids of the Gulf, being found on almost all rocky shores of this region.

The sand dollars avoid heavy surf, and are seldom found alive without diving or waiting for the very lowest tides of the year. When one is fortunate enough to find a bed of these animals, he will often see hundreds or even thousands of them lying flat on the bottom, or standing slightly on edge with two-thirds or so of the disk buried in the sand, the exposed portion angled in the direction of the current. *Encope grandis* is probably the most common sand dollar of the Gulf, but in the northern extremes its abundance is rivaled by *Mellita longifissa* and *Encope micropora*. The sand dollars and heart urchins are found only on sandy substrates. Hedgpeth (1969), Caso (1961), and Clark (1940) stated that the familiar west coast bat star *Patiria minata* occurs around the tip of Baja California and north to La Paz, but we have not collected it in the Gulf

Fig. 12.1
Patiria miniata
(from Allen, 1969)

Because of numerous recent name changes, partial synonomies are given for the echinoderms in the species descriptions.

KEY TO THE COMMON CLASSES OF INTERTIDAL ECHINODERMS IN THE GULF OF CALIFORNIA

1 Animal with 4 or more radiating arms 2
-Animal without arms 3

2(1) Arms with ambulacral grooves; animal without a distinct central disk (figs. 12.1 to 12.12) .... ***Asteroidea*, the sea stars**
-Arms without ambulacral grooves; animal with a distinct central disk (figs. 12.13 to 12.22) .... ***Ophiuroidea*, the brittle stars**

3(1) Body covered with stiff, movable spines; without tentacles around the mouth; body with a rigid external skeleton (figs. 12.23 to 12.38) .... ***Echinoidea*, the sand dollars and sea urchins**
-Body without stiff spines; mouth surrounded by tentacles; body flaccid, without rigid skeleton (figs. 12.39 to 12.42) .... ***Holothurioidea*, the sea cucumbers**

CLASS ASTEROIDEA

Preservation: Sea stars may be dried, or preserved in 10 percent buffered formalin or 70 percent alcohol. Formalin usually keeps the color better than alcohol, but if its buffering quality is not retained the calcareous plates of the sea star will begin deteriorating. Carefully covering the sea star with fine sand during preservation will usually result in the spines of the dorsal surface remaining erect.

Taxonomy: The sea stars are free-living, pentaradial echinoderms that crawl about the substrate by use of the tube feet. The central disk bears five arms, or rays, in most species. In the Gulf, *Heliaster* and *Acanthaster* are the only common multirayed starfish (with 10 to 42 arms). The central disk of asteroids is not marked off from the arms as in the ophiuroids (brittle stars), but grades into them. The taxonomy of this group is based largely on the morphology of the rays, calcareous body plates, pedicellariae, and other external features. The majority of the species are distinct and easily differentiated. The systematic scheme used in this presentation is based on H. L. Clark (1913), L. Lyman (1955), and D. Nichols (1962).

KEY TO THE COMMON INTERTIDAL ASTEROIDS OF THE GULF OF CALIFORNIA

1 Starfish with more than 10 arms 2
-Starfish with 5 to 6 arms 4

2(1) 10 to 17 arms; body and arms with long, pointed spines; usually with multiple madreporites .... ***Acanthaster ellisii***
-Usually more than 18 arms; body and arms without long, pointed spines; single madreporite 3

3(2) 10 to 24 arms; diameter of central disk less than length of individual arm (fig. 12.9) .... ***Heliaster kubinijii***
-30 to 42 arms; diameter of central disk equal to or greater than length of individual arm (fig. 12.10) .... ***Heliaster microbrachius***

4(1) Spines wanting or, if present, only on marginal arm plates; aboral surface of disk and arms covered with granular plates or compacted paxillae (star-shaped pedestals) 5
-Spines present on both disk and arms (spines may be blunt or pointed) ... 10

5(4) Large spines present, but only on marginal arm plates 6
-Large spines absent, small spines only on ambulacra 8

6(5) Large rectangular marginal arm plates present; aboral surface sunken below level of marginal plates (fig. 12.5) ...... ***Astropecten armatus***
-Small marginal arm plates present; aboral surface not sunken as above ... 7

7(6) Lateral margins of arms with paired black spots; several large spines on aboral margin between arm bases; body mosaic tan and black, usually with orange markings (fig. 12.4) ... ***Luidia phragma***
-Lateral margins of arms without paired black spots; without large spines on aboral margin as above; body dark gray to black (fig. 12.3) .... ***Luidia columbia***

8(5) Aboral surface of disk and arms with alternating plates, and spherical granules arranged irregularly; arms very rounded in shape, normally 6 present, but highly variable; mottled gray and red color .... ***Linckia columbia***
-Aboral surface of disk and arms with plates and granules arranged in more definite rows; arms more flattened in shape, always 5; color variable **9**
9(8) Distinct aboral plates; lemon-colored body with white ambulacral spines; a single row of small flat spines present on each side of the ambulacral grooves (fig. 12.6) .... ***Pharia pyramidata***
-Indistinct aboral plates; grayish body, with reddish ambulacral spines; double rows of small flat spines on each side of the ambulacral grooves (figs. 12.7, 12.8) .... ***Phataria unifascialis***
10(4) Body spines blunt; sea star *usually* over 4 inches in diameter **11**
-Body spines sharp; sea star usually less than 4 inches in diameter **13**
11(10) Body cushion shaped and inflated; spines massive, blunt-tipped; color variable **12**
-Body not cushion shaped as above; sea star with a small central disk and very long arms, covered with small blunt spines; pink to orange, with large brown blotches .... ***Mithrodia bradleyi***
12(11) Spines of aboral surface small (1 to 5 mm) and immobile: arms distinctly set off from disk; spines of oral surface small and indistinct; body greenish red with bright orange spines ...... ***Oreaster occidentalis***
-Spines of aboral surface large (8 to 10 mm) and mobile; arms not distinctly set off from disk; spines of oral surface large and distinct; body lavender with purple to black spines (fig. 12.2) .... ***Nidorellia armata***
13(10) Spines long and narrow; body soft and flexible with wide ambulacral grooves; spines bright red-orange at the tip (fig. 12.11) .... ***Astrometis sertulifera***
-Spines very short and stubby; body much more rigid with normal ambulacral grooves; spines may or may not be orange-tipped **14**
14(13) Sea star red to orange, with dark red or black arm tips; body spines very small, often inconspicuous (fig. 12.1) ...... ***Othilia tenuispina***
-Sea star brown or purple, without dark arm tips; body spines larger, conspicuous .... ***Echinaster tenuispina***

SYSTEMATICS AND SPECIES DESCRIPTIONS

Asteroidea

SUBCLASS EUASTEROIDEA

Order Spinulosa

Family Acanthasteridae

Acanthaster ellisii (Gray)

A thick, soft-bodied multirayed sea star, easily recognized by its thin gray skin, long pointed spines on the body and arms, and the numerous small spines that cover the aboral surface. This species usually has more than one madreporite and 10 to 13 arms. *Acanthaster* is largely subtidal, occasionally brought up by the shrimp trawlers, but also occurs in the intertidal. Its depth range is reported as shore to 6 fathoms. It ranges from the Gulf of California to Peru and the Galapagos Islands. This species has been previously reported as *Echinaster ellisii*. We have collected *A. ellisii* intertidally at Isla San José and La Paz, but we have no records from the northern Gulf.

Family Echinasteridae

Echinaster tenuispina **Verrill**

An attractive, small sea star, usually about 60 to 70 mm from arm-tip to arm-tip. The body is mottled brown and purple, and the sharp spines that cover the aboral surface are orange tipped. Each arm usually has nine longitudinal rows of these short, pointed spines. At first glance, this creature may be mistaken for a miniature version of the common west coast *Pisaster brevispinus*. The general body shape and size is similar to that of *Othilia tenuispina* (with which it was confused either anatomically or systematically by Fisher, 1911, as well as by Steinbeck and Ricketts), although the latter has black spots at the tips of the arms. *Echinaster tenuispina* ranges from southern California to the upper Gulf of Panamá, although it is rarely encountered either in the field or in the literature. This apparently is the sea star mentioned by Steinbeck and Ricketts (p. 381) as "*Othilia tenuispina* (Verrill) 1871 as *Echinaster.*" We have collected *Echinaster* infrequently throughout the Gulf.

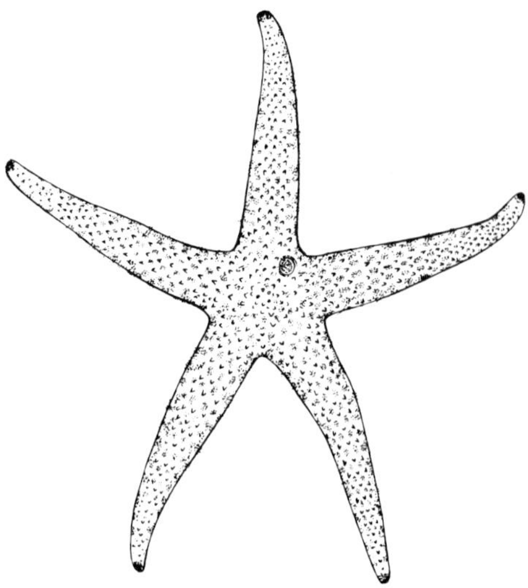

Fig. 12.2 *Othilia tenuispina*

Othilia tenuispina Verrill

A small, yellow-brown to brick-red sea star of the low intertidal rocky coasts of the upper Gulf. The tips of the arms tend to curl up slightly, and are marked with a dark red to black spot. In large specimens the arm span reaches 100 mm. Investigations by Miller (1967) indicate that individuals of this species can move at a rate of 6 inches per minte, in any direction, and can right themselves in about 4 minutes. *Othilia* is photopositive, usually being found on the tops of rocks during the day, rather than hiding beneath them as do most echinoderms. This species preys on various bivalves, especially mussels and gastropods, but can be forced to feed on a number of small invertebrates in the aquarium. It is very common from El Golfo de Santa Clara to Bahía Kino, but we have found it only once as far south as Guaymas.

Family Mithrodiidae

Mithrodia bradleyi Verrill

This is a large and very long-armed sea star, with striking orange-pink body color and large brown blotches. The disk is somewhat small and constricted. Both surfaces are covered with short blunt spines. We have encountered this species at both Guaymas and Isla San José in the Gulf, both times in the subtidal. It has previously been reported from rocky shores at low tide zone to 13 fathoms in the Panamic region. Its range is listed by other authors as Panamá and the Galapagos to the Gulf of California, and also in Hawaii. It has not been reported north of Guaymas in the Gulf.

Order Phanerozonia

Family Oreasteridae

Oreaster occidentalis Verrill

An inflated sea star, with a red-orange to reddish green aboral surface. The spines and marginal plates are bright red. The tube feet lack suckers, and the pedicellariae are sessile (not on movable stalks). The tube feet are usually hidden well within the deep ambulacral grooves. The aboral surface is covered with numerous, blunt, immobile spines. The commensal polychaete worm, *Podarke pugettensis* Johnson, has been reported from the ambulacral grooves of this species. This is the same worm that is so common in the bat star *Patiria miniata* on the west coast of the United States and Baja California. *Oreaster occidentalis* ranges from the northern Gulf of California to Peru and the Galapagos Islands, and is restricted to the lowest low tide zone and the subtidal, being most frequently encountered when one dives over rocks in 1 to 3 fathoms of water.

Nidorellia armata (Gray)

A beautiful lavender to red-orange sea star with heavy marginal plates, and large movable spines on the dorsal surface that break off easily when handled. The aboral spines are usually purple with dark tips. The pedicellariae are sessile, and the tube feet lack suckers. This species occurs in the rocky low intertidal and subtidal from the upper Gulf to Peru and the Galapagos Islands, and is most often encountered when diving (Lyman, 1902, reported it from southern California). It has previously been reported as *Pentaceros armatus*. Dried specimens have deflated bodies and turn a dirty chocolate brown.

Family Luidiidae

Luidia columbia (Gray)

A dark gray to black sea star, similar to *Luidia phragma* but two or three times larger, and without the distinctive black arm blotches of the latter. The tube feet are without suckers, and the pedicellariae are sessile. It inhabits sandy and muddy bottoms of the low intertidal and subtidal, where it is often found buried under a few inches of the substrate. This sea star can grow to be a real giant, occasionally obtaining an arm-tip to arm-tip diameter of 500 mm, although most specimens range from 50 to 250 mm. *Luidia columbia* ranges from the Gulf of California to Peru and the Galapagos Islands. It is one of the commonest invertebrates picked up in otter trawls of shrimp boats throughout the Gulf. These trawls are usually fished at 20 to 40 fathoms. This species has been reported as *Petalaster columbia* by Fell (1963) in his recent article on asteroid phylogeny.

Luidia phragma H. L. Clark

An attractive and common sea star of sandy and muddy bottoms of open coasts and bays. The aboral surface is orange to gray, with black spots and paired black patches on the arms. The pedicellariae are sessile, and the tube feet lack suckers. The arm span of this species reaches about 225 mm, but smaller individuals are more commonly found. *Luidia phragma* ranges from the northern Gulf to Chile. It is often picked up in the nets of the shrimp trawlers. According to Fell's work (1963), this species would now be *Alternaster phragma*.

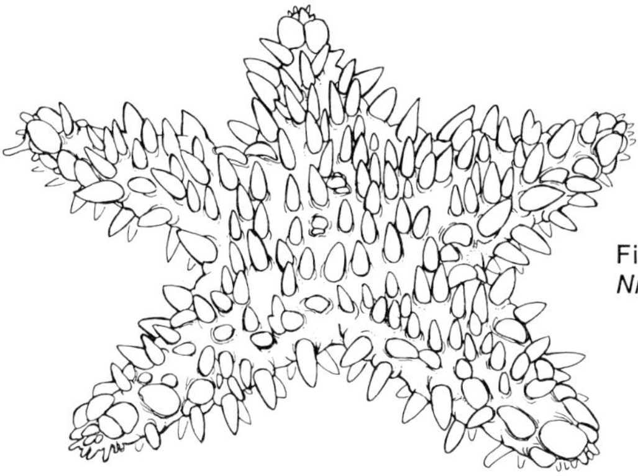

Fig. 12.3
Nidorellia armata

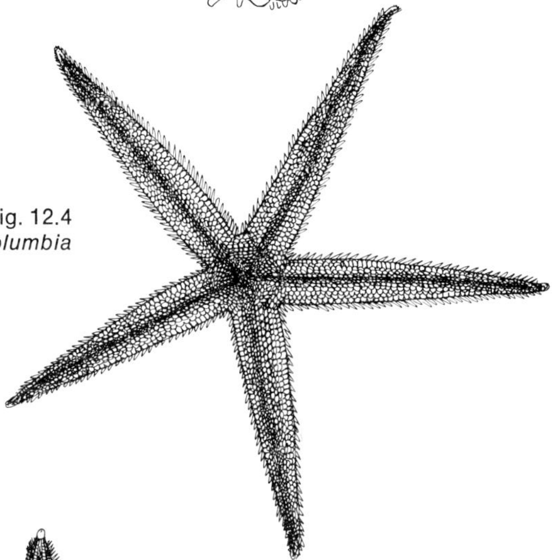

Fig. 12.4
Luidia columbia

Fig. 12.5
Luidia phragma

Family Astropectinidae

Astropecten armatus Gray

Students of the southern California seashore will recognize this common sand star. *Astropecten* burrows just under the sand or mud-sand substrate in the low intertidal and subtidal and has been brought up from depths as great as 83 fathoms. The aboral surface is a dull orange to gray-brown or gray-violet, and is sunk below the level of the large, rectangular, spiny plates that border the arms. The tube feet are without suckers, and the pedicellariae are sessile. The arm span of this species reaches 360 mm. *Astropecten* ranges from southern California and the upper Gulf to Ecuador and turns up consistently in deep water trawls and subtidal sandy beaches throughout the Gulf. It has been previously reported as *Astropecten erinaceus, Astropecten braziliensis armatus,* and *Astropecten orstedii.*

Family Ophidiasteridae

Pharia pyramidata (Gray)

This colorful starfish is found on rocks and reefs in the low intertidal throughout the Gulf. It is purplish brown to dull green-brown, and has a large, red, irregular madreporite (to 0.5 inch in diameter). The arms are lined with longitudinal bands of orange or yellow spots. The tube feet are without suckers, and the pedicellariae are sessile. The arm span reaches 300 mm. Once the color has been bleached out of these sea stars by preservatives, it may resemble preserved specimens of *Phataria unifacialis.* It may be distinguished by its wider ambulacral grooves, which tend to remain open when preserved, exposing the tube feet, whereas in *Phataria* the grooves usually close tight upon preservation, making the tube feet impossible to see. Another feature distinguishing these two stars in their preserved state is the single row of stubby, flattened spines along each side of the ambulacral groove in *Pharia,* whereas *Phataria* has a double row of these stout, flat spines along each side of the ambulacral groove. *Pharia* is undoubtedly one of the most beautiful sea stars of the eastern Pacific, ranging from the northern Gulf to Peru and the Galapagos Islands. We have found it in great abundance on the low intertidal and subtidal rocks of Bahía San Carlos, north of Guaymas. It has previously been reported as *Ophidiaster pyramidatus.*

Phataria unifascialis (Gray)

A very common starfish of the lower Gulf. It is large and narrow-rayed, with the aboral surface dull brown to orange, often with traces of ultramarine and with reddish ambulacral spines. The arms are tan and have numerous, lateral pores imbedded in an orange streak. The tube feet lack suckers, and the pedicellariae are sessile. The species ranges from the lower Gulf to Peru, inhabiting rocky shores in the low intertidal and subtidal to about 60 feet. It has been previously reported as *Linckia unifascialis* (see *Pharia pyramidata,* above).

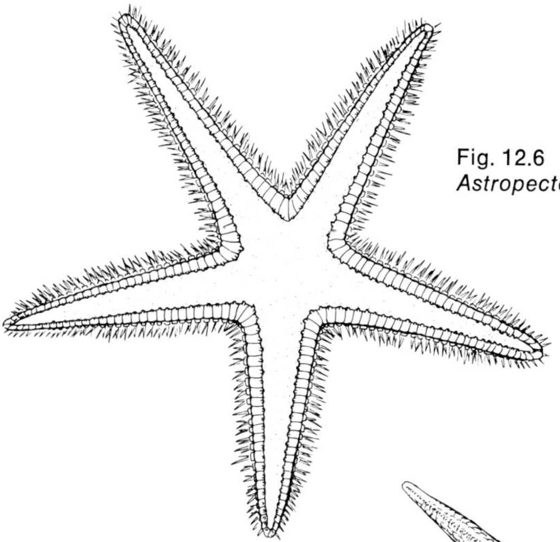

Fig. 12.6
Astropecten armatus

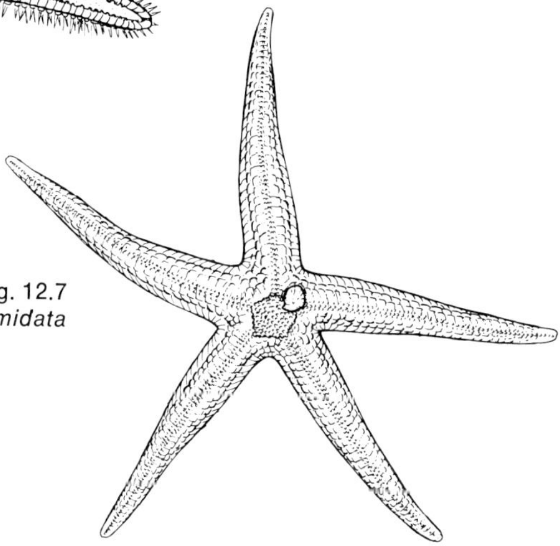

Fig. 12.7
Pharia pyramidata

Fig. 12.8
Phataria unifascialis

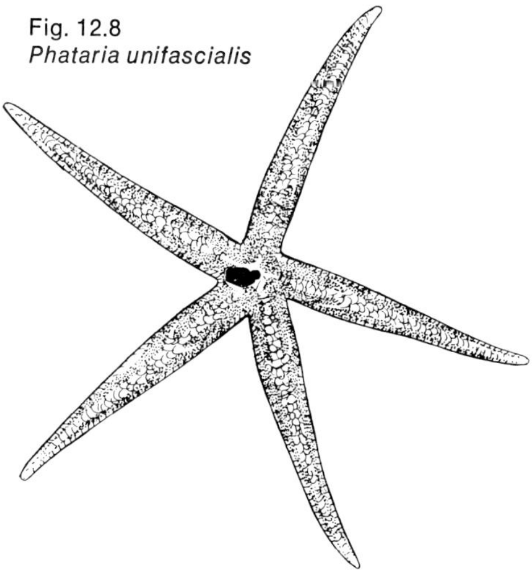

Fig. 12.9
Phataria unifascialis
(side view of arm base)

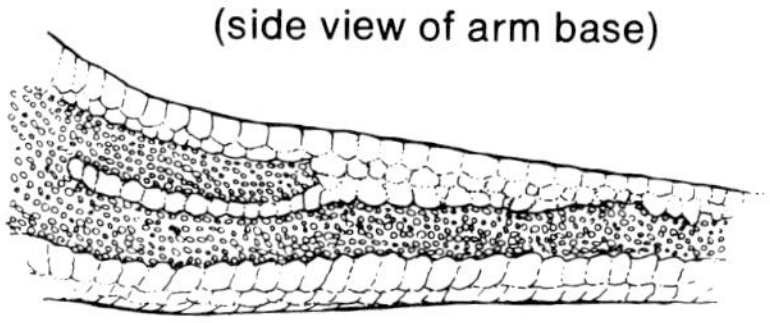

Linckia columbiae Gray

This is one of the most variable starfish known, both in habitat and morphology. It resembles both *Pharia* and *Phataria* but differs in having more rounded arms and a more irregular arrangement of plates and granules on the aboral surface. It is found from the rocky midintertidal, under stones and in crevices in the low intertidal, to flat sandy bottoms of the subtidal. The aboral surface is mottled gray and red and has irregular bumps over the entire surface. This species rarely possesses any symmetry, due to autotomy and species variability. It may be found with one to nine arms of any length, one to seven mouths, one to four anuses, and one to five madreporites (usually two). The best specimens are undoubtedly found subtidally, where the majority of individuals probably have five or six arms of nearly equal length. According to Hedgpeth, Ricketts, and Calvin (1969), Dr. S. P. Monks examined over 400 specimens of this species and found only four symmetrical individuals, stating that no two were alike. The range of *Linckia columbia* is given as southern California and the upper Gulf, to Peru and the Galapagos Islands. We have collected this species from the intertidal and subtidal at Puertecitos. This sea star seems to prefer the rocky low intertidal and subtidal to depths of at least 50 fathoms. Like all members of the order Phanerozonia, it lacks suckers on the tube feet and possesses sessile pedicellariae. This species has been reported previously as *Phataria columbiae*.

Order Forcipulata

Family Heliasteridae

Heliaster kubinijii Xantus

This sun star is probably the most common asteroid of the Gulf intertidal, and one of the dominant animals of the rocky intertidal zone throughout this region. It usually is found with 19 to 25 rays, young individuals beginning life with five but adding additional ones rapidly as they mature. New arms are always added between the existing ones, in the inter-radii. The tube feet have suckers, and the pedicellariae are stalked. The aboral surface is pinkish lavender, with black-and-green mottling, younger specimens being darker than the pinkish adults. *Heliaster* may be seen crawling about on and under rocks in the mid- and low intertidal, younger individuals being more common in the mid- and occasionally high intertidal zones. It seems to show a preference for large, smooth, basalt-type rocks. In the very low intertidal, it is largely replaced by *Othilia tenuispina* in the northern Gulf and by *Phataria unifascialis* in the southern Gulf. Its feeding habits are similar to those of the familiar west coast *Pisaster*. *Heliaster* prefers the sessile mussels (*Brachiodontes multiformis, Cardita affinis,* etc.) and barnacles (*Chthamalus anisopoma)* for food. It has also been seen feeding on a variety of snails, including *Tegula, Cerithidea mazatlanica, Nassarius,* and *Turbo,* the chitons *Stenoplax conspicua* and *Chiton virgulatus,* small crabs, and the colonial sea anemone *Palythoa.* In the aquarium *Heliaster* can be forced, through starvation, to feed on algae and a great variety of small invertebrates. Tanzer (1967) reported the occurrence of a commensal Ophelid polychaete in the ambulacral and oral areas of *Heliaster.* The worm was brown, small (8 to 30 mm long), and had 30 to 50 body segments. Glidden (1967) showed that these same polychaetes are apparently attracted to the sun star by a chemical transmitted through the water, and that these worms may be host-specific for *Heliaster*. Glidden reported the worm as *Polyophalmus* although Jerry Kudenov, author of the polychaete section of this handbook, informs me they are more likely members of the genus *Ophiodromus.*

The range of *Heliaster kubinijii* is given by Caso (1961) as Cape Mendocino, California, to Nicaragua, and throughout the Gulf. We have collected it in abundance throughout the northern Gulf, and in fewer numbers at Guaymas and Mazatlán. We feel the occurrence of this species north of Mexico is highly questionable. Caso also described a variety of this species, *H. kubinijii* var. *nigra* Caso, but did not give any details of its alleged distribution, merely stating it occurs in the Gulf of California. She stated the variety *nigra* differs from the former by having only 19 arms and a nearly black aboral surface, as opposed to what she describes as a dark red color for *H. kubinijii.*

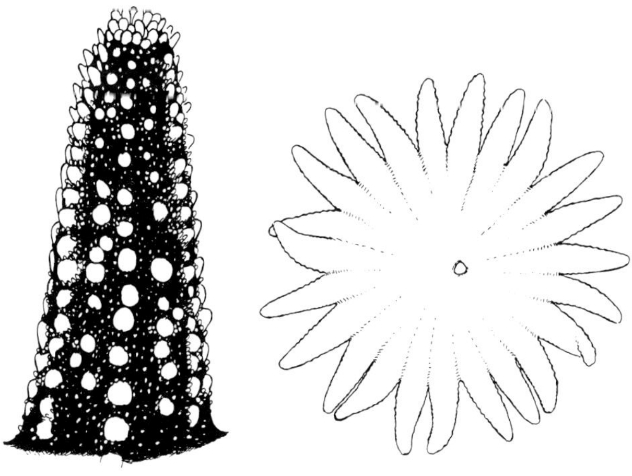

Fig. 12.10 *Heliaster kubinijii* (body and one arm)

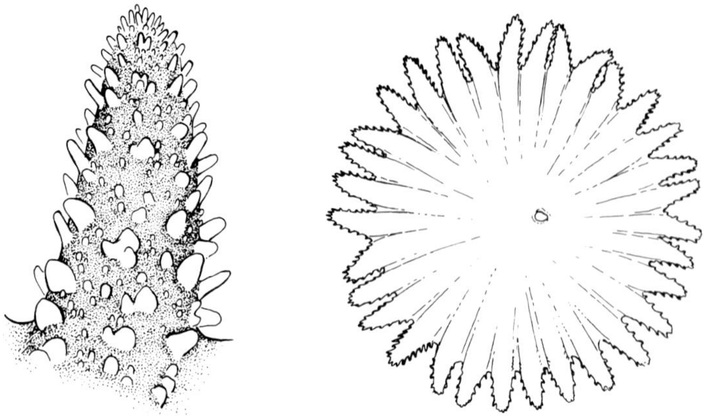

Fig. 12.11 *Heliaster microbrachius* (body and one arm)

Heliaster microbrachius Xantus

This sun star is a close relative of *H. kubinijii,* and the two are apparently sympatric in the regions of Guaymas, Mazatlán, and further south. *Heliaster microbrachius* is a more compact, short-rayed version of *H. kubinijii.* It has 30 to 40 arms (as opposed to 25 or fewer in *H. kubinijii*) and is distributed from Guaymas as far south as Chile. This is a truly Panamic form that apparently can't stand the cold winter temperatures above Isla de la Angel Guarda. Its aboral surface is darker than that of *H. kubinijii,* and the many small bumps that cover the surface of both species are much smaller and more uniform in size in *H. kubinijii.* We have collected it in abundance at Guaymas and Mazatlán. Caso (1961) described a variety of this species also, *H. microbrachius* var. *polybrachius* Caso, as differing by having the free portion of the radius 26 to 29 percent of the total radius, as opposed to 16 to 22 percent in *H. microbrachius,* and in some other minor morphological differences.

Although we have collected both *H. kubinijii* and *H. microbrachius* from the regions of Guaymas and Mazatlán, we have never found both occurring in the same place at the same time. This may be just chance and the result of random collecting, or maybe some interesting factors of population dynamics and interspecific competition are taking place. Data from a study of this genus in the Western Hemisphere could provide someone with a good Ph.D. dissertation. There is, in fact, a third and possibly a fourth species of *Heliaster* on the west coast of Mexico and Central America. Caso reports *H. heliantus* (Lamarck) distributed from Acapulco to Valparaiso, Chile, and Lyman (1902) reported *H. cumingi* (Gray) from Peru, and Hood, Chatham, and Galapagos islands.

Family Asteriidae

Astrometis sertulifera Xantus

A large, soft, highly flexible and spiny sea star of the lowest intertidal and subtidal. The slimy rays are constricted basally by several millimeters, and may number five to six. The aboral surface is somewhat slimy, green to brown with orange and blue spines, tipped with red. The long tube feet are white to yellow, and possess suckers. The pedicellariae are stalked. In the low intertidal, this sea star usually hides under large rocks in a sandy substrate, but subtidally it may be seen crawling about on exposed sand or rock surfaces. *Astrometis* is a voracious predator of gastropods, pelecypods, and barnacles. It is one of the few species of asteroids that easily and readily sheds its arms when handled. The arm span reaches 180 mm.

This species ranges from Oregon and Washington to the Gulf of California and Mazatlán, being fairly common on the open coast, where it often is encountered in great numbers subtidally, although it is rarely seen in the Gulf or above Santa Barbara on the California coast. There is a record of it as far north as Vancouver, British Columbia, but Fisher (1928) suggested this specimen may have been carried to Canada on the bottom of a lumber schooner. This sea star cannot be dried, or it will disintegrate, and it must be preserved in alcohol.

Fig. 12.12 *Astrometis sertulifera*

CLASS OPHIUROIDEA

Preservation: The brittle stars can be preserved in 40 to 50 percent alcohol or a 5 percent buffered formalin solution. They must be handled with great care, as they break off their long, delicate arms readily by autotomy if handled roughly. Specimens must be relaxed in magnesium sulfate, fresh water, or other narcotizing agents prior to preservation. After preservation the larger species can be air dried, but they will remain quite fragile.

Taxonomy: The ophiuroids are pentaradial echinoderms with long slender, flexible arms used for walking, as opposed to the tube feet used by the asteroids. The skeleton is greatly reduced and fragmented, which allows for the flexibility of the arms. The central disk is sharply marked off from the rays, and may be round or pentagonal. The aboral surface of the disk, in most species, bears exposed, paired radial shields at the base of each arm. The arms of brittle stars, unlike those of asteroids, lack ambulacral grooves on the oral surface. Ophiuroids can be found under rocks or in other protective habitats, usually where the substrate is sandy or muddy.

The taxonomy of this group is based largely on spine and skeletal plate morphology. The systematic treatment of this presentation is based on Caso (1961) and agrees with Hyman (1955) and Nichols (1962). One brittle star not treated in the keys, *Ophiolepis variegata* (Lutken), may be encountered in the northern Gulf below 10 to 15 fathoms and is occasionally brought up in shrimp trawls. It is a rather stiff species, the arms displaying very little flexibility. The arm spines are short, and the arms are banded with blue and black, with a large white spot covering several plates of the disk on many specimens.

KEY TO THE COMMON INTERTIDAL OPHIUROIDEA OF THE GULF OF CALIFORNIA

1 Arms branch (figs. 12.21, 12.22) .... ***Astrocaneum spinosum***
-Arms do not branch 2

2(1) Arm spines very short, forming an acute angle with the arm axis, and lying flat against the arm, pointing toward the tip 3
-Arm spines long, perpendicular to arm axis 4

3(2) Arms uniformly slate-colored; aboral arm plates fragmented and broken up into subplates; 9 radial spines (figs. 12.13, 12.14) .... ***Ophioderma teres***
-Arms banded or spotted with white, at least distally; aboral arm plates entire (fig. 12.16); 11 radial spines .... ***Ophioderma panamense***

4(2) Animal with 5 rays; arm span greater than 25 mm; 6 or more oral papillae per jaw 5
-Animal with 6 rays (occasionally 5); size minute, arm span less than 20 mm; 2 to 5 oral papillae per jaw 8

5(4) Spines of disk and arms are themselves serrated or spiculated (fig. 12.19); oral surface of arms banded with tan and orange (fig. 12.18); disk covered with minute spicules, appearing fuzzy or glasslike .... ***Ophiothrix spiculata***
-Spines of disk and arms not spiculated; oral surface of arms not banded as above; disk not as above **6**
6(5) Arms unicolored, not banded .... ***Ophiocoma aethiops***
-Arms banded **7**
7(6) Diameter of disk 18 to 25 mm; animal brown with a white blotch at the base of each arm spine; aboral surface of disk granulated (fig. 12.17) .... ***Ophiocoma alexandri***
-Diameter of disk less than 16 mm; animal without a white blotch at the base of each arm spine; surface of disk not granulated, but covered with minute scales .... ***Ophionereis annulata***
8(4) 5 to 6 spines on each side of arm plate (fig. 12.20); always with 6 rays; radial shields large, joined in pairs distally; 4 to 5 oral papillae per jaw .... ***Ophiactis savignyi***
-4 spines on each side of arm plate; with 5 or 6 rays; radial shields small, not joined in pairs; 2 oral papillae per jaw .... ***Ophiactis simplex***

SYSTEMATICS AND SPECIES DESCRIPTIONS

Ophiuroidea

Order Ophiurae

Family Ophiodermatidae

Ophioderma teres (Lyman)

This species (and the one following) has short arm spines, lying flat against the arms. The arm spines are arranged in vertical rows all along the lateral margin of the arm. *Ophioderma teres* is a large, solid slate-colored brittle star (resembling *O. esmarki* of the California coast), with a disk diameter of 20 to 26 mm. The upper arm plates are fragmented and broken up into numerous small subplates. This species ranges from the upper Gulf to Panamá, reappearing again in the Galapagos Islands. It is found under large stones in the low intertidal. *O. teres* has been reported previously as *Ophiura teres*.

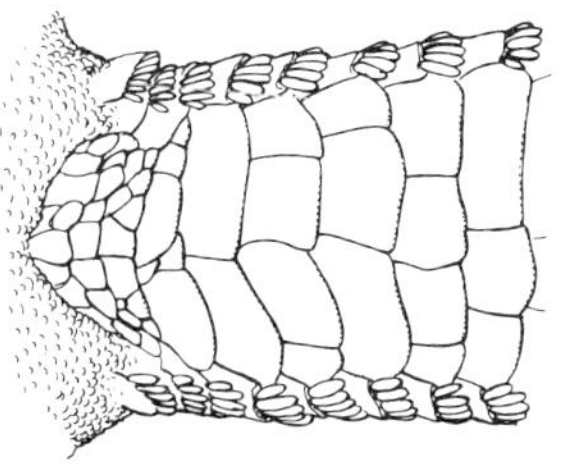

Fig. 12.13 *Ophioderma teres* (close-up of arm showing dorsal plates)

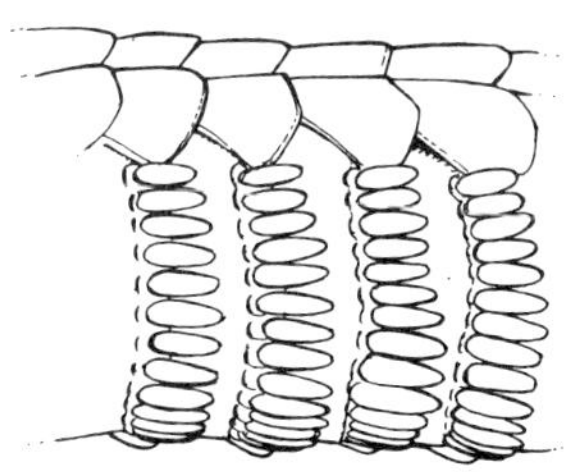

Fig. 12.14 *Ophioderma teres* (close-up of arm showing side spines)

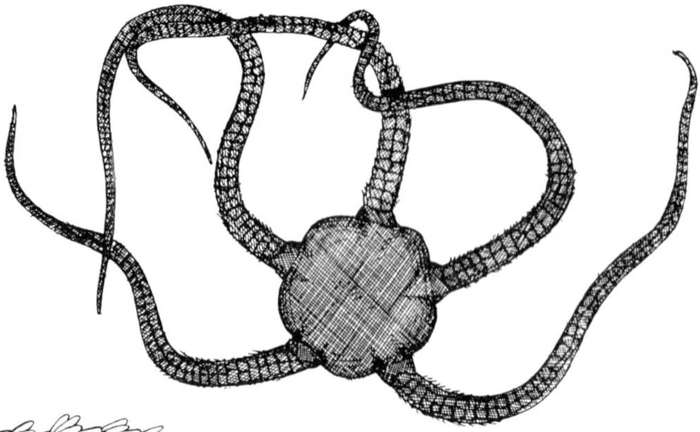

Fig. 12.15
Ophioderma panamense

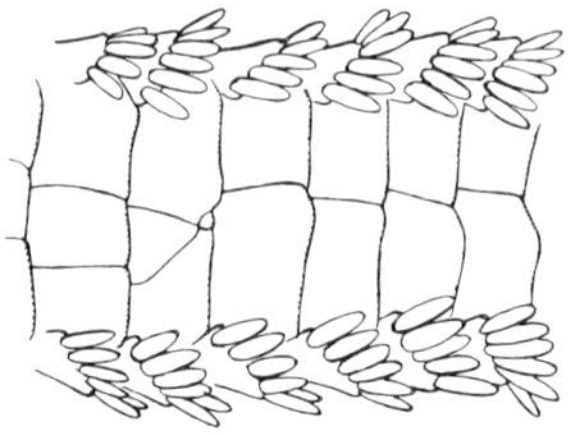

Fig. 12.16
Ophioderma panamense
(close-up of arm showing dorsal plates)

Ophioderma panamense Lutken

A large brittle star, at first resembling *Ophioderma teres,* but with paired white spots on the arm tips and without the aboral arm plates fragmented. The color varies from pale brown to black. The arm spines are short, lying close to the arm. The disk is granulated, with a greatly thickened epidermis, and reaches 25 mm in diameter, while the arm span may occasionally exceed 200 mm. This species ranges from southern California and the upper Gulf to Peru and the Galapagos Islands, occurring in the rocky mid- and low intertidal.

Family Ophiocomidae

Ophiocoma aethiops Lutken

One of the largest and most striking brittle stars of the Gulf, with a disk diameter to 40 mm and an arm span of 340 to 440 mm (nearly 20 inches). The arm spines are long and perpendicular to the arm axis. The aboral surface is purple-brown, unicolored, and the disk is covered with minute granules. The oral shields are much longer than wide. This species is common on rocky shores, under rocks in sand or muddy sand of the mid- and low intertidal zones from the upper Gulf to Panamá and the Galapagos Islands. To the student as yet uninitiated to the Gulf fauna, this may well be the most impressive animal he encounters during his first time to the seashore. Nowhere in western America, north of Mexico, will one encounter a robust, spiny brittle star the size of this giant.

Ophiocoma alexandri Lyman

Another large brittle star, common in the rocky midintertidal from the upper Gulf to Panamá and the Galapagos Islands and also possibly occurring in the Caribbean region. The arm spines are perpendicular to the arm axis, and the arms are banded with gray and white on the aboral surface. The disk is light brown or pale green, usually with a white blotch at the base of each arm spine, covered with minute granules. The disk reaches 30 to 35 mm in diameter, and the arm span reaches approximately 330 mm. The oral shields are about as long as wide. This species is more slender than *Ophiocoma aethiops,* and much lighter in color.

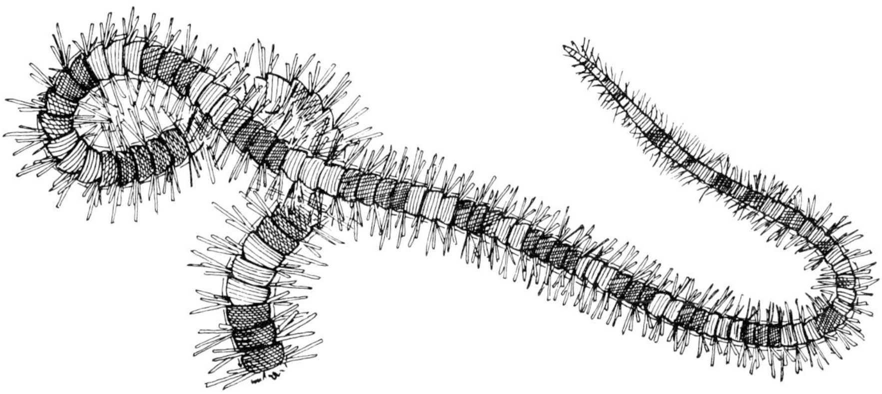

Fig. 12.17 *Ophiocoma alexandri* (arm)

Family Ophiochitonidae

Ophionereis annulata (LeConte)

A small- to medium-sized species of brittle star, with the arm spines perpendicular to the arm axis. This species is black and white, with a disk diameter of 10 to 17 mm, and an arm span of 50 to 100 mm. The arms are usually pale, with dark bands completely encircling them. The surface of the disk is covered with minute scales, not granulated. There are 9 to 10 oral papillae. *Ophionereis annulata* ranges from southern California and the upper Gulf, to Ecuador and the Galapagos Islands. It occurs in the midtidal zone and occasionally in the low intertidal. It has been reported previously as *Ophiolepis annulata.* Hedgpeth (1969) reported this species to be the southern counterpart of the similar appearing *Amphiodia occidentalis,* which ranges from Alaska to central California.

Family Ophiotrichidae

Ophiothrix spiculata LeConte

This attractive little brittle star should be another "familiar face" to California students. It has the appearance of being covered with glass-like, fuzzy spines. The arm spines are perpendicular to the arm axis, and they (as well as the disk spines) bear minute spinulets or serrations. The disk is more or less pentagonal. The color of this species varies from blue or gray-green to a beautiful violet, and the arms are nearly always banded with orange, at least on the oral surface. When the individual is gravid, conspicuous egg masses bulge out between the rays. The disk lacks oral papillae and attains a diameter of 10 to 15 mm, while the arm span of a large individual often exceeds 100 mm. This brittle star is found in the mid- and low intertidal, under rocks and in large sponges, or clinging to algae. It ranges from Monterey, California, and the upper Gulf to Peru and the Galapagos Islands, and is abundant throughout its range from the midintertidal to at least 100 fathoms. It previously has been reported as *Ophiothrix dumosa*.

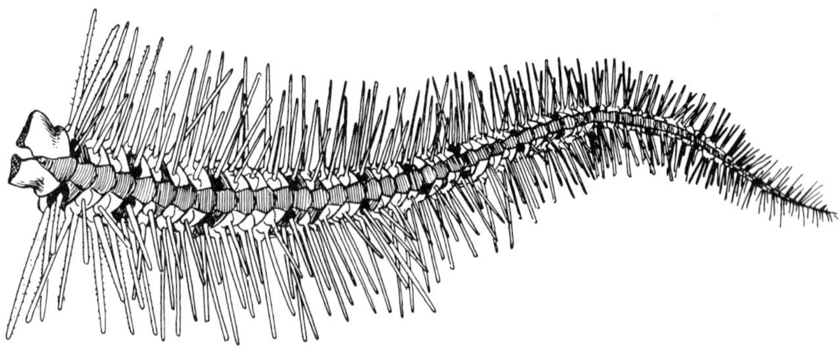

Fig. 12.18 *Ophiothrix spiculata* (arm)

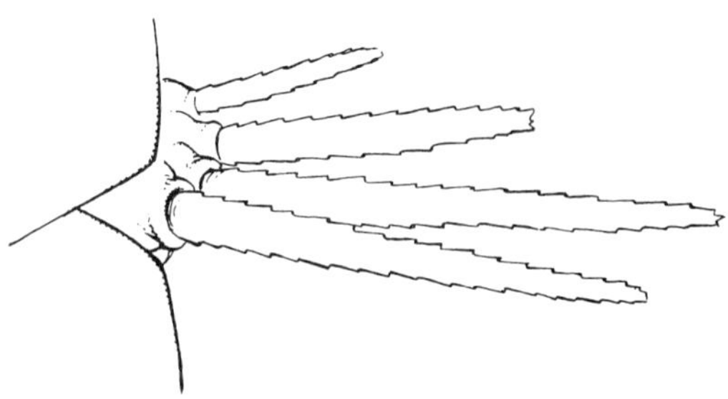

Fig. 12.19 *Ophiothrix spiculata* (close-up of arm spines)

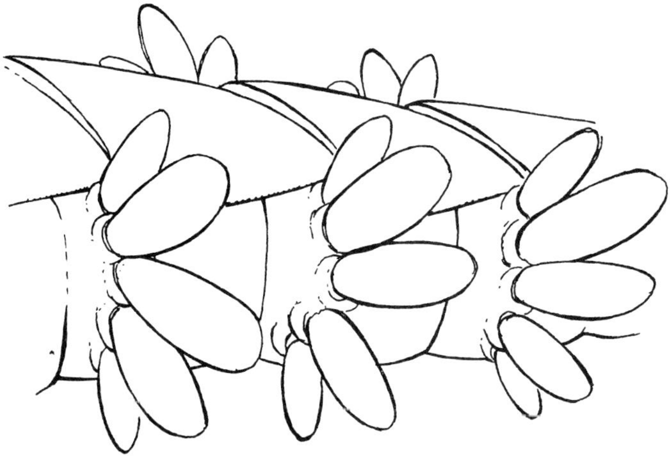

Fig. 12.20 *Ophiactis savignyi*
(close-up of arm showing side spines)

Family Ophiactidae

Ophiactis savignyi (Muller and Troschel)

A small, six-armed brittle star (occasionally seven-armed), with arm spines perpendicular to the arm axis. The predominant color is yellow-green to pale green. It is found in the mid- and low intertidal, under rocks and in mangrove swamps from San Pedro, California, and the upper Gulf to Panamá. It is reported by Clark (1940) and Caso (1961) as tropicopolitan. We have collected it at Puerto Lobos, Puerto Peñasco, and Guaymas. It previously has been reported as *Ophiolepis savignyi.*

Ophiactis simplex (LeConte)

A small five- or six-armed brittle star, similar in appearance to the above species. The arms are marked with light and dark green or occasionally brown bands. It has arm spines perpendicular to the arm axis. *Ophiactis simplex* may be distinguished from *O. savignyi* by the presence of four smooth spines on each arm plate, while in the latter the spines of each arm plate are rough and number five to six. We have found this species to be common throughout the Gulf and have collected it in great abundance from the pores of large sponges (*Geodia, Leucosolenia,* and *Leucetta*). We have found it living in the thick mats of algae on rocks at Puerto Peñasco, Bahía Kino, Guaymas, and Mazatlán. This species ranges from southern California and the upper Gulf to Peru. It previously has been called *O. arenosa.*

Order Euryalae

Family Gorgoncephalidae

Astrocaneum spinosum (Lyman)

This is the Gulf basket star. The body is a glossy tan, and the arms have a single, aboral row of white spots, circled by black rings, extending down the midline. The arms branch dichotomously near the disk but alternate distally in a panicle fashion. Large individuals will have a disk diameter of 35 to 45 mm. Living baskets, when placed in the aquarium, are beautiful sights. The long, knotted arms unfold to form a giant flower-like creature, whose tendrils wave slowly in the water. This species ranges from the upper Gulf to Panamá. In the Gulf it probably is never seen in water shallower than three or four fathoms. It previously has been reported as *Astrophyton spinosum.* The best place to look for basket stars is in the catch of the shrimp trawlers, where they are often brought up in small numbers, wrapped tightly around gorgonian coral or other objects, including the shrimp net itself.

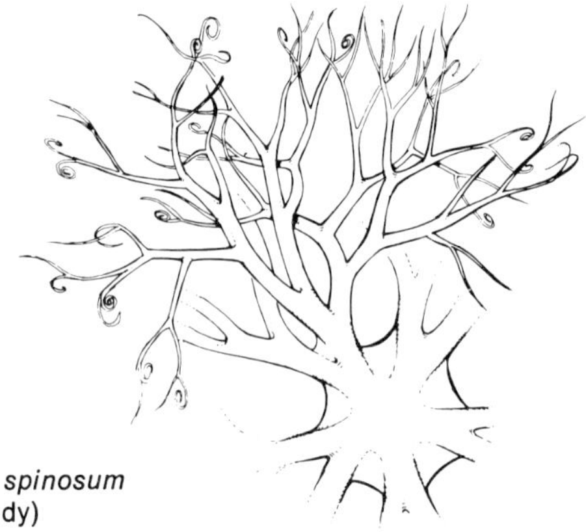

Fig. 12.21
Astrocaneum spinosum
(portion of body)

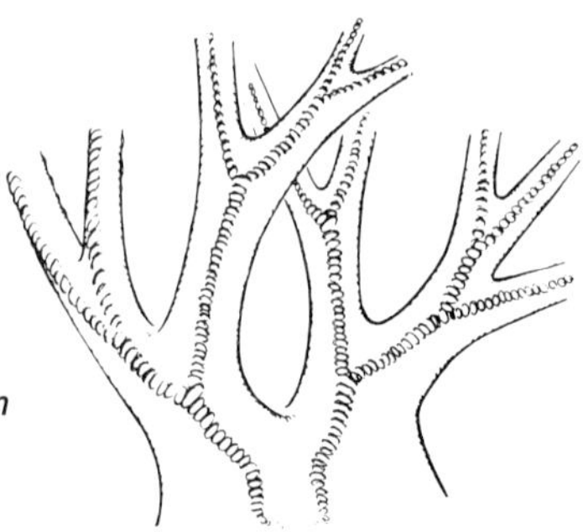

Fig. 12.22
Astrocaneum spinosum
(oral view of arm)

CLASS ECHINOIDEA

Preservation: The sea urchins and sand dollars may be dried, or preserved in 70 percent alcohol. The fine-sand technique described for the asteroids also works well with echinoids.

Taxonomy: Echinoids are globose, ovate, or discoidal echinoderms that include sand dollars, regular sea urchins, and heart urchins. The Gulf fauna of this class includes representatives of all three of these groups. The regular urchins are found in rocky areas, while the heart urchins and sand dollars usually may be found on sandy or muddy tidal flats. The taxonomy of this group is based primarily on the position and morphology of the spines, test plates, pedicellariae, mouth parts (Aristotle's lantern), and other external features. References in the key to the *internal lunule* of sand dollars refer to that lunule whose margins are entirely surrounded by the test, as opposed to the lunules whose distal end is open, forming a "slit" in the test margin. The systematics of this class have undergone numerous revisions by various authors over the last 60 years. The taxonomic scheme followed in our treatment is based on Clark (1912, 1914, 1917, and 1948) and Caso (1961).

KEY TO THE COMMON INTERTIDAL ECHINOIDS OF THE GULF OF CALIFORNIA

1 Body ovate or globose in shape (urchins) **7**
 -Body discoidal, or flattened in shape (sand dollars) **2**

2(1) Test with holes, slits, or notches (lunules) **3**
 -Lunules absent **5**

3(2) Lunules narrow and slitlike; test extremely flattened and fragile; maximum disk diameter about 75 mm (fig. 12.36) .... ***Mellita longifissa***
 -Lunules rounded, not slitlike; test thicker and more robust; maximum disk diameter about 130 mm **4**

4(3) Posterior lunule large and confined within the margin, 5 other lunules all open at the margin (fig. 12.34) .... ***Encope grandis***
 -All lunules confined within margin of test (fig. 12.35) .... ***Encope micropora***

5(2) Test circular or oval, with interradial margins convex (fig. 12.33) .... ***Clypeaster rotundus***
 -Test more or less pentagonal, with interradial margins concave or straight across **6**

6(5) Diameter of test may reach 200 mm; color pale pink to red .... ***Clypeaster europacificus***
 -Diameter of test rarely over 130 mm; color dark purple to black (fig. 12.32) .... ***Clypeaster testudinarius***

7(1) Body globose (regular urchins) **8**
 -Body ovate to spherical (heart urchins) **15**

8(7) Test armed with rows of blunt, club-shaped spines, usually covered with encrusting organisms (figs. 12.23, 12.24) .... ***Eucidaris thouarsii***
 -Test armed with sharp pointed spines; rarely, if ever, with encrusting organisms **9**

9(8) Aboral surface with 5 spineless rows running down from aboral pole (figs. 12.25, 12.26) .... ***Arbacia incisa***
-Aboral surface without 5 spineless zones **10**
10(9) Spines unicolored, not banded **11**
-Spines banded with 2 or more colors **13**
11(10) Spines with numerous barbs, projecting toward the tip of each spine; length of longest spines greater than 2 times height of test .... ***Diadema mexicanum***
-Spines without barbs as above; length of longest spines less than 2 times height of test **12**
12(11) Spines short, equal to or shorter than height of test; test and spines pale straw or greenish with faint rose or lavender tint; test dome-shaped (figs. 12.27, 12.28) .... ***Lytechinus pictus***
-Spines long, equal to or 1.5 times test height; test and spines purple to dark lavender; test globular (figs. 12.29, 12.30) .... ***Echinometra vanbrunti***
13(10) Spines banded near base only; spines not annulated; urchin lavender with 5 large, pale interradial spots .... ***Astropyga pulvinata***
-Spines banded near tip, annulated; urchin unicolored, not spotted as above **14**
14(13) Short claviform spinelets present on uppermost interambulacral plates; peristomial plates around mouth bear spines as well as pedicellariae (fig. 12.31) .... ***Centrostephanus coronatus***
-Without claviform spinelets on uppermost interambulacral plates; peristomial plates around mouth bear pedicellariae only, without spines .... ***Diadema mexicanum***
15(7) Urchin with numerous long dorsal spines (subequal to test width) color purple; body strongly ovate (fig. 12.28) .... ***Lovenia cordiformis***
-Urchin with very short, close-cropped spines; color tan to dark brown; body slightly ovate to spherical **16**
16(15) Test elongate-oval (width equals .77 to .84 percent of length); aboral surface with a well-defined, 5-pointed star pattern; test and spines gray to white .... ***Brissus latecarinatus***
-Test more circular (width equals .82 to .93 percent of length) aboral surface with a poorly defined 5-pointed star pattern; test and spines almost always uniformly dull brown **17**
17(16) Length 15 to 60 mm; test extremely rounded; occasionally with white spines (fig. 12.37) .... ***Agassizia scrobiculata***
-Length 75 to 150 mm; test ovate; never with white spines ... ***Meoma grandis***

SYSTEMATICS AND SPECIES DESCRIPTIONS

Echinoidea

SUBCLASS ENDOCYCLICA or REGULARIA (regular sea urchins)

Order Cidaroidea

Family Cidaridae

Eucidaris thouarsii (Valenciennes)

One of the most common echinoids of the Gulf and Panamic region in general. This sea urchin has ten vertical rows of five to eight clubbed spines each. The test diameter reaches 65 mm. The primary (large) and

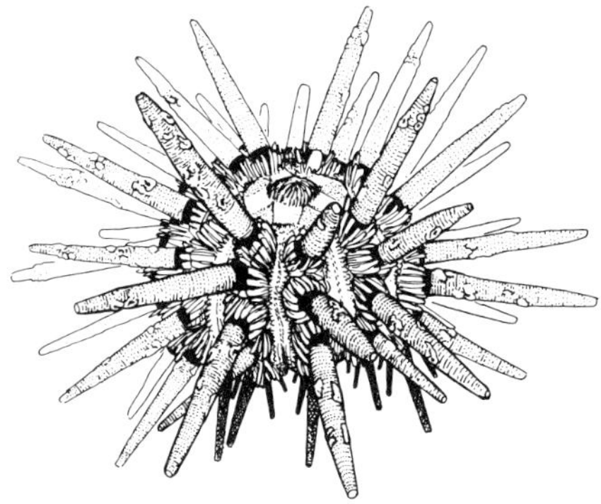

Fig. 12.23 *Eucidaris thouarsii*

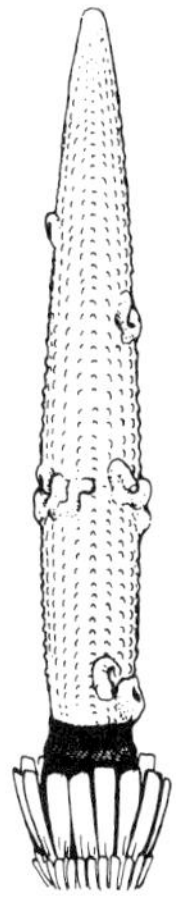

Fig. 12.24 *Eucidaris thouarsii* (single spine)

secondary (small) spines differ in shape, the secondaries being short and flat, truncate or chisel-tipped, while the primaries are large and round, tapering gradually toward the blunt tip. The brownish purple spines are nearly always coated with white bryozoan colonies, sponges, and calcareous algae. This species occurs on rocks and reefs in the mid- and low tide zone, often exposed to violent surf. It also ranges subtidally to depths of near 83 fathoms. It also has been reported as *Cidaris thouarsii*. *Eucidaris thouarsii* is the most characteristic urchin of tropical and subtropical west America, ranging from the upper Gulf of California to Ecuador and the Galapagos Islands. Clark (1923) reported this urchin as a possible enemy of pearl shell growers. Some workers give Agassiz and Desor as the authors of this species. The correct authorship may be open to debate, as both names appeared in the same year, 1846.

Order Aulodonta

Family Diadematidae

***Diadema mexicanum* A. Agassiz**

A large purple, black, or brown sea urchin with long slender spines. The spines are covered with a toxic mucoid substance that is quite painful, although relatively harmless otherwise. The spines break off readily and are difficult to remove because they possess recurved spinelets. The test diameter reaches 80 mm, and the dorsal spines reach 50 mm in length. Due to the fragility of its spines, this species can rarely be preserved with all the spines intact. Fully mature adults are uniformly black-purple, while younger specimens have banded spines, usually white and purple. This species is restricted to the low tide zone and subtidal and occurs from the mid-Gulf of California (Bahía de los Angeles and Guaymas) to Colombia and the Galapagos Islands. It also has been reported as *Centrechinus mexicanus.* Caso (1961) reported *Diadema mexicanum* from Puget Sound, Washington, but this record is doubtful, since this species appears to be strictly Panamic. In the Gulf it probably occurs in greatest numbers between Loreto and Cabo San Lucas, on the Baja California peninsula.

Astropyga pulvinata (Lamarck)

This urchin is highly variable in color (violet, red, yellow, or brown), usually brown-violet with reddish violet markings. It is easily recognized by the five large, pale, triangular blotches present in the interradial area of the aboral surface, and the rather flattened test. The test diameter reaches 120 mm. The species is largely subtidal and somewhat rare in the Gulf, ranging from the mid-Gulf to Ecuador and the Galapagos Islands. It also has been reported as *Cidarites pulvinata, Astropyga venusta,* and *Astropyga depressa.*

Centrostephanus coronatus (Verrill)

This species is often mistaken for *Diadema mexicanum* and avoided, but *Centrostephanus coronatus* is not poisonous. The test diameter reaches 63 mm, and the largest spines are approximately twice the width of the test. The spines have an annulated texture and are fragile and easily broken. The uppermost interambulacral plates possess short, claviform spinelets, with bright red or purple tips. Young individuals have the spines banded with purplish brown and yellow. Large adults may be without banding. This species has been reported as *Echinodiadema coronata.* It occurs in the low tide zones of Newport Beach and the upper Gulf, south to the Galapagos Islands, and has been reported from depths of up to 350 feet.

Order Stirodonta

Family Arbaciidae

Arbacia incisa (A. Agassiz)

One of the more common urchins of the Gulf, with the test diameter reaching 40 mm, although the average size is closer to 20 mm. Young individuals are red, becoming violet to black in adults. The interambulacral areas of the aboral surface are without spines and usually bear red spots on the plates. This species has short dorsal and ventral spines, but long lateral spines. *Arbacia incisa* prefers surf-protected rocky shores and is found in the midintertidal from southern California and the upper Gulf of California to Peru and the Galapagos Islands. It has been reported as *Echinocidaris incisa* and *Arbacia stellata.*

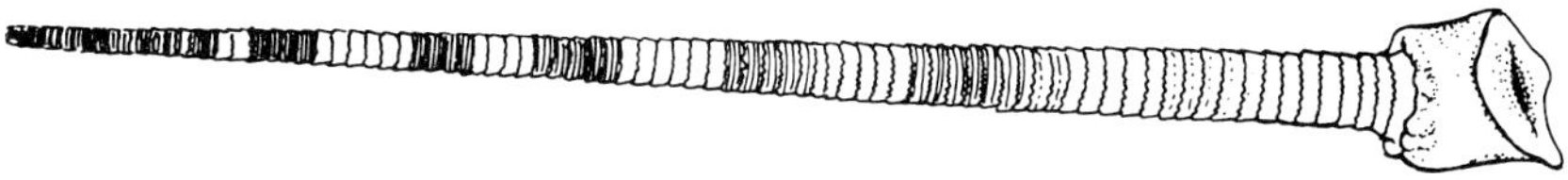

Fig. 12.25 *Centrostephanus coronatus* (single spine)

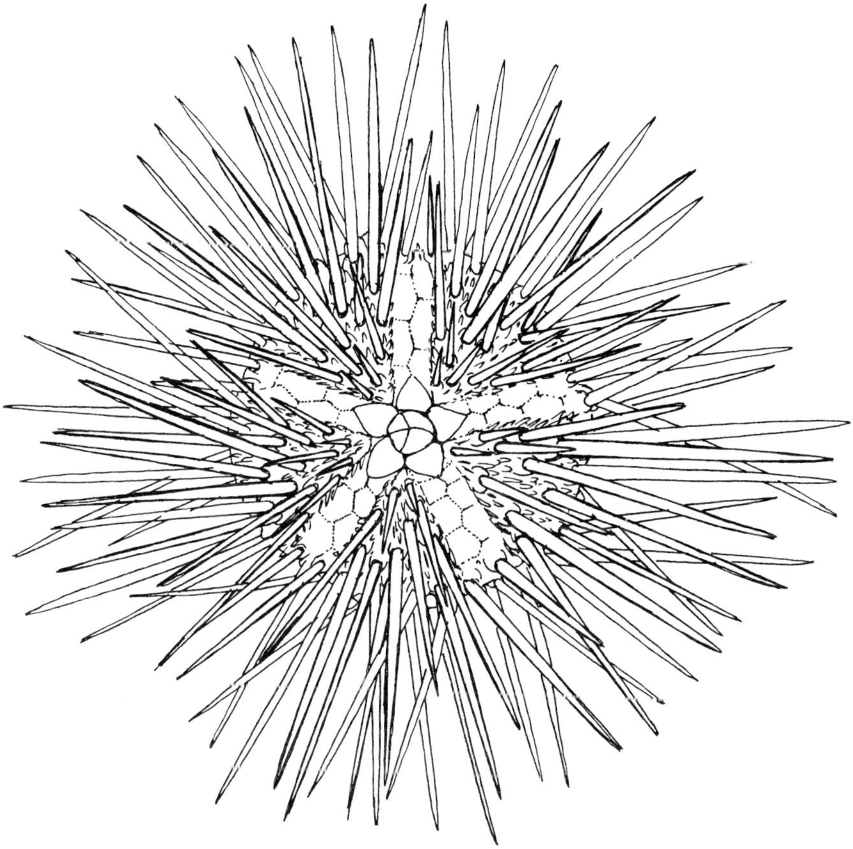

Fig. 12.26 *Arbacia incisa*

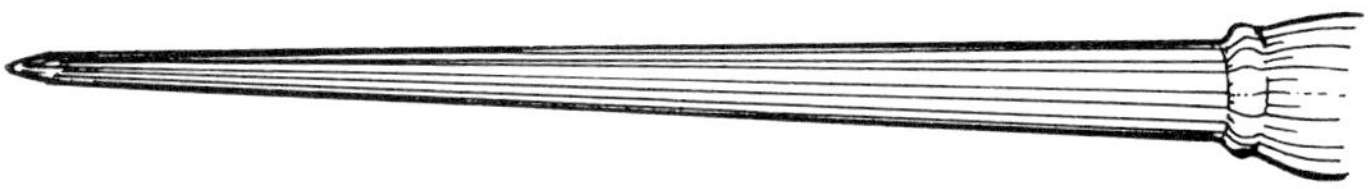

Fig. 12.27 *Arbacia incisa* (single spine)

Order Camarodonta

Family Toxopheustidae

Lytechinus pictus (Verrill)

This short-spined urchin has a test diameter reaching nearly 40 mm. The spines are short, thick and blunt. The color is a straw brown to gray, occasionally with a rosy hue. The basal half of the spines usually remains white, regardless of the distal coloring. In California this species is found in the mid- and low intertidal, but in the Gulf it is probably restricted to the very lowest minus tides and the subtidal. Its range is from central California and the northern Gulf to Ecuador. This species is nearly identical to *Lytechinus anamensus* H. L. Clark, differing mainly in its shorter spines and larger test, the test diameter of *L. anamensus* being only 10 to 20 mm. Hedgpeth (1969) reported *L. anamensus* as occurring in the Gulf, but we have no other reference to it from Gulf waters. Clark (1940) stated these may be ecological types of a single species.

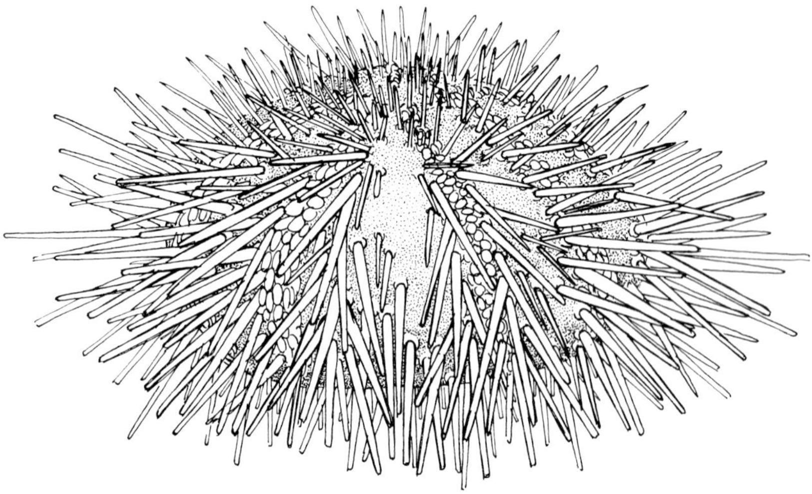

Fig. 12.28 *Lytechinus pictus*

Fig. 12.29 *Lytechinus pictus* (single spine)

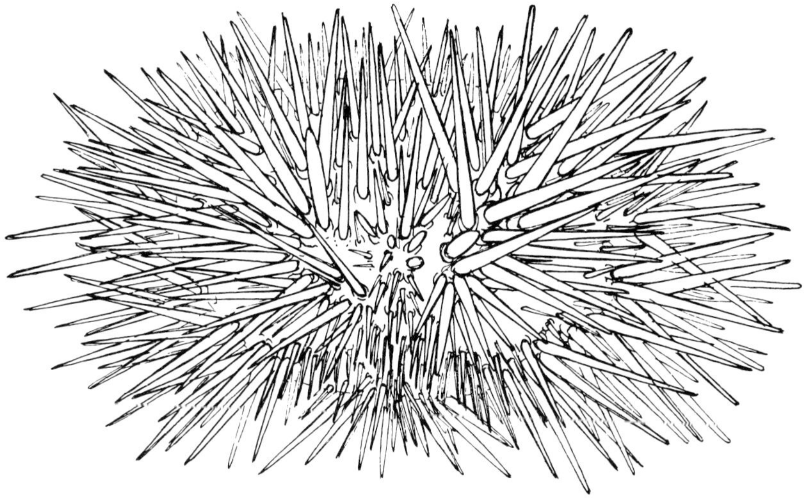

Fig. 12.30 *Echinometra vanbrunti*

Fig. 12.31 *Echinometra vanbrunti* (single spine)

Family Echinometridae

Echinometra vanbrunti A. Agassiz

This is one of the most common sea urchins of the Gulf. Young individuals are nearly circular, with the test diameter being 5 to 10 mm, while adults are more oval, with a test length reaching 70 mm. In the adult the test width equals about 90 percent of its length. This purple urchin prefers crevices and cavities in rocks and reefs, where it can avoid the full impact of the surf, although it occasionally is found on exposed rock surfaces where the surf is calm. *Echinometra vanbrunti* occurs in the mid- and low intertidal zones from the upper Gulf to Peru and the Galapagos Islands. As in the California species of purple urchin, *Strongylocentrotus purpuratus, E. vanbrunti* often is found singly, each urchin occupying its private pocket in soft rock, eroded by the rotating action of its body and spines. However, at Bahía Kino we have seen many handfuls of these urchins congregated on the sand under a single rock (December, 1970), and at Cabo San Lucas we've seen them so thick on the rocks as to obscure the substrate entirely (December, 1971). This species also has been reported as *Heliocidaris stenopora.*

SUBCLASS EXOCYCLICA or IRREGULARIA
(sand dollars and heart urchins)

Family Clypeastridae

Clypeaster europacificus H. L. Clark

A large sand dollar, reaching 200 mm in length. The width usually equals 95 percent of the length, making the test nearly circular. The body margin is easily broken, which leads to large numbers of deformed adults with strange looking tests. This species is pale red-violet when young and violet to brown when adult. *Clypeaster europacificus* is largely subtidal, as are all sand dollars, and it ranges from the mid-Gulf to Ecuador and the Galapagos Islands. This and other species of sand dollars and heart urchins are rarely collected alive, although the naked tests are quite commonly found washed up on the expanse of a sand flat or estuary. One is most likely to encounter the living animals during the lowest of low tides, when large beds of hundreds or even thousands of them may be partially exposed in lagoons and large bays, where water circulates over a sandy bottom.

Clypeaster testudinarus Gray

This sand dollar attains a test diameter of 130 mm, with a width of 87 to 95 percent of the length. Young are lavender to brown, and the adults dark purple to black, much darker overall than the above species. It also has been reported as *Clypeaster speciosus*. We have collected it only at Guaymas. Records indicate it has a range similar to the above species in the Gulf, but it probably does not occur as far south as Ecuador.

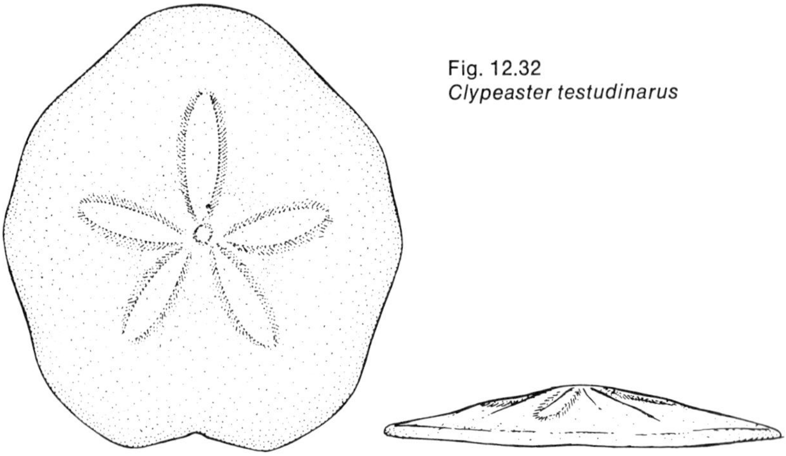

Fig. 12.32
Clypeaster testudinarus

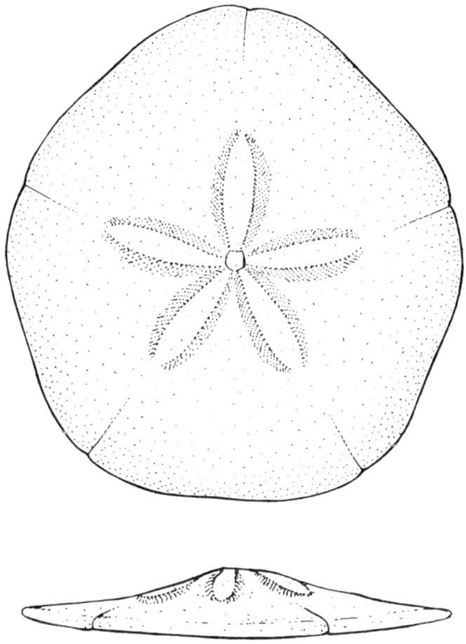

Fig. 12.33 *Clypeaster rotundus*

Clypeaster rotundus A. Agassiz

A nearly circular sand dollar, the width being 90 to 95 percent of the length, which reaches 175 mm. The spines are black, dark purple, or dark brown. This species is often called the "sea biscuit." Like most sand dollars, it has a prominent star pattern on its aboral surface, formed by the wide ambulacra. It ranges from the upper Gulf to Ecuador and the Galapagos Islands. The species is nearly an identical twin to the common, but smaller, *Dendraster excentricus,* which ranges from Alaska to northern Baja California.

Family Scutellidae

Encope grandis L. Agassiz

The giant keyhole sand dollar is abundant in the Gulf. It has a very coarse and heavy test, with a large unpaired lunule (the "keyhole.") The test is nearly circular, up to 120 mm in length. The posterior, unpaired lunule varies from a nearly circular hole to a long, narrow slit. This species is dull purple to black or dark brown. *Encope grandis* is probably endemic in the Gulf of California. Burns (1967) reported the odd occurrence of hermit crabs in the shells of *Cerithidea mazatlanica,* wedged into the anal lunule of individuals of this species at Puerto Peñasco. This and the two following species of *Encope* are quite often infected with a

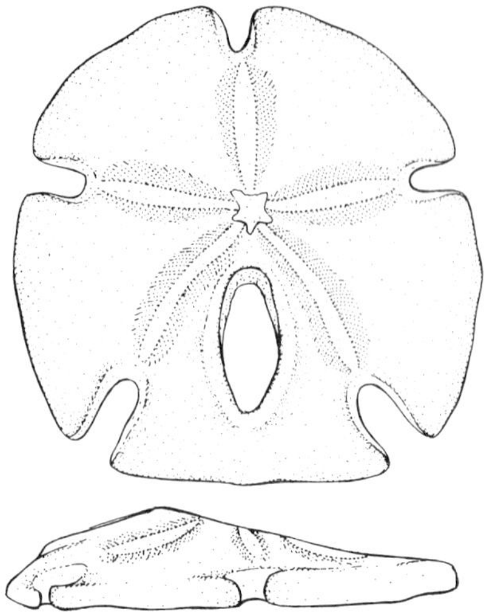

Fig. 12.34
Encope grandis

commensal pea crab of the genus *Dissodactylus,* which also inhabits the anal lunule. These commensal crustaceans feed on anal wastes and food being carried to the mouth in the food grooves of the sand dollar. We have also found a small porcelain crab and hermit crabs in the shells of various mud snails on *E. grandis* in the Puerto Peñasco area.

Encope micropora L. Agassiz

The test of this species may reach 130 mm, but usually small individuals are found with the diameter ranging from 10 to 50 mm. *Encope micropora* is circular, often with rounded lobes between the lunules, and may be various shades of brown to purple or black. The lunules are usually more or less circular and are normally all enclosed. The species ranges from the upper Gulf of California to Chile and the Galapagos Islands. *E. micropora* is often called *E. californica,* but the former name has priority. This is the *E. californica* of Steinbeck and Ricketts (pp. 186, 193, 400).

Mellita longifissa Michelin

This small sand dollar is purple to gray when alive, but has a prominent green-gray pattern when preserved. It is circular, with a test diameter reaching 75 mm, but individuals are normally from 20 to 40 mm in diameter. The lunules are unique, in that they are long, narrow slits, the width being only 5 to 10 percent of the length. This species ranges from the upper Gulf of California to Panama, and is especially common north of Bahía Kino in the Gulf.

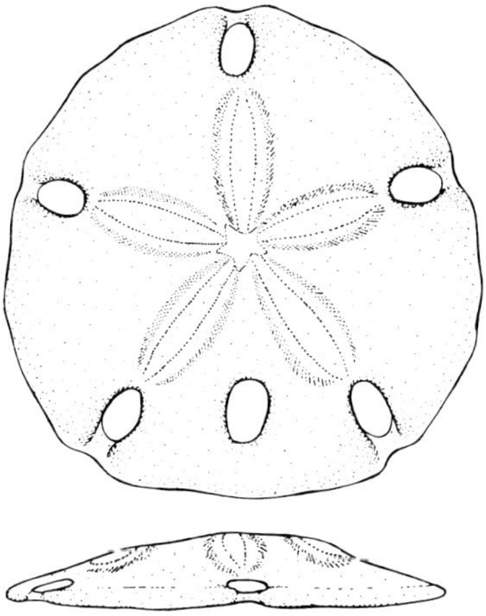

Fig. 12.35
Encope micropora

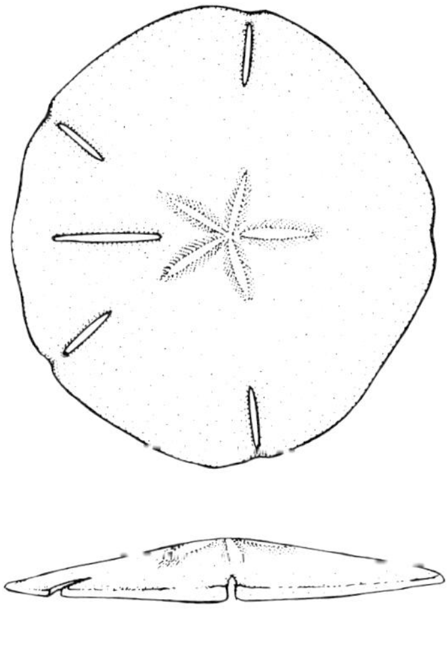

Fig. 12.36
Mellita longifissa

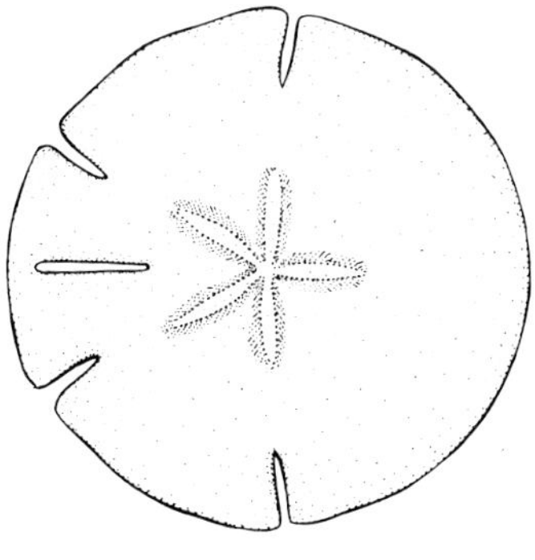

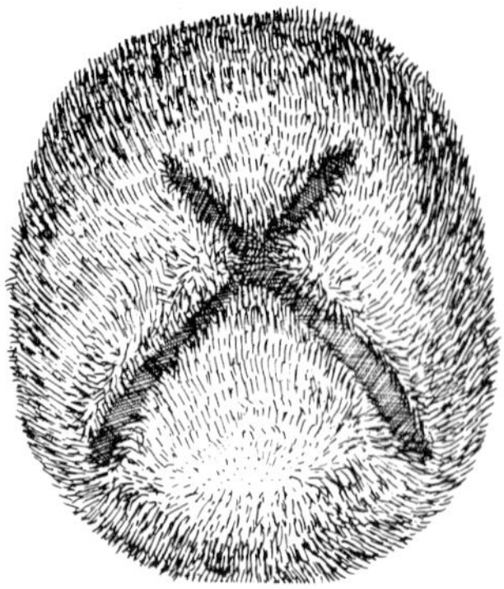

Fig. 12.37
Agassizia scrobiculata

Family Hemiasteridae

Agassizia scrobiculata Valenciennes

The normal size for this species of heart urchin is 19 to 25 mm. However, individuals up to 55 mm in length have been reported. *Agassizia scrobiculata* is a pale brown urchin, occasionally with white spines, largely subtidal, occurring from the upper Gulf of California to Peru and the Galapagos Islands. The width is about 80 to 90 percent of the length. This may be the most common heart urchin of the northern Gulf.

Family Spatangidae

Brissus latecarinatus (Leske)

This heart urchin is usually about 40 mm long and 30 mm wide; however, a giant of 166 mm has been reported. It has been listed previously as *Spatangus brissus* and *Brissus carinatus*. The width of the test equals about 77 to 85 percent of the length, and it has a normal, well-defined, star-shaped, petaloid pattern on the oral surface. *Brissus latecarinatus* ranges from the Gulf of California to Panamá and across most of the Indo-Pacific region. Unlike the other heart urchins, it seems to prefer the intertidal habitat over the subtidal.

Meoma grandis Gray

A brown to violet or nearly black heart urchin with a strong test. One of the largest heart urchins in the Gulf, with its length reaching 150 mm in large individuals and averaging 70 to 80 mm in smaller specimens. The width equals about 90 to 92 percent of the length. This giant heart urchin is found intertidally and subtidally, from the upper Gulf to Ecuador and the Galapagos Islands.

Lovenia cordiformis A. Agassiz

This unique heart urchin is the only species in the Gulf with long aboral spines and a nearly consistent, deep purple color. *Lovenia* is a large urchin, reaching 75 mm in length and 50 mm in width. The test is more oblong than the other heart urchins of the Gulf and has an odd-shaped petaloid pattern. Its color occasionally varies to a gray-brown (rarely), and the dorsal spines are often banded with a variety of colors. This species, like the majority of heart urchins, burrows in the sand or sandy mud of the lowest tide zones and subtidal. *Lovenia* is reported from southern California (and Hawaii) and the upper Gulf to the Galapagos Islands. It is commonly referred to as the "sea porcupine." The extremely long dorsal spines (up to 100 mm long) are said to contain a toxic substance. When disturbed they become erect, forming a defense mechanism to be avoided by barefoot beachcombers. We have not collected this species in the Gulf, although it is reported by other authors from Bahía de los Angeles, Bahía Santa Ynez, and Isla Espíritu Santo.

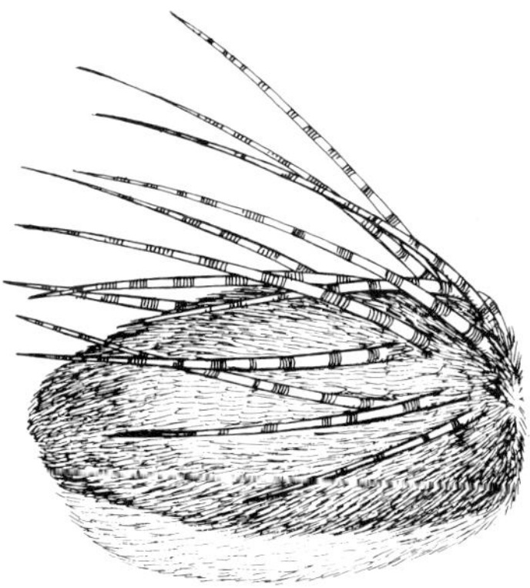

Fig. 12.38
Lovenia cordiformis

CLASS HOLOTHURIOIDEA

Preservation: Sea cucumbers always should be preserved in alcohol, as formalin tends to erode the calcareous plates of the epidermis, upon which the systematics of this class are based and identification of its members ultimately depend. Inol or glycerine added to the alcohol will often increase its color retention ability. It is difficult to preserve holothurians without rigorous narcotizing techniques, such as relaxing with Epsom salts (magnesium sulfate) over long periods of time. Properly preserved specimens must have the buccal tentacles extended. An alternate method of relaxing, one we have found to be successful in about 80 percent of the cases, is to allow the animal to die naturally in seawater, either by diluting it slowly with freshwater over a period of 5 to 8 hours, or by simply allowing the seawater to stand until it becomes putrid and the animal dies. This usually prevents eversion of the digestive tract, and the animal dies with the buccal tentacles still extended. This method has proven successful for numerous invertebrates, including polychaete annelids, sipunculids, ophiuroids, coelenterates, octopuses, nemerteans, and flatworms. Another trick you might want to try with the sea cucumbers is to simply grab the base of their tentacles when they are fully extended with a strong pair of forceps or pliers (we have used large forceps or hemostats), and then hold the animal in formalin for a few minutes until it is dead. Once dead, with the tentacles extended, it can be washed and placed in 70 percent alcohol.

Taxonomy: The sea cucumbers differ from all other echinoderms one may encounter in the Gulf in that they have lost their radial symmetry and reverted to the ancestral, bilateral symmetry. They are also soft, the skeleton being reduced to many small, calcareous ossicles imbedded in the body wall. The mouth is surrounded by feeding tentacles, and the body has numerous rows of tube feet, sometimes restricted to the ventral surface and sometimes scattered all over the body surface. The systematic scheme used in this handbook follows that of Nichols (1962) and Deichmann (1941, 1958). Two circumstances make the writing of a nontechnical field key to the Holothuroidea difficult. First, the taxonomy of the class is based largely on the ossicles confined to the thick, muscular body wall and other internal structures, and, second, the vast majority of species are variable in color, while greatly resembling one another in general body form. For this reason, only those species of sea cucumbers collected by the author are included in the following key. It is hoped these represent the more common species of the Gulf intertidal.

KEY TO THE COMMON INTERTIDAL HOLOTHURIANS OF THE GULF OF CALIFORNIA

1 Body length less than 2 cm; dorsal surface covered by imbricating scales; ventral surface modified into a creeping sole .... ***Psolidium dorsipes***

-Body length greater than 3 cm; dorsal surface not covered by imbricating scales; ventral surface may or may not form a flat, creeping sole **2**

2(1) Tube feet absent; small vermiform animals; usually found burrowing in sand **3**

-Tube feet present; large nonvermiform animals; usually found clinging to rocks **4**

3(2) Body pink to white, with white papillae; 12 buccal tentacles (fig. 12.41) .... ***Chiridota aponocrita***

-Body white; 10 buccal tentacles .... ***Epitomapta tobagae***

4(2) Tube feet occur only on ventral surface; ventral surface strongly flattened; body chocolate brown to red-orange with large orange-tipped dorsal spines .... ***Isostichopus fuscus***

-Tube feet not restricted to ventral surface; ventral surface not strongly flattened; body not colored as above **5**

5(4) Buccal tentacles highly dendritic and not completely retractile; buccal tentacles begin branching near base; 10 buccal tentacles **6**

-Buccal tentacles rarely dendritic, usually peltate (leaf or shieldlike) and completely retractile; buccal tentacles if branched, branch distally only; 15 to 30 buccal tentacles **8**

6(5) Body pure white, gray or black; buccal tentacles same length as body .... ***Cucumaria californica***

-Body usually lavender to purple (occasionally white); buccal tentacles much shorter than body **7**

7(6) Feet restricted to ambulacra (in 5 narrow bands); body reddish lavender; usually assumes an L-shape or lies straight when clinging to rocks, etc. (fig. 12.42) .... ***Pentamera chierchia***

-Feet not restricted to ambulacra, present in interambulacral regions; body white to purple; usually assumes U-shape when clinging to rocks, etc. .... ***Neothyone gibbosa***

8(5) Dorsal surface dark (black, dark gray, brown, or purple) **9**

-Dorsal surface light (white, pale gray, or tan) **10**

9(8) Body bottle-shaped with a long "neck"; paired black blotches dorsally; body papillae large, not yellow-tipped (fig. 12.39) .... ***Brandtothuria impatiens***

-Body not bottle-shaped; without paired black, dorsal blotches; body papillae minute and yellow-tipped (fig. 12.40) .... ***Selenkothura lubrica***

10(8) Body cylindrical, not flattened dorso-ventrally; usually with paired black spots dorsally .... ***Brandtothuria arenicola***

-Body noticeably flattened dorso-ventrally; without paired black spots dorsally .... ***Fossothuria rigida***

Order Aspidochirota

Family Stichopodidae

Isostichopus fuscus (Ludwig)

This is a large, robust sea cucumber, chocolate-brown to orange-brown, with large orange papillae on the dorsal surface. It attains a length of 20 cm. Its body is heavy and rigid, with the skin being smooth and tough to the touch. The ventral surface forms a strongly flattened sole, with three rows of close set tube feet. The mouth is circular and ventrally located. *Isostichopus fuscus* is found in sheltered areas where the wave shock is not too great, in the low intertidal and subtidal, where its attractive body may be seen slowly moving across the surface of submerged rocks and sand bottoms. For some reason this species does not preserve well, and pickled specimens generally do not resemble the living cucumber at all. It ranges from the upper Gulf of California to Ecuador and the Galapagos Islands. It also has been reported as *Stichopus fuscus*.

Family Holothuriidae

Brandtothuria arenicola (Semper)

A slender, spindle-shaped sea cucumber attaining a length of 20 cm, although most specimens average 10 to 15 cm in length. The feet are arranged in five indistinct bands. The dorsal surface is slightly papilliform, and the ventral surface is convex, not flattened. This species is white to gray with dark spots, usually in two dorsal rows (some individuals may be black or red, possibly from pigments in the substrate). It may be found buried under the sand or sandy mud in the rocky mid- and low intertidal zones, from the upper Gulf of California to Ecuador and the Galapagos Islands. *Brandtothuria arenicola* is nearly circumtropical in distribution. It has been reported previously as *Holothuria arenicola* and *H. monsuni*.

Brandtothuria impatiens (Forskal)

This sea cucumber reaches 15 cm in length and is usually distinctly bottle-shaped, with a long "neck." The tube feet are on the ends of long papillae, randomly spaced or in indistinct rows. It is mottled gray or brown (occasionally reddish), with paired black blotches posteriorly that merge together anteriorly to form black bands. The surface is rough and sandy to the touch. This cucumber ranges from the upper Gulf of California to Colombia and the Galapagos Islands and is nearly circum-

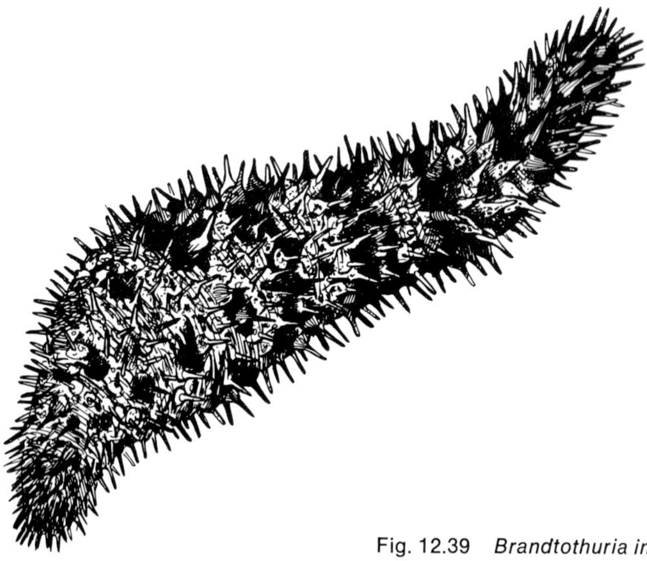

Fig. 12.39 *Brandtothuria impatiens*

tropical. It is commonly found wedged between large rocks, or partly buried in the substrate where rocks are present. It also has been reported as *Fistularia impatiens* and *Holothuria impatiens*.

Selenkothuria lubrica (Selenka)

One of the most common cucumbers of the Gulf, this species attains a maximum length of about 15 cm, although most individuals average 5 to 12 cm. It is often called the "sulfur cucumber" because the tips of the feet and body papillae are usually a sulfur-yellow color. The ground color is slate gray to mottled gray or brown, or a dark purplish black. The lighter specimens often have two rows of dark, dorsal blotches. The tube feet form a rather indistinct ventral sole. *Selenkothuria lubrica* is found under rocks in the mid- and low intertidal zones, where large numbers of individuals are often clustered. The gills are often seen protruding from a rock crevice and quickly retract at the least provocation. This very common species has, in the past, carried the names *Holothuria lubricata* and *H. kapiolaniae*. It ranges from the upper Gulf of California to Ecuador and the Galapagos Islands.

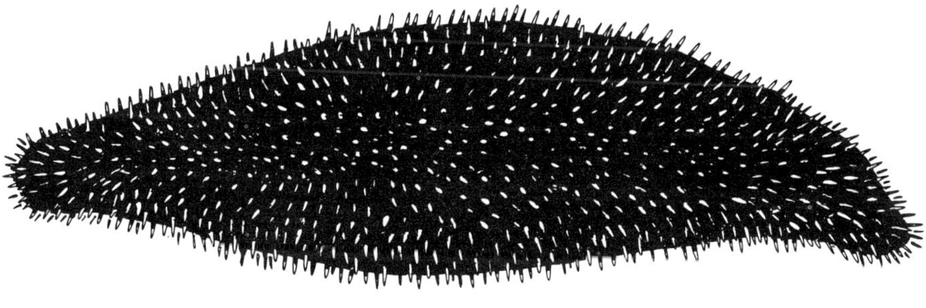

Fig. 12.40 *Selenkothuria lubrica*

Fossothuria rigida (Selenka)

A white to gray species of sea cucumber, usually with two rows of dark blotches on the dorsal surface. This species buries itself just under the surface of the sand in shallow water of the low tide zone and subtidally. The ventral surface is slightly flattened. *Fossothuria rigida* ranges from the upper Gulf of California and the west coast of Baja California to the Galapagos Islands and is also found in Africa and the Red Sea. It has been described also as *Stichopus rigidus, Holothuria rigida, Cystipus pleuripus, Holothuria pleuripus, Mulleria aegyptiana,* and *Holothuria aegyptiana*.

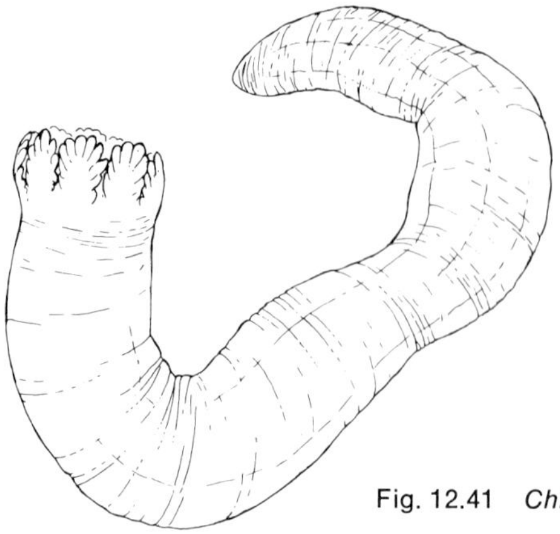

Fig. 12.41 *Chiridota aponocrita*

Order Apoda

Family Chiridotidae

Chiridota aponocrita

This is a small, white to pinkish red sea cucumber, with many minute papillae. It lacks visible tube feet and is quite vermiform in shape. *Chiridota aponocrita* is found buried in the sand, usually under rocks, in the mid- and low intertidal zone. It is common in the northern Gulf, but uncommon south of Bahía Kino.

Epitomapta tabogae Heding

Another small, pale sea cucumber. It is white, about 6 cm long, and somewhat vermiform. Tube feet are lacking, and the buccal tentacles number ten. *Epitomapta tabogae* burrows in the sand in the low intertidal zone throughout the Gulf.

Order Dendrochirota

Family Psolidae

Psolidium dorsipes Ludwig

A very small sea cucumber with a flattened, creeping sole. The length of this species rarely exceeds 1 or 2 cm. The sole is set off by three crowded bands of tube feet. The dorsal surface is covered by imbricating scales, about 10 to 12 from anterior to posterior. The dorsal scales bear several papillae (feet) each. This species is mainly subtidal and ranges from the Gulf of California to Panamá. It also has been reported from the Strait of Magellan.

Family Cucumariidae

Cucumaria californica Semper

A small, soft, slippery sea cucumber, attaining a maximum length of only about 10 cm. It has ten soft, bushy tentacles, approximately the same length as the body. The feet are large, soft, and completely retractable and are restricted to the five ambulacra. This species is red to gray or black (or occasionally white). It ranges from the upper Gulf and west coast of Baja California to Peru, occurring in the rocky mid- and low intertidal.

Pentamera chierchia (Ludwig)

A small sea cucumber, only 3 to 4 cm long. It has numerous tube feet in five narrow bands. It is black or dark brown to reddish purple and is one of the most common cucumbers of the Panamic region. Individuals of this species often assume an L-shape when attached to the rocks of the mid- and low intertidal zones. *Pentamera chierchia* ranges from the upper Gulf of California to Ecuador and the Galapagos Islands. It also has been reported as *Cucumaria chierchia.*

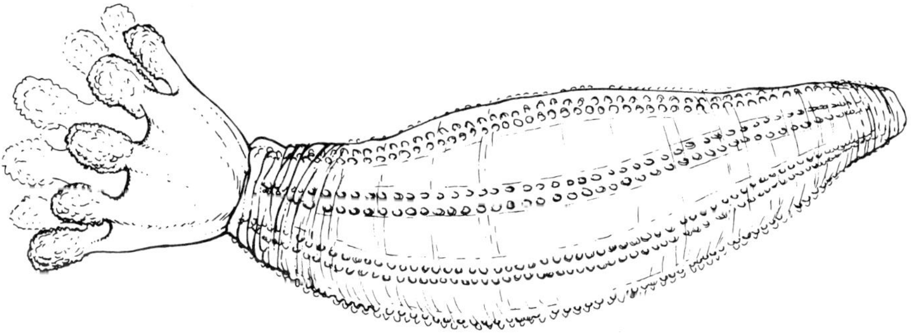

Fig. 12.42 *Pentamera chierchia*
(preserved specimen with tentacles extremely distended)

Neothyone gibbosa Deichmann

A small- to medium-sized cucumber, with a white to purple body. It has long tube feet and a rigid body wall. This species often assumes a U-shape when attached to the rocks of the mid- and low intertidal zones. *Neothyone gibbosa* occurs throughout the Gulf, and also has been reported from Peru. We have found this cucumber in great numbers at Guaymas and Bahía Kino.

Hemichordata (Acorn worms)

CLASS ENTEROPNEUSTA (acorn worms)

Preservation: The acorn worms should be carefully narcotized to prevent extreme contraction. Knudsen recommends allowing the worms to fully expand in fresh seawater, then adding alcohol, drop by drop, until a 10 percent solution is attained. The worms are then allowed to set for 4 to 6 hours before preservation in 5 percent formalin or 70 percent alcohol.

Taxonomy: The phylum Hemichordata is generally divided into three classes: Pterobranchiata, Graptolita, and Enteroptneusta. The pterobranches are rare, benthic, deep-water marine forms; the graptolites are extinct forms, known mainly from the structure of their tubes; while the enteropneusts (acorn worms) are free-living, wormlike creatures of the marine intertidal and subtidal. The acorn worms do not differ a great deal among themselves, so have not been placed in categories higher than the family level. There are three families: Ptychoderidae, Harrimaniidae, and Spengelidae. The systematics of the enteropneusts are based largely on the internal anatomy and its corresponding external morphology. The systematics used in this handbook are from Hyman (1959).

Enteropneusts of the Gulf: There are at least three species of acorn worms occurring in the Gulf of California intertidal region. Dr. T. H. Bullock reports (personal correspondence) the following: *Ptychodera flava* Eschscholtz, of the family Ptychoderidae, collected by E. F. Ricketts at Espíritu Santo Island in 1940; *Saccoglossus sonorensis* Rao, Ritter and Bullock (M. S.) of the family Harrimaniidae, from Bahía Cholla; and *Balanoglossus sonorensis* Rao, Ritter and Bullock (M. S.), of the family Harrimaniidae, taken by Steve Glassell in 1935 at Bahía Cholla, by Ricketts in 1940 on the west side of the Gulf south of Santa Rosalía, and in Sonora (Estero de la Luna), and by Dr. Bullock in 1949 at Bahía Cholla. *Ptychodera flava* is a well-known species from the Indo-Pacific region and throughout the tropics. The species of *Saccoglossus* and *Balanoglossus* are both new and are being described at this writing by Bullock et al. (see reference section).

Saccoglossus can be differentiated from *Ptychodera* and *Balanoglossus* by the length of the proboscis, which is very long in *Saccoglossus,* more than six times longer than wide and two to three times the length of the collar, which is the ring-shaped region immediately behind the proboscis. There are, however, no obvious and reliable external characters with which to differentiate *Ptychodera* and *Balanoglossus*. Dr. Bullock has informed me that the following characteristics are *usually* reliable: *Balanoglossus* is most likely to be in mud and to show spiral castings (like those of *Saccoglossus,* only larger), while *Ptychodera* seems to prefer a coarse sand, frequently made up of fragmented coral; *Ptychodera* rarely burrows to a depth greater than 3 to 5 inches, whereas *Balanoglossus* is frequently found at depths of 6, 8, or 10 inches (because of its habit of burrowing at these depths, *Balanoglossus* is frequently recovered only in fragments); *Ptychodera* is commonly 10 cm or less in length, whereas *Balanoglossus* is frequently several times that length.

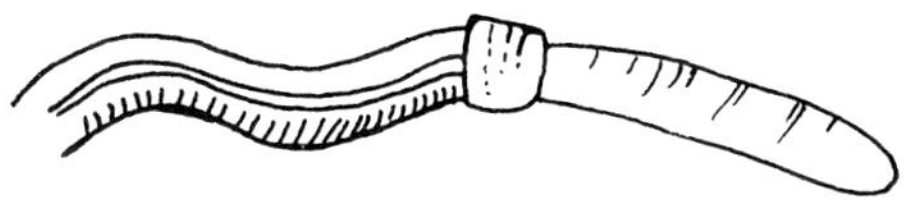

Fig. 13.1 *Saccoglossus* sp.

Chordata (Ascidians)

CLASS ASCIDIACEA

Preservation: For general preservation, where expanded individuals are not required, 5 percent formalin or 70 percent alcohol are good preservatives. If the specimens are going to be kept for any length of time, alcohol is preferable, even though it will bleach the color from the animal. For purposes of laboratory study and specific identifications, individuals with fully expanded siphons are desirable, requiring slow relaxation prior to preservation. Relaxing can be done with magnesium sulfate or chloral hydrate. Bianco (1899) recommended adding small quantities of chrom-acetic fixative to the container of seawater the expanded specimens were in. This diffuses into the water, killing the animals slowly over a period of several hours.

Taxonomy: The subphylum Urochordata is divided into three classes: Larvacea, Thaliacea, and Ascidiacea. The first two classes are pelagic, free-living marine forms, while the ascidians are sedentary benthic forms. The intertidal ascidians may be loosely grouped into two categories, solitary and compound. The solitary (or simple) ascidians develop only from eggs and do not bud. Solitary ascidians are usually saclike animals, often transparent, with two large siphons. Compound (or colonial) ascidians can reproduce by budding, to form aggregations of a few to many individuals, all imbedded in a common gelatinous matrix. Colonial ascidians often appear as a mass of colored jelly with clusters of individuals, usually of a different color, arranged in a petaloid fashion throughout the matrix. The taxonomy of the ascidians is based on a number of characters, of which the most important are growth form and anatomy of the branchial sac, alimentary system, and gonads. Specific identification is a difficult and time consuming task, and may require examination of the endostyle, branchial tentacles, folds of the branchial sac, etc.

There is no general agreement on how to taxonomically arrange the higher taxa of ascidians. Various schemes of classification have been suggested, based on different aspects of the animal's anatomy. Borradaile et al. (1961) discuss three principal classifications, those of Lahille, Seeliger, and Berrill. However, the classification of this group at the generic level is claimed by Van Name (1945) to be quite stable, and he states, "More than 90 percent of the known ascidians fit perfectly into a comparatively small number of genera . . . that are selfevidently natural groups . . . " This puts the tunicates in a comfortably workable category, which may be due in part to the fact that no fossil ascidians are known, thus placing the burden of natural phylogenies on the 2,000 or so extant forms.

Ascidians of the Gulf: Specific determination of ascidians requires rather exacting microscopic dissections and is rarely attempted by anyone other than a specialist. In addition to this, the ascidians tend to exhibit a great deal of variability in growth form and color, while different species with similar growth patterns often look very much alike. For these reasons no key has yet been attempted for this group. The species list below represents only specimens collected by the author from the region of Puerto Peñasco in the northern Gulf. These certainly represent only a fraction of the total Gulf intertidal ascidian fauna, but should, on the other hand, include many of the most commonly encountered forms in the northern Gulf. Identifications were made by Dr. D. Abbott, and many of the following comments are derived from personal correspondence with him.

SYSTEMATICS AND SPECIES DESCRIPTIONS

Ascidians

CLASS ASCIDIACEA

Order Enterogona

Suborder Aplousobranchia

Family Polyclinidae

Aplidium Savigny

This common and widespread genus is often still referred to as *Amaroucium*, a name no longer valid. All species of *Aplidium* have the zooids divided into a thorax, abdomen, and postabdomen. The atrial aperture bears a small process, or languet. The stomach has a series of longitudinal folds on the wall. There are at least three species of this genus common to the northern Gulf intertidal. "Species A" forms a massive, fleshy colony, with the zooids bearing 15 to 18 rows of stigmata, and the stomach wall thrown into ten weakly developed folds. This species resembles the typical "sea-pork" so often found washed ashore, or dragged up from the bottom in shrimp nets. The color is usually pale to vivid orange. "Species B" consists of a compact cluster of slender lobes, each lobe completely encrusted with sand. The lobes are joined only basally, but they arise close to one another. They are about all the same length, so the whole cluster forms a very compact mass, resembling a zoanthid anemone colony. The zooids have ten to 11 rows of stigmata and only five to seven folds on the stomach wall. "Species C" is a massive, fleshy colony, resembling "sea-pork," with the zooids bearing nine to ten rows of stigmata, and the stomach folded into 20 to 25 regular and marked folds.

Synoicum Phipps

A species of this genus found commonly on tidal flats consists of a cluster of globular lobes, 1 to 3 cm or more in diameter. The zooids have the body divided into thorax, abdomen, and postabdomen (like in *Aplidium*). The postabdomen is long and broadly attached to the posterior end of the abdomen. The stomach wall is smooth.

Polyclinum Savigny

Polyclinum laxum Van Name is a common inhabitant of the Puerto Peñasco region. It generally forms a smooth, raised colony, dark brown to black, with yellow to white individuals arranged in oval or circular groups. The colonies grow to several inches across, and are more or less hemispherical, or rounded and depressed. A slice taken across the colony will reveal the zooids are all near the outer surface, and that most of the colony is a more or less transparent, stiff, jellylike tunic. The zooids have their body divided into a thorax, abdomen, and postabdomen. The thorax is the largest of the three divisions. The postabdomen is quite short and joined to the abdomen by only a narrow neck. Steinbeck and Ricketts reported this species (as *Polyclinum* sp., page 565) from Isla Espíritu Santo and El Mogote. They described it as small, circular, brown colonies and reported a description given them by Van Name in a personal communication as, "Forms large rounded colonies of soft consistency, not raised on a pedicel."

Family Didemnidae

Diplosoma MacDonald

A species of ascidian, probably in this genus, is common on the rocks of the Puerto Peñasco region. It is a solid black, encrusting form. Dr. Abbott informs me this species is related to *Didemnum* and its allies, but lacks spicules.

Family Polycitoridae

Archidistoma Garstang

This genus was formerly called *Eudistoma,* but the latter genus was recently largely placed in the genus *Archidistoma. Archidistoma pachecae* (Van Name) is a common encrusting tunicate found throughout the Gulf. It is most easily recognized by the more or less colorless tunic that bears dark lavender, purple, or almost black zooids. The zooids are sometimes arranged geometrically, other times irregularly. Steinbeck and Ricketts described this species (as *Eudistoma* sp., page 567), as a soft, fleshy encrustation, dark brown or purplish in preservation. They collected it at Isla Espíritu Santo, Punta Marcial, Puerto Escondido, Bahía Concepción, Bahía de los Angeles, Puerto Refugio, and the south end of Isla Tiburón.

Suborder Phlebobrachia

Family Ascidiidae

Ascidia Linnaeus

Members of this genus are mostly solitary tunicates, found attached to the undersides of rocks. *Ascidia interrupta* Heller is a common species found throughout the Gulf. It has a transparent body with orange-tipped siphons. Dr. D. Abbott informs me that he has been calling this west coast form *A. interrupta,* although there are a few minor differences from the Caribbean form. He considers these differences too trivial and undependable to substantiate specific separation of the two forms. I have found this species to be common in the rocky intertidal, at least as far south as Guaymas and La Paz.

In addition to the above species, Dr. P. Pickens (1970) reported the following species from the Puerto Peñasco region (identified by Dr. Abbott): *Cystodytes, Botryllus, Sumplegma,* and *Styela.* Steinbeck and Ricketts reported the following species: *Amaroucium californicum* Ritter and Forsyth, from Bajio Plumo, Isla Espíritu Santo, La Paz area, Bahía de los Angeles, Puerto Refugio (Isla Angel de la Guarda) and Isla Tiburón; *Didemnum carnulentum* Ritter and Forsyth, from Cabo San Lucas, La Paz, Islas Coronado, Bahía Concepción, Bahía de los Angeles, Bahía San Francisquito, and Puerto Refugio; *Trididemnum opacum* Ritter, from Cabo San Lucas; *Cystodytes dellechiajei* (Della Valle) from Isla Espíritu Santo, Puerto Escondido, Bahía Concepción, Bahía de los Angeles, Puerto Refugio, and Isla Tiburón; *Clavelinia* sp. from La Paz; *Botrylloides diegensis* Ritter and Forsyth from La Paz, Isla Espíritu Santo, and Bahía San Francisquito; and *Pyura* sp., a solitary tunicate, from Isla Espíritu Santo, Islas Coronado, and Puerto Escondido.

Appendix

I. APPROXIMATE LATITUDES OF SOME SELECTED LOCALITIES IN THE GULF OF CALIFORNIA

Puerto Peñasco	31° 20′ N
San Felipe	31° 03′ N
Desemboque	30° 30′ N
Puertecitos	30° 25′ N
Puerto Libertad (Cabo Lobos)	29° 50′ N
Bahía San Luis Gonzaga	29° 48′ N
Isla Angel de la Guarda (Puerto Refugio)	29° 33′ N
Bahía de los Angeles	28° 55′ N
Bahía Kino	28° 45′ N
Isla San Lorenzo Sur	28° 40′ N
Guaymas	27° 55′ N
Bahía Concepción	26° 43′ N
Topolobampo	25° 35′ N
Boca del Rio	25° 20′ N
Isla Las Animas	25° 05′ N
La Paz	24° 10′ N
Mazatlán	23° 15′ N
Cabo San Lucas	22° 50′ N

II. MARINE STATIONS IN THE GULF OF CALIFORNIA

1. Vermilion Sea Field Station at Bahía de los Angeles. Operated by the San Diego Society of Natural History.
2. University of Arizona-University of Sonora Cooperative Marine Station at Puerto Peñasco, Sonora. Operated jointly by the Universities of Sonora and Arizona.
3. Las Cruces Biological Station at Las Cruces, Baja California Sur. Operated by Immaculate Heart College of Los Angeles.
4. Escuela de Ciencias Marítimas y Technología de Alimentos de Guaymas. (Marine laboratories of the Instituto Tecnologico de Monterrey, Nuevo Leon, Mexico.)

III. DOMINANT INVERTEBRATES OF SOME ACCESSIBLE AREAS OF THE SONORAN-SINALOAN COAST

What follows is a list of the invertebrate animals that I have found to be most commonly encountered, year round, at the easily reached collecting sites along the east coast of the Gulf of California. The lists were compiled during the years 1969 to 1972 and represent animals found consistently all four years. The lists do not by any means represent *all* the animals occurring in these areas, but are meant to give the student traveling to these regions (all easily accessible by good roads) an idea of what conspicuous and common forms he may expect to encounter. The molluscs of these same areas are treated separately in the following Section IV of the Appendix.

Puerto Peñasco-Bahía de Adair-Bahía Cholla Region

Sponges:
- *Geodia mesotriaena*
- *Pseudosuberites pseudos*
- *Hymeniacidon sinapium*
- *Leucetta losangelensis*

Hydroids:
- *Aglaophenia* sp.

Sea anemones:
- *Bunodactis mexicana*
- *Bundosoma californica*
- *Phyllactis cocinnata*
- *Anthothoe carcinophila*
- *Palythoa ignota*
- *Zoanthus danae*

Corals:
- *Porites californica*

Flatworms:
- *Stylochoplana* sp.
- *Pseudoceros mexicanus*
- *Pseudoceros bajae*

Nemertines:
- *Baseodiscus delineatus*

Peanut worms:
- *Themiste lissum*
- *Phascolosoma perlucens*

Barnacles:
- *Tetraclita squamosa*
- *Chthamalus anisopoma*
- *Balanus tintinnabulum*

Isopods:
- *Ligia occidentalis*

Shrimp:
- *Alpheus californiensis*
- *Synalpheus* sp.
- *Lysmata californica*
- *Callianassa* sp.
- *Upogebia* sp.

Brachyuran crabs:
- *Callinectes bellicosus*
- *Uca crenulata*
- *Eriphia squamata*
- *Ala cornuta*
- *Pilumnus limosus*
- *Pilumnus townsendi*
- *Pilumnus gonzalensis*
- *Leptodius occidentalis*
- *Eurytium albidigitum*

Porcelain crabs:
- *Petrolisthes gracilis*
- *Petrolisthes armatus*
- *Pachycheles setimanus*
- *Porcellana cancrisocialis*

Hermit crabs:
- *Clibanarius digueti*
- *Petrochirus californiensis*

Sea stars:
- *Heliaster kubinijii*
- *Othilia tenuispina*

Brittle stars:
Ophioderma panamense
Ophiothrix spiculata
Ophiocoma aethiops
Ophiocoma alexandri
Ophionereis annulata
Sea urchins:
Echinometra vanbrunti
Sand dollars:
Encope grandis
Sea cucumbers:
Neothyone gibbosa
Selenkothuria lubrica
Brandtothuria sp.

Bahía Kino Region

Sponges:
Leucosolenia sp.
Sea anemones:
Bundosoma californica
Flatworms:
Pseudoceros bajae
Peanut worms:
Phascolosoma perlucens
Barnacles:
Tetraclita squamosa
Chthamalus anisopoma
Nemertines:
Baseodiscus punneti
Isopods:
Ligia occidentalis
Shrimp:
Alpheus sp.
Synalpheus sp.
Brachyuran crabs:
Ala cornuta
Pilumnus gonzalensis
Pilumnus townsendi
Tetragrapsus jouyi
Porcelain crabs:
Petrolisthes crenulatus
Pachycheles setimanus
Hermit crabs:
Paguristes anahuacus
Sea stars:
Heliaster kubinijii
Brittle stars:
Ophiothrix spiculata
Ophioderma panamense
Ophiocoma alexandri
Ophionereis annulata
Ophiocoma aethiops
Sea urchins:
Echinometra vanbrunti
Sea cucumbers:
Brandtothuria arenicola
Brandtothuria impatiens
Chiridota aponocrita

Guaymas Region

Sponges:
Geodia mesotriaena
Leucetta sp.
Hydroids:
Aglaophenia sp.
Sea anemones:
Phyllactis sp.
Bunodactis sp.
Bundosoma sp.
Corals:
Porites californica
Nemertines:
Baseodiscus punnitti
Peanut worms:
Themiste lissum
Phascolosoma perlucens
Barnacles:
Balanus concavus

Balanus anisopoma
Tetraclita squamosa
Isopods:
Ligia occidentalis
Shrimp:
Alpheus sp.
Synalpheus sp.
Lysmata californica
Brachyuran crabs:
Callinectes bellicosus
Leptodius occidentalis
Ala cornuta
Pilumnus limosus
Tetragrapsus jouyi
Ocypode occidentalis
Pilumnus gonzalensis
Porcelain crabs:
Petrolisthes hirtipes
Petrolisthes gracilis
Petrolisthes crenulatus
Pachycheles setimanus
Sea stars:
Pharia pyramidata
Phateria unifascialis
Heliaster kubinijii
Brittle stars:
Ophioderma panamense
Ophiocoma aethiops
Ophiocoma alexandri
Ophiothrix spiculata
Ophiactis simplex
Sea urchins:
Echinometra vanbrunti
Eucidaris thouarsii
Sea cucumbers:
Selenkothuria lubrica
Pentamera chierchia
Brandtothuria sp.

Topolobampo Region

Barnacles:
Balanus trigonus
Tetraclita squamosa
Shrimp:
Alpheus sp.
Synalpheus sp.
Brachyuran crabs:
Eriphia squamata
Leptodius occidentalis
Eurytium affine
Panopeus purpureus
Pachygrapsus transversus
Uca crenulata
Uca princeps
Brittle stars:
Ophiactis simplex

Yavaros Region

Barnacles:
Balanus amphitrite
Isopods:
Ligia occidentalis
Brachyuran crabs:
Uca princeps
Leptodius occidentalis
Tetragrapsus jouyi
Eurytium affine
Panopeus purpureus
Ocypode occidentalis

Huatabampito Region

Balanus tintinnabulum
Callinectes bellicosus

Boca del Rio-La Gloria Region

Ocypode occidentalis
Uca princeps

IV. COMMON MOLLUSCS OF SOME ACCESSIBLE AREAS OF THE SONORAN-SINALOAN COAST

By Roy S. Houston

The following lists constitute the common molluscan fauna found to occur during the years 1970 to 1972 in the following areas. The lists are based on collections made by Richard C. Brusca.

Puerto Peñasco-Bahía de Adair-Bahía Cholla region

Gastropoda:
Collisella dalliana
Collisella stanfordiana
Tegula rugosa
Turbo fluctuosus
Littorina aspera
Nerita funiculata
Cerithium stercusmuscarum
Cerithidea mazatlanica
Calyptraea mamillaris
Crucibulum spinosum
Serpulorbis margaritaceus
Hexaplex erythrostomus
Acanthina angelica
Thais biserialis
Morula ferruginosa
Nassarius moestus
Columbella fuscata
Columbella strombiformis
Anachis varia
Anachis coronata
Mitrella guttata
Olivella dama
Crassispira pluto
Conus perplexus

Pelecypoda:
Arca pacifica
Noetia reversa
Glycymeris gigantia
Leptopecten tumbezensis
Anomia adamas
Pinna rugosa
Lithophaga aristata
Lithophaga spatiosa
Ostrea conchaphilia
Modiolus capax
Dosinia dunkeri
Mytella guyanensis
Tivela planulata
Pitar lupanaria
Chione amathusia
Protothaca grata
Donax punctatostriatus
Tagelus affinis

Scaphapoda:
Dentalium oerstedii

Polyplacophora:
Chiton virgulatus
Chiton stokesii
Acanthochitona exquisita
Stenoplax conspicua
Callistochiton gabbi

Opisthobranchiata:
Chromodoris banksi
Chromodoris norrisi
Chromodoris sedna
Hypselodoris californiensis
Navanax inermis

Bahía Kino region

Gastropoda:
Acmaea filosa
Diodora inaequalis
Tegula ligulata
Turbo fluctuosus
Cerithidea mazatlanica

Cerithium maculosum
Cerithium sculptum
Acanthina angelica
Mitrella guttata
Columbella fuscata

Pelecypoda:
Arcopsis solida
Ostrea sp.
Cardita affinis

Polyplacophora:
Acanthochitona angelium
Chiton virgulatus
Ischnochiton sp.
Stenoplax conspicua

Guaymas region

Gastropoda:
Diodora inaequalis
Collisella atrata
Collisella mitella
Collisella stanfordiana
Tegula ligulata
Tegula rugosa
Turbo fluctuosus
Nerita scabricosta
Turritella gonostoma
Cerithium maculosum
Cerithium sculptum
Cerithium stercusmuscarum
Cerithidea mazatlanica
Crucibulum scutellatum
Trivia californiana
Trivia solandri
Strombus gracilior
Muricanthus nigritus
Anachis nigricans
Columbella fuscata
Mitrella guttata
Nassarius luteostoma
Nassarius versicolor

Pelecypoda:
Barbatia reeveana
Anadara sp.
Arcopsis solida
Lithophaga sp.
Mytella guyanensis
Pteria sterna
Ostrea sp.
Cardita affinis
Trachycardium panamense
Dosinia dunkeri
Chione sp.
Tagelus sp.

Polyplacophora:
Chiton virgulatus
Ischnochition sp.

Topolobampo region

Gastropoda:
Diodora inaequalis
Nerita funiculata
Nerita scabricosta
Cerithium stercusmuscarum
Crucibulum spinosum

Pelecypoda:
Arcopsis solida
Pecten sp.
Cardita affinis
Anomalocardia subimbricata
Protothaca grata

Huatabampito region

Gastropoda:
Turritella gonostoma
Architectonia nobilis
Calyptraea mamillaris

Pelecypoda:
Trachycardium panamense
Pitar sp.
Chione californiensis
Tivela planulata
Tellina sp.
Donax carinatus

Boca del Rio-La Gloria region

Gastropoda:
- *Turritella leucostoma**
- *Cerithidea mazatlanica**
- *Calyptraea mamillaris*
- *Crucibulum spinosum**

Pelecypoda:
- *Anadaria mazatlanica**
- *Anadaris tuberculosa*
- *Leptopectin tumbezensis**
- *Anomia peruviana**
- *Polymesoda maritima*
- *Trachycardium procerum**
- *Pitar lupanaria**
- *Chione amuthusia*
- *Mulinia coloradensis**
- *Rangia mendica*
- *Tellina ecuadoriana**
- *Donax punctatostriatus*

*Indicates possible beach drift.

Yavaros region

Gastropoda:
- *Nerita funiculata*
- *Nerita scabricosta*
- *Cerithium stercusmuscarum*
- *Cerithidea mazatlanica*
- *Nassarius versicolor*
- *Fusinus dupetitthouarsi*

Pelecypoda:
- *Anadara* sp.
- *Mytella guyanensis*
- *Pitar concinnus*
- *Chione californiensis*
- *Chione gnidia*
- *Protothaca grata*

Glossary

NOTE: Definition of terms used in the keys or species descriptions, but not included in the glossary, may be found in the introductory section of their respective chapters.

— A —

Aboral. The region or end of an animal's body opposite its mouth.

Acontia. Coelenterates: internal, threadlike processes constituting a continuation of the middle lobe of a free-edged mesentery. The acontia are sites of enzyme production and intracellular digestion and absorption. They also may bear nematocysts.

Ala. A winglike projection or structure.

Allopatric. Referring to the geographical relationship of different populations of animals or plants, usually of two different species or subspecies, wherein the populations do not occur together; having different areas of distribution.

Ambulacral groove. Asteroids: a furrow along the oral surface of each ray of a sea star, sheltering the tube feet.

Ambulatory. Used for walking.

Amorphous. Shapeless, or lacking symmetry.

Anastomose. To branch and intertwine; interconnecting branches.

Annulated. Bearing ringlike segments, or bands.

Anterior. At or near the head or front of the body.

Anterolateral. The region where the front and sides join; as, on a crab, the angular or rounded meeting of the front and side of the carapace. (=anterio-lateral)

Asconoid. Sponges: the simplest and probably most primitive body plan, wherein the external ostia of the sponge body open directly into the spongocoel (via the incurrent canals), and little or no folding of the body wall has taken place.

Asymmetrical. Having no symmetry.

Athecate. Hydrozoa: when the hydranth or gastrozooid is *not* encased in a chitinous outer covering (the perisarc); gymnoblastic hydroids are athecate.

Autotomy. The process whereby an animal can "voluntarily" break off certain appendages, or break its body apart at certain locations; self-amputation.

— B —

Biramous. Consisting of or divided into two branches.

Blastostyle. Cnidaria: a modified zooid (gonozooid) from which the medusoids are budded; may or may not be protected by a surrounding theca.

Bouin's solution. A preservative/fixative made of 75 parts picric acid, 25 parts formalin, and five parts acetic acid.

Branchial. Of or pertaining to the gills.

Buccal. Of or pertaining to the mouth.

— C —

Calcareous. Composed of or containing large amounts of calcium carbonate ($CaCO_3$).

Calyptoblastea. Cnidaria; the order of thecate hydroids having each hydranth protected by a cuplike hydrotheca.

Campanulate. Bell-shaped.

Capitate tentacle. Cnidaria: short tentacles with a terminal knob, which is usually studded with nematocysts.

Carapace. Crustacea: the exoskeletal, dorso-lateral shield covering the cephalothorax of some crustaceans.

Caudal. Referring to the tail or posterior region of an animal.

Cephalic. Of or pertaining to the head or head region.

Cephalothorax. A body region combining the head and thorax as one.

Chelae. Crustacea: in strict sense, the first pair of thoracic appendages of decapod crustaceans, which are modified to form a pincer; in common usage, the term has come to mean the "hand region" or terminal part of the cheliped. (sing., chela)

Cheliped. Crustacea: the entire first thoracic leg of an arthropod (especially Crustacea) that is modified to form a pincer. (pl., chelipeds)

Chloretone. A chemical used as a relaxing agent for invertebrates (made by the Parke-Davis drug firm).

Circumtropical. Referring to a geographical distribution spread around the globe in the tropical regions.

Cirrus. A small, slender, and usually flexible appendage, generally threadlike. (pl., cirri).

Clavate. Club-shaped, gradually thickened near one end.

Clove oil. A chemical used as a relaxing agent for invertebrates and vertebrates.

Commensal. Members of two different species living in close association, with little or no mutual influence.

Cortex. An outer layer or outer region.

Cosmopolitan. Of world-wide distribution.

— D —

Dactylozooid. Cnidaria: the long nematocyst-bearing tentacles of some hydrozoans, such as the Portuguese Man-of-War, *Physalia* sp. The dactylozooid is a specific type of polyp used for defense and food capture. Hyman in 1940 described it as being derived from the gastrozooid (feeding zooid) by reduction or loss of mouth, tentacles, and gastric cavity. (= machozooid or protective polyp)

Denticulate. With teeth or small pointed projections. (dentate)

Depressed. Flattened.

Discoidal. Disklike.

Distal. Away from, especially from the place of attachment of an appendage, or an entire animal.

Dorsal. Referring to the back or back side.

— E —

Encrusting. Forming a crustlike layer on another object.

Endemic. Something that is constantly present in, and restricted to, a specific locality.

Endocoel. Cnidaria: referring to the paired septa of sea anemones and corals, the space between the members of a pair being the endocoel.

Epidermis. The outermost layer of cells of a plant or animal. In most invertebrates the epidermis is one cell layer thick.

Epsom Salts. A crude compound of magnesium sulfate. May be used for narcotizing and relaxing various invertebrates; $MgSO_4 \cdot 7H_2O$.

Euryhaline. Able to tolerate wide fluctuations in salinity, or osmotic pressures in general.

Eurythermic. Able to tolerate wide fluctuations in temperatures.

Eversible. Capable of turning inside out, or everting.

Exocoel. Cnidaria; referring to the region of paired septa of sea anemones and corals, the space between the pairs of septa being the exocoel.

Exoskeleton. Any invertebrate skeleton existing as the outermost covering of the body; principally in arthropods.

— F —

FAA. Formal-Acetic-Alcohol; 10 parts commercial formaldehyde, 50 parts alcohol, 2 parts acetic acid, and 38 parts water.

Flaccid. Soft; limp; flabby.

Filiform. Filamentous or threadlike.

Foliose. Leafy or leafy appearing. (= foliaceous)

Formaldehyde. A colorless, water-soluble, toxic gas (HCHO) derived by the oxidation of methyl alcohol. U.S.P. formaldehyde solution is 38 to 40 percent dissolved gas in water, or 38 to 40 percent saturation.

Formalin. An aqueous solution of 38 to 40 percent formaldehyde. (= commercial formaldehyde)

Fusiform. Spindle-shaped.

— G —

Gastrozooid. Cnidaria: a nutritive polyp. (= hydranth)

Globose. Spherical or globular.

Gonangium. Cnidaria: term applied to the reproductive polyp, consisting of the gonotheca together with its enclosed blastostyle and gonophores.

Gonotheca. Cnidaria: the vaselike case of periderm that surrounds the gonophores and blastostyles, the whole reproductive polyp being termed the gonangium.

Gonozooids. Cnidaria: modified polyps that bud off medusoids or their morphological equivalents, especially on hydroid colonies.

Gravid. With eggs.

Gymnoblastea. Cnidaria: the order of athecate hydroids. The polyps lack the skeletal cups of other hydroids into which the soft parts can be withdrawn.

— H —

Handbook. In taxonomy, a publication designed primarily as an aid to field and laboratory identification, as opposed to the presentation of new taxonomic conclusions (Mayr, 1969, *Principles of Systematic Zoology*).

Hexagonal. Having six angles and six sides.

Holotype. The single specimen designated or indicated as "the type" by the original author at the time of the publication of the original species description.

Hydrotheca. Cnidaria: the vaselike case of periderm (perisarc) that surrounds the feeding polyp of a hydroid, the entire feeding zooid being the hydranth or gastrozooid.

— I —

Imbricated. Overlapping in a regular order, like shingles.

Introvert. A closed tubular pocket cabable of being unrolled and extended inside out.

Infraspecific. Within the species; as, infraspecific variation.

— K —

Key. A tabulation of diagnostic characters for specific animals or animal groups (or plants) facilitating rapid identification.

Key character. In taxonomy, a character of special utility in a key or diagnosis of an animal (or plant).

— L —

Lamella. A thin, flat plate or scale. (pl., lamellae).

Lamellar. Bearing lamellae.

Lamellate. Consisting of any number of lamellae, in any type of arrangement.

Lateral. Referring to the side of an animal (or plant).

Leuconoid. Sponges: the most complex and probably most advanced body plan, wherein the sponge wall, the incurrent canals, and the radial canals have all undergone folding, giving the sponge a complex, asymmetrical body. The spongocoel has been reduced in these forms to numerous small water channels, and there are usually several to many oscula.

Local population. Individuals in a population of a given, specific locality which potentially form a single interbreeding community.

Lunule. Crescent-shaped, or approximating a crescent shape. In sand dollars, the elongate oval holes in the test or body.

— M —

Madreporite. Echinoderms: The button-shaped structure on the aboral surface of certain echinoderms that serves as the external connection of the water-vascular system.

Magnesium sulfate. A chemical often used for relaxing invertebrates prior to preservation; $MgSO_4$; available in drug stores.

Manus. Hand.

Median. Situated in or towards the plane which divides a bilaterally symmetrical organism or organ into right and left halves; centrally located.

Medusoid. Cnidaria: referring to an animal having the medusa form, which is bell or umbrella-shaped, like the jellyfish.

Megascleres. Sponges: the larger spicule types, comprising the major structural elements of most sponge skeletons.

Menthol. A chemical often used to relax or narcotize invertebrates; a colorless, crystalline, slightly water-soluble alcohol obtained from peppermint oil or synthesized; $CH_3C_6H_9(C_3H_7)OH$; available in drug stores.

Micron. One thousandth of a millimeter. ($= \gamma$)

Microscleres. Sponges; the smaller spicules of a sponge skeleton.

Monaxon spicule. Sponges: Spicules formed by growth in one or both directions along a *single* axis, which may be straight or curved; an unbranched spicule.

Monograph. In taxonomy, an exhaustive treatment of a higher taxon in terms of all available information pertinent to the group.

—N—

Narcotize. To subject an animal to a narcotizing agent, which is a chemical that induces torpor or relaxation of the individual.

Nematophore. Cnidaria: the small, highly modified zooid (dactylozooid) of protective function which is characteristic of the family Plumulariidae, a group of hydroids.

Niche. The precise group of environmental factors into which a species fits, or which is required by a species.

Nicotine. A chemical that may be used as a narcotizing agent for invertebrates; a colorless, oily, water-soluble, highly toxic liquid alkaloid, $C_{10}H_{14}N_2$.

—O—

Obovate. Inversely ovate.

Ocelli. Simple eyes (sing., ocellus).

Ommatidium. A compound eye.

Oral. Of or pertaining to the mouth.

Oral disk. Cnidaria: the flat surface surrounding the mouth of anthozoans, which bears the tentacles.

Oral papillae. Echinoderms: nipple-shaped projections around the mouth of ophiuroids.

Orifice. A mouthlike opening or aperture.

Osculum. Sponges: the opening to the spongocoel where the water leaves the body of the sponge.

Ostium. Sponges: the opening through which water initially enters the body.

Ovate. Egg-shaped.

Ovigerous. With eggs.

—P—

Papillate. Bearing many papillae. (= papillose)

Papilla. Small, nipplelike process or projection.

Pectinate. Comblike in appearance or structure.

Pedicellariae. Echinoderms: small, specialized jawlike appendages found on asteroids and echnoids that are used for protection and to capture small animals. Pedicellariae may be stalked or sessile. Stalked pedicellariae (as in the order Forcipulata of the asteroids) have the jawlike apparatus surmounted upon a short fleshy stalk, while the sessile pedicellariae (as in the order Phanerozonia of the asteroids) do not have the jaws mounted at the ends of flexible stalks.

Pelagic. Animals living in the open sea, as opposed to bottom dwellers. Pelagic animals may be free-swimming (nekton) or planktonic and may live at any depth, although deepwater forms are usually referred to as bathypelagic; some authors restrict the term "pelagic" to those animals living at or near the sea surface, for example to a depth of 100 fathoms or so.

Peltate. Shield- or leaf-shaped; having the stalk attached to the central portion of a flattened end.

Phenotype. The complete appearance of an individual as based on the totality of its characteristics. The phenotype is the result of the interaction between the genotype (the animal's heredity) and the environment.

Photonegative. Showing a negative reaction to light. An animal that attempts to maintain a position away from the sunlight is photonegative.

Photopositive. Showing a positive reaction to light. An animal that has a preference for light and will attempt to maintain a position where he receives a certain degree of sunlight is photopositive.

Pinnate. Resembling a feather; having side branches arranged on either side of a common stalk.

Plankton. Small plant and animal life floating in the upper layer of a body of water.

Podia. Echinoderms: tube feet.

Polymorphism. The simultaneous occurrence of several different phenotypes in a population.

Polyp. Cnidaria: an individual having a hollow cylindrical body, closed and attached at one end and open and free at the other, which bears the mouth. Examples of the polypoid form include sea anemones and individual members of a coral or hydroid colony.

Posterior. Toward the back or tail; opposite the head or anterior end.

Priority. The principle of taxonomy wherein of the two or more competing names for a single taxon, the name given that taxon is the one with priority, which is in almost all cases the name first validly published.

Proboscis. A long, flexible, anterior snout.

Proximal. Near or toward the point of origin, especially in regard to place of attachment of an appendage or an entire animal.

— R —

Race. In taxonomy, a subspecies.

Rachis. Cnidaria: the main axis or stem of a gorgonian coral.

Radial shield. Echinoderms: a calcareous plate at the base of each arm in Ophiuroids.

Ramus. A branch of something, as a ramus or branch of a thoracic appendage. (pl., rami)

Reticulate. Radiating ribs and concentric ridges criss-crossing at right angles, giving the surface a cross-hatched or netlike appearance.

Robust. Strong; healthy appearing; vigorous; well-formed.

Rostrum. Any snoutlike extension of the head of an animal.

Rugae. Folds or wrinkles.

Ruff. Resembling the old-time neckpieces or collars of lace (ruffs) worn in the 16th and 17th centuries. These were gathered and drawn into regular folds, and anything consisting of a series of regular folds is said to be ruffled; the circular disk around the mouth of some sea anemones is often incised and thrown into a series of rounded rays and is called a ruff.

— S —

Sedentary. Remaining stationary much of the time, but not necessarily permanently attached to the substrate.

Series. In taxonomy, the sample of animals (or plants) which a taxonomist collects in the field or has available for study.

Serrate. A margin that is notched or toothed.

Sessile. Permanently attached to the substrate.

Setae. Hairlike filaments.

Setaceous. Bearing setae.

Siphonoglyph. Cnidaria: In polyps the oral disk often has a ciliated groove, the siphonoglyph, at one or both ends of its elongated mouth. It is by means of these structures that water currents are directed into the interior.

Species. Groups of actually or potentially interbreeding natural populations which are reproductively isolated from other such groups (Mayr, 1969, *Principles of Systematic Zoology*).

Spongin. The flexible substance making up the skeleton of some sponges, composed of a fibrous protein related to keratin and collagen.

Steno. A prefix meaning narrow limits of tolerance or preference. Thus a stenohaline animal can withstand only a narrow fluctuation of salinities (or osmotic change) and a stenothermal animal can withstand only a narrow fluctuation in temperature.

Striations. Narrow lines, grooves, or ridges.

Sub. A prefix meaning near or almost; subequal means nearly equal though not necessarily exactly equal in all dimensions.

Sympatry. The occurrence of two or more populations in the same area.

Synconoid. Sponges: A secondary or intermediate degree of body wall folding wherein the size of the spongocoel has been decreased from that of the ascon types, and the choanocytes no longer line the spongocoel but are confined to specialized evaginations called radial canals or flagellated canals. The corresponding invagenations on the outside of the sponge are the incurrent canals. The two canal systems are connected by openings called prosopyles.

Synonym. In nomenclature, each of two or more different names for the same taxon.

Systematics. The science dealing with the diversity of organisms, generally involving taxonomy.

— T —

Taxon. A taxonomic group that is sufficiently distinct to be worthy of being distinguished by name and to be ranked in a definite category (Mayr, 1969, *Principles of Systematic Zoology*).

Taxonomic category. Designates rank or level in a hierarchic classification. It is a class, the members of which are all taxa assigned a given rank (Mayr).

Taxonomic character. Any attribute of a member of a taxon by which it differs or may differ from a member of a different taxon (Mayr).

Taxonomy. The theory and practice of classifying organisms.

Tentacle. Short or long, fleshy, flexible appendage on an animal, usually restricted to the anterior region.

Tetraxon spicule. Sponges: Sponge spicules that have four rays, each pointed in a different direction. The rays may be equal or one may be longer than the others.

Thecate. Cnidaria: hydroids that possess a protective, cuplike hydrotheca surrounding the hydranth; the Calyptoblastea are thecate hydroids.

Thigmotropism. An innate tendency to seek enclosing contact with a solid or rigid surface, as in a burrow. (An animal that is thigmotactic is one that displays thigmotropism.)

Tropicopolitan. Referring to a widespread geographical occurrence, encompassing the tropical regions throughout the world.

Truncate. Abruptly cut off, or cut short; an abrupt, flat ending.

Tube feet. Echinoderms: hollow extensile appendages connected to the water vascular system. In starfish and sea urchins they usually end in suckers and are locomotor, while in brittle stars and crinoids suckers are absent. In holothurians the ones around the mouth are used for feeding, and in crinoids they are ciliated and form part of the food-collecting mechanism.

Tubercle. A small knobby protuberance.

Turbinate. Whorled, usually in an inverse conical shape.

Type. The zoological (or botanical) object which serves as the base for the name of a taxon; the actual animal on which the name is based.

Type locality. The locality at which a type is collected (whether it be the holotype, lectotype, neotype, etc.).

— V —

Valve. One of the two halves of a clam shell.

Ventral. Bottom or belly surface; abdominal; opposite from dorsal.

Vermiform. Wormlike; long and slender like a worm.

Vernacular name. The colloquial designation of a taxon; the common name applied to an animal or group of animals, as opposed to its scientific name.

Verrucae. Cnidaria: protuberances along the column of some sea anemones, usually in longitudinal rows between the septal attachments. They are hollow projections of the body wall provided with a special musculature and having a concavity lined by a glandular epidermis, by which means they can attach bits of shell and sand to their bodies, presumably for some protective purpose.

— Z —

Zooid. The individual members of a colony or compound organism, having more or less an independent life of its own, i.e., the zooids of hydroid colonies, corals, bryozoans, ascidians, etc.

References

Introductory Material

ABBOTT, R. TUCKER
1958. Seashells of North America. Golden Press, New York. 280 p.

ALLEN, C. R.
1957. San Andreas Fault Zone in San Gorgonio Pass, southern California. Bull. Geol. Soc. America 68:315–350.

ALLEN, RICHARD K.
1969. Common intertidal invertebrates of southern California. Peek Publications, Palo Alto, California. 170 p.

ANDEL, T. H. VAN AND G. G. SHOR, JR., EDITORS
1964. Marine geology of the Gulf of California. Am. Assoc. Petroleum Geologists, Memoir 3.

ANDERSON, C. A., ET AL.
1950. The 1940 E. W. Scripps cruise to the Gulf of California. Geol. Soc. Am. Memoir 43.

ANDERSON, D. L.
1971. The San Andreas Fault. Scien. Amer., November, 53–68.

ARNAL, ROBERT E.
1957. Limnology, sedimentation, and microorganisms of the Salton Sea, California. Univ. So. Calif. Doctoral Dissertation, 222 p.

ARNOLD, AUGUSTA FOOTE
1916. The sea-beach at ebb-tide: A guide to the study of the seaweeds and the lower animal life found between tide-marks. Century, New York. 490 p.

AXELROD, DANIEL I.
1948. Climate and evolution in western North America during middle Pliocene time. Evolution 2:127–144.

BARNES, ROBERT D.
1969. Invertebrate Zoology. 2nd ed. W. B. Saunders Co., Philadelphia. 743 p.

BARRETT, E. C.
1957. Baja California, 1535–1956; a bibliography of historical, geographical, and scientific literature relating to the peninsula of Baja California and to the adjacent islands in the Gulf of California and the Pacific Ocean. Bennett and Marshall, Los Angeles.

BARTSCH, P.
1921. Ocean currents, the prime factor in the distribution of marine molluscs on the west coast of America. Bernice P. Bishop Mus. Spec. Pub. 7:505–526.

BERRY, S. S.
1956. A tidal flat on the Vermilion Sea. Journ. Conch. 24(3):81–84.
BLAKE, W. P.
1907. Lake Cahuilla: the ancient lake of the Colorado Desert. Nat. Geog. Mag. 18:830.
1914. The Cahuilla Basin and Desert of the Colorado. *In:* The Salton Sea, by W. T. MacDougal, Carnegie Inst., Wash. Publ. 193:1–12.
BORRADAILE, L. A., AND F. A. POTTS
1961. The Invertebrata. Cambridge Univ. Press. 820 p.
BROWN, JOHN S.
1922. Fault features of Salton Basin, California. Journ. Geol., 30:217–226.
1923. The Salton Sea region, California. U.S. Geol. Surv. Water Supply Paper 497. 436 p.
BRUSCA, RICHARD C.
1969. A distributional list of the Opisthobranchiate molluscs from the Pacific coast of North America, with some notes on faunal provinces. MS.
BYRNE, J. V., AND K. O. EMERY
1960. Sediments of the Gulf of California. Geol. Soc. Am. Bull. 71(7):983–1010.
CANNON, R.
1966. The Sea of Cortez. Lane Magazine and Book Co., Menlo Park, California. 283 p.
CLARK, B. L.
1921. The marine Tertiary of the west coast of the United States: its sequence, paleogeography, and the problems of correlation. Journ. Geol. 29:588–614.
CLEMONS, ELIZABETH
1964. Tidepools and beaches. Knopf, New York. 84 p.
COCKERELL, T. D. A.
1945. The Colorado Desert of California: its origin and biota. Kansas Acad. Sci. Trans. 48:1–39.
1946. The age of Lake Cahuilla. Science 103:235.
DALL, W. H.
1909. Report on a collection of shells from Peru, with a summary of the littoral marine Mollusca of the Peruvian zoological province. Proc. U.S. Nat. Mus. 37(1704):147–294.
DARLINGTON, PHILIP J.
1957. Zoogeography: the geographical distribution of animals. John Wiley, New York. 675 p.
DAWSON, E. YALE
1960. A review of the ecology, distribution and affinities of the benthic flora. *In* Symposium: The biogeography of Baja California and adjacent seas, Part II. Marine Biotas. Systematic Zoology 9(3):93–99.

DEEVEY, E. S.
1949. The biogeography of the Pleistocene. Bull. Geol. Soc. Am. 60:1315–1416.

DIBBLEE, T. W.
1954. Geology of the Imperial Valley region, California. Calif. State Div. Mines Bull. 170:21–28.

DURHAM, J. W.
1950. Cenozoic marine climates of the Pacific Coast. Bull. Geol. Soc. Am. 61:1243–1264.

DURHAM, J. W. AND E. C. ALLISON
1960. The geologic history of Baja California and its marine faunas. Systematic Zool. 9(2):47–91.

DURHAM, J. W., A. R. V. ARELLANO, AND J. H. PECK, JR.
1955. Evidence for no Cenozoic Isthmus of Tehuantepec seaways. Bull. Geol. Soc. Am. 66:977–992.

EKMAN, SVEN
1953. Zoogeography of the sea. Sidgwick and Jackson, London. 417 p.

EMERSON, W. K.
1958. Results of the *Puritan*-American Museum of Natural History Expedition to western Mexico. 1. General account. Am. Mus. Natur. Hist. Novit. 1894:1–25.

EMERY, K. O.
1960. The sea off southern California, a modern habitat of petroleum. John Wiley, New York. 366 p.

FARMER, WESLEY M.
1968. Tidepool animals from the Gulf of California. Wesword Co., San Diego, California. 68 p.

GARTH, J. S.
1955. The case for a warm-temperate marine fauna on the west coast of North America. *In* Essays in the natural sciences in honor of Captain Allan Hancock:19–27. Univ. So. Calif. Press.
1961. Distribution and affinities of the Brachyuran Crustacea. Systematic Zool. 9(3):105–123.

GAYMAN, W.
1968. Barrier formation in the Gulf of California. Unpublished Ph.D. dissertation, Univ. Calif., San Diego. (Scripps Institution of Oceanography).

GORSLINE, D. S.
1967. Sedimentologic studies of the Colorado Delta. Univ. So. Calif. Dep. Geol. Submitted to Geography Branch, Office of Naval Research, in partial fulfillment of the contractual requirements of Contract Nonr 228 (29), NR 388–079.

HAMILTON, E. L.
1956. Sunken islands of the mid-Pacific Mountains. Geol. Soc. Am. Memoir 64:1–97.

HAMILTON, W.
1961. Origin of the Gulf of California. Geol. Soc. Am. Bull. 72:1307–1318.
HASTINGS, J. R.
1964a. Climatological data for Baja California. Univ. Ariz. Institute of Atmospheric Physics, Technical Reports on the Meteorology and Climatology of Arid Regions 14.
1964b. Climatological data for Sonora and northern Sinaloa. Univ. Ariz. Institute of Atmospheric Physics, Technical Reports on the Meteorology and Climatology of Arid Regions 15.
HEDGPETH, JOEL W.
1953. An introduction to the zoogeography of the northwestern Gulf of Mexico with reference to the invertebrate fauna. Publ. Inst. Marine Sci. Univ. Texas 3:107–224.
1961. Common seashore life of southern California. Naturegraph Ocean Guidebooks. 65 p.
1962. Introduction to seashore life of the San Francisco Bay region and the coast of northern California. Univ. Calif. Press Natural History Guides 9. 136 p.
HEDGPETH, JOEL W., E. F. RICKETTS, AND J. CALVIN
1969. Between Pacific tides. Stanford University Press. 614 p.
HERTLEIN, L. G., AND E. K. FORDAN
1927. Paleontology of the Miocene of Lower California. Proc. Calif. Acad. Sci., ser. 4, 16:605–646.
HERTLEIN, L. G., AND W. K. EMERSON
1956. Marine Pleistocene invertebrates from near Puerto Penasco, Sonora, Mexico. Trans. San Diego Soc. Natur. Hist. 12:154–176.
HERZEN, R. P. VON
1963. Geothermal heat flow in the Gulfs of California and Aden. Scripps Inst. Oceanogr. Contr., n.s., 1504:343–344.
HILL, M. L.
1954. Tectonics of faulting in southern California. *In* Geology of Southern California, Calif. Dept. Natur. Resources, Div. Mines, Bull. 170, pt. 2, Ch. 4:5–14.
HOLBROOK, G. F.
1927. Probable future stages of Salton Sea. U.S. Geol. Surv., 27 p. (Typed report).
HUBBS, CARL L.
1948. Changes in the fish fauna of western North America correlated with changes in ocean temperature. Sears Found. Journ. Marine Res. 7:459–482.
1960. The marine vertebrates of the outer coast. *In* Symposium: The biogeography of Baja California and adjacent seas, Part II, Marine biotas. Systematic Zool. 9(3):134–147.

INTERNATIONAL BOUNDARY AND WATER COMMISSION, UNITED STATES AND MEXICO
1960. Western water bulletin — flow of the Colorado River and other western boundary streams and related data.
1961. Western water bulletin — flow of the Colorado River and other western boundary streams and related data.
1962. Western water bulletin — flow of the Colorado River and other western boundary streams and related data.

IVES, R. L.
1951. High sea-levels of the Sonoran shore. Amer. Journ. Sci. 249:215–223.
1959. Shell dunes of the Sonoran shore. Amer. Journ. Sci. 257(6):449–457.

JOHNSON, M. E., AND H. J. SNOOK
1967. Seashore animals of the Pacific coast. Dover Publ., New York. 659 p. (facsimile of original 1927 edition)

KEEN, A. MYRA
1937. An abridged check list and bibliography of west North American marine mollusca. Stanford Univ. Press. 87 p.
1956. Supplement to "An abridged check list and bibliography of west North American marine mollusca." (Papers published during the years 1937 to 1956). Mimeographed. Distributed by John Q. Burch and the author. 13 p.
1958. Sea shells of tropical west America. Stanford Univ. Press. 625 p.
1963. Marine molluscan genera of western North America. Stanford Univ. Press. 126 p.

KEEN, A. MYRA AND DON L. FRIZZELL
1939. Illustrated key to the west North American pelecypod genera. Stanford Univ. Press. 32 p.

KEEN, A. MYRA, AND JOHN C. PEARSON
1952. Illustrated key to west North American gastropod genera. Stanford Univ. Press. 39 p.

KING, P. B.
1958. Evolution of modern surface features of western North America. *In* C. L. Hubbs, ed., Zoogeography. Am. Assoc. Advanc. Sci. Publ. 51:3–60.

LARSON, R. L., H. W. MENARD, AND S. M. SMITH
1968. Gulf of California. A result of ocean-floor spreading and transform faulting. Science 161(3843):781–784.

LAWSON, L. M.
1916. The Yuma project silt problem. Reclamation Record 7:358–359.
1950. River and tide effects, lower Colorado River. Civil Eng. 20:389–391.

LIGHT, S. F.
1941. Laboratory and field text in invertebrate zoology. Berkeley: Assoc. Students Store. 232 p.
LIGHT, S. F., ET AL.
1954. Intertidal invertebrates of the central California coast. S. F. Light's "Laboratory and field text in invertebrate zoology" revised by Ralph I. Smith, Frank A. Pitelka, Donald P. Abbott, and Frances M. Weesner. Univ. Calif. Press. 446 p.
LONGWELL, C. R.
1954. History of the Lower Colorado River and the Imperial Depression. Calif. State Div. Mines Bull. 170:53–56.
LOWE, H. N.
1934. On the Sonoran side of the Gulf. Nautilus 48(1):1–46.
MACDOUGAL, D. T.
1906. The delta of the Rio Colorado. Am. Geogr. Soc. Bull. 38:1–16.
1914. The Salton Sea. Carnegie Inst. Wash. Publ. 193. 182 p.
MACGINITIE, G. E., AND NETTIE MACGINITIE
1968. Natural history of marine animals, second edition. McGraw-Hill, New York. 524 p.
MACMILLAN, D. H.
1966. Tides. American Elsevier Publ. Co., New York. 240 p.
MATTHEWS, J. B., D. A. THOMSON, AND S. R. BROWNING
1967. Tide calendar for Puerto Penasco, Mexico. Univ. Ariz., Tucson.
MATTHEWS, J. B.
1968. The tides of Puerto Penasco, Gulf of California. Typescript paper presented to Am. Geophys. Union, 49th Ann. Meeting, Washington, D.C.
n.d. (MS) Cooscillating tides of the Gulf of California.
MCLEAN, JAMES H.
1961. Marine molluscs from Los Angeles Bay, Gulf of California. Trans. San Diego Soc. Natur. Hist. 12(28):449–476.
1969. Marine shells of southern California. Los Angeles Co. Mus. Natur. Hist., Sci. Ser. 24, Zool. No. 11. 104 p.
MENARD, H. W.
1964. Marine geology of the Pacific. McGraw-Hill, New York. 270 p.
MINA, U. F.
1956. Bosquejo geologico de la parte sur de la peninsula de Baja California. Twentieth International Geol. Congr., Excursion A–7:1–79.
NELSON, EDWARD W.
1921. Lower California and its natural resources. Mem. Nat. Acad. Sci. 16:1–194.
NEWELL, I. M.
1948. Marine molluscan provinces of western North America: a critique and a new analysis. Proc. Am. Philos. Soc. 92:155–166.

NOBLE, L. F.
1954. The San Andreas fault zone from Soledad Pass to Cajon Pass, California. *In* Geology of southern California. Calif. Dept. Natur. Resources, Div. Mines Bull. 170, pt. 2, Ch. 4:37–48.

PARKER, R. H.
1963. Zoogeography and ecology of macro-invertebrates, particularly mollusks, in the Gulf of California and the continental slope off Mexico. Vidensk. Medd. fra Dansk Naturhist. Foren. 126:1–178.

PHLEGER, F. B., AND G. C. EWING
1962. Sedimentology and oceanography of coastal lagoons in Baja California, Mexico. Geol. Soc. Am. Bull. 73(2):145–181.

PICKENS, PETER E.
1970. A field guide to some of the marine invertebrates found in the northern Gulf of California. Ditto manuscript available from the Department of Biological Sciences, Univ. Ariz., Tucson.

PRECIADO, FRANCISCO
(1540) Francisco Preciado's account of the voyage of Francisco de Ulloa to the Gulf of California. *In:* Ramusio, Navigationi et Viaggi, 1st ed., 3:339–394. Venice, 1556.

REED, R. D.
1932. Geology of California. Am. Assoc. Petroleum Geologists, Tulsa. 355 p.

REISH, DONALD J.
1968. Marine life of Alamitos Bay. Printed by Seaside Printing Co., Long Beach, and distributed by Forty-Niner Shops, Inc. Calif. State College at Long Beach, Long Beach, California.

RODEN, G. I.
1964. Oceanographic aspects of Gulf of California. *In* T. H. van Andel and G. G. Shore, Jr., eds., Marine geology of the Gulf of California. Am. Assoc. Petroleum Geologists Mem. 3:50–58.

RODEN, G. I., AND G. W. GROVES
1959. Recent oceanographic investigations in the Gulf of California. Journ. Marine Research 18:10–35.

ROUND, F. E.
1967. The phytoplankton of the Gulf of California. Part I, Its composition, distribution and contribution to the sediments. Journ. Exp. Marine Biol. and Ecol. 1(1):76–97.

RUSNAK, G. A., AND R. L. FISHER
1964. Structural history and evolution of Gulf of California. *In:* T. H. van Andel and G. G. Shore, eds., Marine geology of the Gulf of California. Am. Assoc. Petroleum Geologists Mem. 3:144–156.

SCHENCK, H. G. AND A. MYRA KEEN
1936. Marine molluscan provinces of western North America. Proc. Am. Philos. Soc. 76:921–938.
1937. An index-method for comparing molluscan faunules. Proc. Am. Philos. Soc. 77:162–168.
1940. Biometrical analysis of molluscan assemblages. Soc. Biogeographique Mem. 7:379–392.
SCHMITT, WALDO L.
1940. The stomatopods of the west coast of America. Allan Hancock Pac. Exped. 5(4):129–225.
1965. Crustaceans. Univ. Mich. Press, Ann Arbor. 204 p.
SOULE, J. D.
1960. The distribution and affinities of the littoral marine Bryozoa (Ectoprocta). *In* Symposium: The biogeography of Baja California and adjacent seas, Part II. Marine Biotas. Systematic Zool. 9(3):100–104.
STEINBECK, JOHN, AND E. F. RICKETTS
1941. Sea of Cortez. A leisurely journal of travel and research. Viking Press, New York. 598 p.
SVERDRUP, H. U., M. W. JOHNSON, AND R. H. FLEMING
1942. The oceans: their physics, chemistry and biology. Prentice-Hall, Inc., New York. 1087 p.
SYKES, GODFREY
1914. Geographical features of the Cahuilla Basin. *In:* The Salton Sea, by D. T. MacDougal. Carnegie Inst. Wash. Publ. 193:13–20.
1926. The delta and estuary of the Colorado River. Geog. Rev. 16:232–255.
1937. The Colorado Delta. Carnegie Inst. Wash. Publ. 460. 193 p.
THOMPSON, ROBERT W.
1968. Tidal flat sedimentation on the Colorado River delta, northwestern Gulf of California. Geol. Soc. Am. Mem. 107.
THOMSON, D. A., J. B. MATTHEWS, AND S. R. BROWNING
1968. 1971 tide calendars for the northern Gulf of California. Univ. Ariz., Tucson.
THOMSON, D. A., ET AL.
1969. Environmental impact of brine effluents on Gulf of California. U.S. Dept. Interior, Research and Progress Report No. 387.
TOWNSEND, C. H.
1901. Dredging and other records of the U.S. Fish Commission steamer *Albatross,* with bibliography relative to the work of the vessel. Report of Commissioner, U.S. Commission of Fish and Fisheries, Pt. XXVI, for the year ending June, 1900. 175 p.
1916. Scientific results of the expedition to the Gulf of California in charge of C. H. Townsend, by the U.S. Fisheries steamship *Albatross* in 1911. I. Bull. Am. Mus. Natur. Hist. 35(24): 399–476.

U.S. COAST AND GEODETIC SURVEY
1956. Surface water temperatures at tide stations, Pacific coast of North and South America. USCGS, Washington, D.C., Spec. Publ. 31–4. 74 p.

U.S. GEOLOGICAL SURVEY
1954. Compilation of surface waters of the United States through September 1950, Part 9. Colorado River Basin Water-Supply Paper 1313, prepared under the direction of J. V. B. Wells, Chief, Surface Water Branch, USGS, Washington, D.C. 749 p.
1964. Compilation of records of surface waters of the United States, October 1950 to September 1960, Part 9, Colorado River basin, Water-Supply Paper 1733, prepared under the direction of E. L. Hendricks, Chief, Surface Water Branch, USGS, Washington, D.C. 586 p.

VIOLETTE, P. E.
1964. Shelling in the Sea of Cortez. D. S. King Publ., Tucson, Arizona. 95 p.

WALKER, BOYD W.
1960. The distribution and affinities of the marine fish fauna of the Gulf of California. *In* Symposium: The biogeography of Baja California and adjacent Seas, Part II. Marine Biotas. Systematic Zool. 9(3):123–133.

WALKER, BOYD W., EDITOR
1961. The ecology of the Salton Sea, California, in relation to the sport fishery. Calif. Dept. Fish and Game Fish Bull. 113. 204 p.

Porifera

BAKUS, G. J.
1966. Marine poecilosclcridan sponges of the San Juan Archipelago, Washington, Journ. Zool. Soc. Lond. 149:415–531.
1971. Marine Ecology course data sheet. Dept. Biol., Univ. So. California.

CARTER, H. J.
1882. Some sponges from the West Indies and Acapulco in the Liverpool Free Museum described, with general and classificatory remarks. Ann. Mag. Natur. Hist., ser. 5 VI:35–61.

DE LAUBENFELS, M. W.
1926. New sponges from California. Ann. Mag. Natur. Hist., ser. 9 XVII:567–573.
1927. The red sponges of Monterey Peninsula, California. Ann. Mag. Natur. Hist., ser. 9. XIX:258–266.
1930. The sponges of California. Stanford Univ. Bull. V(98):24–29.
1932. The marine and fresh-water sponges of California. Proc. U.S. Nat. Mus. 81(4):1–140.
1935. Some sponges of Lower California (Mexico). Am. Mus. Natur. Hist. Novit. 779. 14 p.

De Laubenfels, M. W. (cont'd.)

1936a. A discussion of the sponge fauna of the Dry Tortugas in particular and the West Indies in general, with material for a revision of the families and orders of the Porifera. Papers from the Tortugas Laboratory, Carnegie Inst., Vol. 30. 225 p.

1936b. A comparison of the shallow-water sponges near the Pacific end of the Panama Canal with those at the Caribbean end. Proc. U.S. Nat. Mus. No. 2993. 83:441–464.

1948. The order Keratosa of the phylum Porifera, a monographic study. Allan Hancock Pub. Occ. Papers 3. 217 p.

Dendy, A., and R. W. H. Row

1913. The classification and phylogeny of the calcareous sponges; with a reference list of all the described species, systematically arranged. Proc. Zool. Soc. London 704–813.

Dickinson, M. G.

1945. Sponges of the Gulf of California. Allan Hancock Pac. Exped. 11(1):1–251.

Hartman, W. D.

1958. A re-examination of Bidder's classification of the calcarea. Systematic Zool. 7(3):97–110.

Hentschel, E.

1923. Porifera. Handbuch der Zoologie von Kükenthal. Berlin and Leipzig. 1:307–418.

Hyatt, A.

1877. A monograph of North American Poriferae, part II. Mem. Boston Soc. Natur. Hist. 2(18):1–481.

Hyman, Libbie H.

1940. The Invertebrates, Vol. I. McGraw-Hill, New York: 284–364.

Lambe, L. M.

1894. Sponges from the western coast of North America. Trans. Roy. Soc. Canada 12:113–138.

Lendenfeld, R. von

1910a. The sponges I. The Geodidae. Mem. Mus. Comp. Zool. 41(1):1–260.

1910b. The sponges II. The Erylidae. Mem. Mus. Comp. Zool. 41(1):265–323.

Levi, C.

1957. Ontogeny and systematics in sponges. Systematic Zool. 6(4):174–183.

Minchen, E. A.

1900. Porifera. *In* E. Roy Lankester, A treatise on zoology. Part II: The Porifera and Coelenterata. Adam and Charles Black, London. 178 p.

Vosmaer, G. C. J.

1882–1886. Porifera (*In* Bronn's Klassen und Ordnung des Tierreichs II)

1885. The sponges of the "Willem Barents" expedition, 1880 and 1881. Bijdr. Dierk. 12:1–47.

VOSMAER, G. C. J. (cont'd.)
1928. Bibliography of sponges, 1551–1913. Edited by G. P. Bidder and C. S. Vosmaer-Roell. Cambridge Univ. Press. 234 p.

WILSON, H. V.
1904. The Sponges. No. 30 of reports on an exploration off the west coasts of Mexico, Ceneral and South America, and off the Galapagos Islands. Mem. Mus. Comp. Zool. 30(1):1-164.

Cnidaria

CARLGREN, OSKAR
1940. Eastern Pacific Expeditions of the New York Zoological Society. XIX. Actiniaria from the Gulf of California. Zoologica 25(2): 211–219.
1949. A survey of the Ptychodactiaria, Corallimorpharia, and Actiniaria. K. Sven. Vet. (4)1(1). 121 p.
1951. The Actinian fauna of the Gulf of California. Proc. U.S. Nat. Mus. 101(3282):415–449.

DURHAM, J. WYATT
1947. Corals from the Gulf of California and the North Pacific coast of America. Geol. Soc. Am. Mem. 20. 68 p.

DURHAM, J. W., AND J. L. BARNARD
1952. Stony corals of the Eastern Pacific collected by the *Velero III* and *Velero IV*. Allan Hancock Pac. Exped. 16(1):1–110.

FISHER, W. K.
1938. Hydrocorals of the north Pacific ocean. Proc. U.S. Nat. Mus. 84:493–554.

FRASER, C. MCLEAN
1911. The hydroids of the west coast of North America. Bull. Lab. Natur. Hist. Univ. Iowa 6(1).
1937. Hydroids of the Pacific coast of Canada and the United States. Univ. Toronto Press. 207 p.
1938a. Hydroids of the 1934 Allan Hancock Pacific Expedition. Allan Hancock Pac. Exped. 4(1):1–104.
1938b. Hydroids of the 1936 and 1937 Allan Hancock Pacific Expeditions. Allan Hancock Pac. Exped. 4(2):107–126.
1938c. Hydroids of the 1932, 1933, and 1935 Allan Hancock Pacific Expeditions. Allan Hancock Pac. Exped. 4(3):129–152.

GRAVIER, C. H.
1904. Recherches sur un cerianthaire pelagique du Golfe de Californie (*Dactylactis benedeni,* n. sp.), Ann. Sci. Natur. Zool., ser. 8, 20:253–292.

HAND, CADET
1954. The sea anemones of central California. Part I. The Corallimorpharian and Athenarian anemones. Wasmann Journ. Biol. 12(3):345–375.

HAND, CADET (cont'd.)

1955a. The sea anemones of central California. Part II. The Endomyarian and Mesomyarian anemones. Wasmann Journ. Biol. 13(1):37–99.

1955b. The sea anemones of central California. Part III. The Acontiarian anemones. Wasmann Journ. Biol. 13(2):189–251.

HYMAN, LIBBIE H.

1940. The Invertebrates: Vol. 1, Protozoa through Ctenophora. McGraw-Hill, New York. p. 364–661.

MCMURRICH, J. P.

1893. Scientific results of explorations by the U.S. Fish Commission steamer *Albatross,* No. XXIII. Report on the Actiniae collected by the United States Fish Commission steamer *Albatross* during the winter of 1887–1888. Proc. U.S. Nat. Mus. 16:119–216.

1904. The Actiniae of the Plate collection: Fauna Chilensis. Abhandlungen zur Kenntniss der Zoologie Chile. Zool. Jahrb. (6)3(2): 215–306.

1910. The Actiniaria of the Siboga Expedition. Part I: Ceriantharia. Siboga-Expeditie Monograph 15a (livr. 53):1–48.

PICKENS, PETER

1970. A field guide to some of the marine invertebrates found in the northern Gulf of California. Ditto manuscript available from the Department of Biology, Univ. Ariz., Tucson.

RUSSELL, F. S.

1953. The medusae of the British Isles: Anthomedusae, Leptomedusae, Limnomedusae, Trachymedusae, and Narcomedusae. Cambridge Univ. Press, Cambridge, England. 530 p.

SQUIRES, DONALD F.

1959. Results of the *Puritan*-American Museum of Natural History Expedition to western Mexico. 7. Corals and coral reefs in the Gulf of California. Bull. Am. Mus. Natur. Hist. 118(7): 367–432.

STEPHENSON, T. A.

1928. The British sea anemones. London: Ray Society. Vol. 1, 148 p.

1935. The British sea anemones. London: Ray Society. Vol. II, 426 p.

TORREY, HARRY B.

1902. The hydroida of the Pacific coast of North America. Univ. of Calif. Publ. Zool. 1:1–104.

TORREY, H. B., AND F. L. KLEEBERGER

1909. Three species of Cerianthus from southern California. Univ. Calif. Publ. Zool. 6(5):115–125.

VERRILL, A. E.

1866. On the corals and polyps of Panama with descriptions of new species. Proc. Boston Soc. Natur. Hist. 10:323–333.

1869. Review of the corals and polyps of the west coast of America. Trans. Conn. Acad. 1(6):377–558.

1869. On the geographic distribution of the polyps of the west coast of America. Trans. Conn. Acad. 1(7):558–570.

Platyhelminthes

BOCK, SIXTEN

1913. Studien uber Polycladen. Zoologiska Bidrag fran Uppsala. 2:31–344.

1925. Planarians, parts I to III and part IV. Vidensk. Medd. fra Dansk naturh. Foren. 79:1–84, 97–184.

BOONE, E. S.

1929. Five new polyclads from the California coast. Ann. Mag. Natur. Hist. (10)3:33–46.

HYMAN, LIBBIE H.

1939. Polyclad flatworms collected on the Presidential Cruise of 1938. Smithsonian Misc. Coll. 98(17):1–13.

1951. The Invertebrates: Vol. II, Platyhelminthes and Rhynchocoela. McGraw-Hill, New York. 52–458.

1953. The polyclad flatworms of the Pacific coast of North America. Bull. Am. Mus. Natur. Hist. 100(2):265–392.

1955. The polyclad flatworms of the Pacific coast of North America; additions and corrections. Am. Mus. Novit. 1704. 11 p.

KARLING, TOR G.

1962. Marine Turbellaria from the Pacific coast of North America. I: Plagiostomidae. Arkiv. f. Zool. (2)15(6):113–141. II: Pseudostomidae and Sylindrostomidae, *ibid.* (10):181–209.

1964. Marine Turbellaria from the Pacific Coast of North America. III: Otopanidae. Arkiv. f. Zool. 16(26):527–541.

MACGINITIE, G. E., AND NETTIE MACGINITIE

1968. Natural history of marine animals, second edition. McGraw-Hill, New York. 524 p.

WOODWORTH, W. MCM.

1894. Report on the Turbellaria (*Albatross* 1891). Bull. Mus. Comp. Zool. 25(4):49–52.

Nemertea

COE, W. R.

1905a. Nemerteans of the west and northwest coasts of America. Bull. Mus. Comp. Zool. 47:1–319.

1905b. Synopsis of North American Invertebrates. XXI, The nemerteans. Amer. Nature 39:425–447.

1940. Revision of the nemertean fauna of the Pacific coasts of North, Central, and northern South America. Allan Hancock Pac. Exped. 2(13):247–323.

1943. Biology of the nemerteans of the Atlantic coast of North America. Trans. Conn. Acad. Arts and Sci. 35:129–328.

1944. Geographical distribution of the nemerteans of the Pacific coast of North America, with descriptions of two new species. Journ. Wash. Acad. Sci. 34(1):27–32.

HYMAN, LIBBIE H.
1951. The Invertebrates: Vol. II, Platyhelminthes and Rhynchocoela. McGraw-Hill, New York. 459–532.

JOUBIN, L.
1905. Note sur quelques Nemertiens receuillis en Basse-Californie par M. Diguet. Bull. Mus. Hist. Natur. Paris 11:309–315.

Polychaeta

BERKELEY, E., AND C. BERKELEY
1939. On a collection of Polychaeta, chiefly from the west coast of Mexico. Ann. Mag. Natur. Hist. London, ser. 11. 3:321–346.
1948. Annelida, Polychaeta errantia. Canadian Pacific Fauna 9b(1). Fish. Res. Bd. Canada, Toronto. 100 p.
1952. Polychaeta Sedentaria. Canadian Pacific Fauna 9b(2). Fish. Res. Bd. Canada, Toronto. 139 p.

BUSH, KATHARINE J.
1904. Tubicolous annelids of the tribes Sabellides and Serpulides from the Pacific Ocean. Harriman Alaska Exped., N.Y. 12:169–355.

DALES, R. P.
1962. The polychaete stomodaeum and the inter-relationships of the families of polychaetes. Proc. Zool. Soc. London 139(3): 309–428.

DAY, J. H.
1967. A monograph on the Polychaeta of South Africa. 1. Errantia. 2. Sedentaria. Brit. Mus. Natur. Hist. London. Publ. 656. 878 p.

FAUCHALD, K.
1968. Onuphidae (Polychaeta) from western Mexico. Allan Hancock Monogr. Mar. Biol. 3. 82 p.
1970. Polychaetous annelids of the families Eunicidae, Lumbrineridae, Iphitimidae, Arabellidae, Lysaretidae, and Dorvilleidae from western Mexico. Allan Hancock Monogr. Mar. Biol. 5. 335 p.

FAUVEL, P.
1923. Polychètes errantes. Faune de France, Paris. Vol. 5. 488 p.
1927. Polychètes sedentaires. Faune de France, Paris. Vol. 16. 494 p.
1932. Annelida Polychaeta of the Indian Museum, Calcutta. Mem. Indian Mus. 12:1–262.

HARTMAN, OLGA
1939. Polychaetous annelids. Pt. 1. Aphroditidae to Pisionidae. Allan Hancock Pacific Exped. 7(1):1–156.
1940. Polychaetous annelids. Pt. 2. Chrysopetalidae to Goniadidae. Allan Hancock Pacific Exped. 7(3):173–287.
1941b. Polychaetous annelids. Pectinariidae, with a review of all species from the Western Hemisphere. Allan Hancock Pacific Exped. 7(5):325–345.

HARTMAN, OLGA (cont'd.)

1944a. Polychaetous annelids. Pt. 5. Eunicea. Allan Hancock Pacific Exped. 10(1):1–237.

1944c. Polychaetous annelids. Pt. 6. Paraonidae, Magelonidae, Longosomidae, Ctenodrilidae and Sabellariidae. Allan Hancock Pacific Exped. 10(3):311–389.

1947a. Polychaetous annelids. Pt. 7. Capitellidae. Allan Hancock Pacific Exped. 10(4):391–481.

1947b. Polychaetous annelids. Pt. 8. Pilargiidae. Allan Hancock Pacific Exped. 10(5):483–523.

1950. Polychaetous annelids. Goniadidae, Glyceridae, Nephtyidae. Allan Hancock Pacific Exped. 15(1):1–181.

1957. Orbiniidae, Apistobranchidae, Paraonidae and Longosomidae. Allan Hancock Pacific Exped. 15(3):211–393.

1968. Atlas of the Errantiate Polychaetous annelids from California. Allan Hancock Misc. Pub. 828 p.

1969. Atlas of the Sedentariate Polychaetous annelids from California. Allan Hancock Misc. Pub. 812 p.

MONRO, C. C. A.

1928. Papers from Dr. Th. Mortensen's Pacific Expedition, 1914–1916. XLV. On the Polychaeta collected by Dr. Th. Mortensen off the coast of Panama. Vidensk. Medd. Dansk. Naturh. Foren. 85:75–103.

REISH, D. J.

1968. A biological survey of Bahia de los Angeles, Gulf of California, Mexico. II. Benthic polychaetous annelids. Trans. San Diego Soc. Natur. Hist. 15(7):67–106.

RIOJA, E.

1941a. Estudios anelidológicos. II. Observaciones acerca de varias especies del genero *Hydroides* Gunnerus (*sensu* Fauvel) de las costas Mexicanas del Pacífico. Inst. Biol. México, An. 12: 161 175.

1941b. Estudios anelidológicos. III. Datos para el conocimiento de la fauna de poliquetos de las costas del Pacífico de México. Inst. Biol. México, An. 12:669–746.

1942a. Estudios anelidológicos. IV. Observaciones sobre especies de Serpulidos de las costas del Pacífico de México, con descripción de una especie nueva del género *Hydroides*. Inst. Biol. México, An. 13:125–135.

1942b. Estudios anelidológicos. V. Observaciones acerca de algunas especies del género *Spirorbis* Daudin, de las costas Mexicanas del Pacífico. Inst. Biol. México, An. 13:137–153.

1947a. Estudios anelidológicos. XVII. Contribución al conocimiento de los anelidos poliquetos de Baja California y Mar de Cortez. Inst. Biol. México, An. 18:197–224.

1947b. Estudios anelidológicos. XVIII. Observaciones y datos sobre algunas anelidos poliquetos del Golfo de California y costas de Baja California. Inst. Biol. México, An. 18:517–526.

RIOJA, E. (cont'd.)
1962. Estudios anelidológicos. XXVI. Algunos anelidos poliquetos de las costas del Pacífico de México. Inst. Biol. México, An. 33:131–229.
TREADWELL, A. L.
1937. The Templeton Crocker Expedition. Polychaetous annelids from the west coast of lower California, the Gulf of California and Clarion Island. Zoologica 22:139–160.
USCHAKOV, P. V.
1955. (Polychaetes from the seas in the far East). Akademia Nauk. SSSR. Opredeliteli po Faune USSR 56. 445 p. (In Russian).
WOODWICK, K. H.
1953. *Polydora nuchalis,* a new species of polychaetous annelid from California. Journ. Wash. Acad. Sci. 43(11):381–383.

Mollusca

ADAMS, ARTHUR, AND L. REEVE
1848–1850. Mollusca. *In* A. Adams, The zoology of the voyage of *H.M.S. Samarang;* under the command of Captain Sir Edward Belcher. London. 87 p.
ALLISON, E. C.
1964. Geology of areas bordering Gulf of California. Marine geology of the Gulf of California — a Symposium, Mem. 3, Am. Assoc. Petrol. Geol. 3–29.
BAKER, FRED, AND G. D. HANNA
1927. Marine Mollusca of the Order Opisthobranchiata. Proc. Calif. Acad. Sci., ser. 4. 16(5):123–135.
BAKER, FRED, G. D. HANNA, AND A. M. STRONG
1928. Some Pyramidellidae from the Gulf of California. Proc. Calif. Acad. Sci., ser. 4 17(7):205–246.
1930a. Some Rissoid Mollusca from the Gulf of California. Proc. Calif. Acad. Sci. 19(4):23–50.
1930b. Some Mollusca of the Family Epitoniidae from the Gulf of California. Proc. Calif. Acad. Sci. 19(5):41–56.
1938a. Some Mollusca of the Families Cerithiopsidae, Cerithiidae and Cyclostrematidae from the Gulf of California and adjacent waters. Proc. Calif. Acad. Sci. 23(15):217–244.
1938b. Columbellidae from western Mexico. Proc. Calif. Acad. Sci. 23(16):245–254.
BARTSCH, PAUL
1950. New West American turrids. Nautilus 63(3):87–97.
BEEMAN, R. D.
1963. Notes on the California species of *Aplysia* (Gastropoda: Opisthobranchia). Veliger 5(4):145–147.
BEQUAERT, J.
1942. Random notes on American Potamididae. Nautilus 56(1): 20–30.

BERGH, RUDOLPH
1894. Die Opisthobranchien. Bull. Mus. Comp. Zool., Harvard 25(10):125–233.

BERRY, S. S.
1912. A review of the cephalopods of western North America. Bull. Bur. Fisheries, Doc. 761. 30:269–336.
1953a. West American razor-clams of the genus *Ensis*. San Diego Soc. Natur. Hist. Trans. 11(15): 393–404.
1953c. Preliminary diagnoses of six West American species of Octopus. Leaflets in Malacol. 1(10):51–58.
1956b. Diagnoses of new Eastern Pacific chitons. Leaflets in Malacol. 1(13):71–74.
1956c. A new West Mexican prosobranch mollusk parasitic on Echinoids. Am. Midland Natur. 56(2):355–357.
1957. Notices of new eastern Pacific Mollusca. I. Leaflets in Malacol. 1(14): 75–82.
1958a. Notices of new eastern Pacific Mollusca. II. Leaflets in Malacol. 1(15):83–90.
1958b. West American molluscan miscellany. II. Leaflets in Malacol. 1(16):91–98.
1959. Notices of new eastern Pacific Mollusca. III. Leaflets in Malacol. 1(18): 108–113.
1960. Notices of new eastern Pacific Mollusca IV. Leaflets in Malacol. 1(19): 115–122.
1963a. Diagnoses of new eastern Pacific chitons II. Leaflets in Malacol. 1(22): 135–138.
1963b. Notices of new eastern Pacific Mollusca V. Leaflets in Malacol. 1(23): 139–146.
1964. Notices of new eastern Pacific Mollusca VI. Leaflets in Malacol. 1(24): 147–154.
1968a. Some unusual mollusks, mainly Panamic. Am. Malacol. Union Ann. Rept. for 1967: 71–72.
1968b. Notices of new eastern Pacific Mollusca VII. Leaflets in Malacol. 1(25): 155–158.
1969. Notices of new eastern Pacific Mollusca VIII. Leaflets in Malacol. 1(26): 159–166.

BERTIN, VICTOR
1878. Revision des tellinides du Museum d'Histoire Naturelle. Nouv. Arch. Mus. Hist. Natur., Paris, ser. 2. 1:201–361.
1881. Revision des donacidees du Museum d'Histoire Naturelle. Nouv. Arch. Mus. Hist. Natur., Paris, ser. 2. 4:57–121.

BERTSCH, HANS
1970a. *Dolabrifera dolabrifera* (Rang, 1828), range extension to the Eastern Pacific. Veliger 13(1):110–111.
1970b. Opisthobranchs from Isla San Francisco, Gulf of California, with the description of a new species. Santa Barbara Mus. Natur. Hist. Contr. Sci. 2. 16 p.

BERTSCH, HANS, AND A. A. SMITH
1970. Observations on opisthobranchs of the Gulf of California. Veliger 13(2):171–174.
BEU, ALAN G.
1970. The Mollusca of the subgenus *Monoplex* (Family Cymatiidae). Trans. Royal Soc. New Zealand, Biol. Sci. 11(17):225–237.
BINNEY, W. G.
1860. Notes on American shells. Proc. Acad. Natur. Sci. Philadelphia 12(6):150–154.
BOONE, L.
1928. Scientific results of the second oceanographic expedition of the "Pawnee," 1926. Mollusks from the Gulf of California and the Perlas Islands. Bull. Bingham Oceangr. Coll. Peabody Mus. Natur. Hist., Yale Univ. 2(5):1–17.
BRANN, DORIS C.
1966. Illustrations to "Catalogue of the collection of Mazatlan shells" by Philip P. Carpenter. Paleontol. Res. Inst., Ithaca, New York. 111 p.
BRATCHER, TWILA, AND R. D. BURCH
1970. Five new species of Terebra from the eastern Pacific. Veliger, 12(3):295–300.
BURCH, J. Q. AND ROSE L. BURCH
1962. New species of Oliva from west Mexico. Nautilus 75(4): 165–166.
1963. Genus *Olivella* in eastern Pacific. Nautilus 77(1):1–8.
1964. A new species of Sinum from the Gulf of California. Nautilus 77(4):109–110.
CAMPBELL, G. B.
1961. Colubrariidae (Gastropoda) of tropical West America, with a new species. Nautilus 74(4):136–142.
1962. A new deep-water *Anadara* from the Gulf of California. Veliger 4(3):152–154.
1964. New terebrid species from the eastern Pacific (Mollusca: Gastropoda). Veliger 6(3):132–138.
CARPENTER, P. P.
1855. List of four hundred and forty species of shells from Mazatlan. Rept. 24th Meeting, Brit. Assoc. Adv. Sci., for 1854 (Trans.): 107–108.
1856a. Descriptions of (supposed) new species and varieties of shells, from the Californian and west Mexican coasts, principally in the collection of Hugh Cuming, Esq. Proc. Zool. Soc. London, for 1855, pt. 23:228–235.
1856b. Notes on the species of Hipponyx inhabiting the American coasts, with descriptions of new species. Proc. Zool. Soc. London, for 1856, pt. 24:3–5.

CARPENTER, P. P. (cont'd.)
1860. Lectures on the shells of the Gulf of California. Ann. Rept. Smithsonian Inst. for 1859: 195–219.
1864a. Review of Prof. C. B. Adams's "Catalogue of the shells of Panama," from the type specimens. Proc. Zool. Soc. London, for 1863: 339–369. (Reprinted, Carpenter, 1872, p. 173–205.)
1864b. Diagnosis of new forms of mollusks collected at Cape St. Lucas, Lower California, by Mr. J. Xantus. Ann. Mag. Natur. Hist., ser. 3. 13(311–315 and 474–479) and 14(45–49).
1865a. Diagnoses of new forms of Mollusca from the west coast of North America, first collected by Colonel E. Jewett: Ann. Mag. Natur. Hist., ser. 3. 15(177–182 and 394–399).
1865b. Diagnoses de mollusques nouveaux provenant de Californie et faisant partie du Musee de l'Institution Smithsonienne. Journ. de Conchyl., ser. 3. 5(2):129–149.
1865c. Diagnoses of new species of mollusks, from the west tropical region of North America, principally collected by the Rev. J. Rowell, of San Francisco. Proc. Zool. Soc. London, for 1865, pt. 1:278–282.

CATE, C. N.
1961. Remarks on a variation in *Cypraea annettae* Dall, 1909. Veliger 4(2):112–114.
1969a. A revision of the eastern Pacific Ovulidae. Veliger 12(1): 95–102.
1969b. The eastern Pacific cowries. Veliger 12(1):103–119.

CATE, JEAN M.
1967. The radulae of nine species of Mitridae. Veliger 10(2): 192–195.

CATE, JEAN, AND CRAWFORD CATE
1962. The type of *Lamellaria sharonae* Willet, 1939 (Gastropoda). Veliger 5(2):91.

CERNOHORSKY, W. O.
1969. List of type specimens of Terebridae in the British Museum (Natural History). Veliger 11(3):210–222.

CHACE, E. P.
1958. A new mollusk from San Felipe, Baja California. Trans. San Diego Soc. Natur. Hist. 12(20):333–334.

CLARKE, A. H., AND R. J. MENZIES
1959. *Neopilina* (*Vema*) *ewingi*, a second living species of the Paleozoic Class Monoplacophora. Science 129(3355): 1026–1027.

COAN, E. V.
1962. Notes on some tropical West American mollusks. Veliger 5(2):92.
1965. Kitchen midden mollusks of San Luis Gonzaga Bay. Veliger 7(4):216–218.

COAN, E. V. (cont'd.)
1968. A biological survey of Bahia de los Angeles, Gulf of California, Mexico, III. Benthic Mollusca. Veliger 15(8):107–132.
1970. The date of publication of Gould's "Description of shells from the Gulf of California." Veliger 13(1):109.

COAN, E. V., AND B. ROTH
1965. A new species of *Persicula* from west Mexico. Veliger 8(2): 67–69.

COLLIER, C. L., AND W. M. FARMER
1964. Additions to the nudibranch fauna of the east Pacific and the Gulf of California. Trans. San Diego Soc. Natur. Hist. 13(19): 377–396.

CONRAD, T. A.
1849. The following new and interesting shells are from the coasts of Lower California and Peru, and were presented to the Academy by Dr. Thomas B. Wilson. Proc. Acad. Natur. Sci. Philadelphia 4:155–156.
1867. Descriptions of new West Coast shells. Amer. Journ. Conch. 3(2):192–193.

CONTRERAS, F.
1932. Datos para el estudio de los Ostines Mexicanos. An. Inst. Biol., Mexico 3(3):193–212.

COOPER, J. G.
1895. Catalogue of marine shells collected chiefly on the eastern shore of Lower California for the California Academy of Sciences during 1891–92. Proc. Calif. Acad. Sci., ser. 2. 5(34–48).

COASTA, O. G.
1844. Catalogo de testacei viventi nel piccolo e grande mare il Taranto redatto sul sistems del Lamarck. Ann. Accad. Sci. 5(2) (1843): 14–66.

CROSSE, H.
1861. Etude sur le genre Cancellaire, suivie du catalogue des especes vivantes et fossiles actuellement connues. Journ. de Conchyl. 9:220–256.

DALL, W. H.
1871. Descriptions of sixty new forms of mollusks from the west coast of America and the north Pacific Ocean, with notes on others already described. Am. Journ. Conch. 7(2):93–160.
1885. Comments, *In:* Notes on the mollusks of the vicinity of San Diego, Calif., and Todos Santos Bay, Lower California, by Charles R. Orcutt. Proc. U.S. Nat. Mus. 8:534–552.
1892. On some types new to the fauna of the Galapagos Islands. Nautilus 5(9):97–99.
1894. Synopsis of the Mactridae of Northwest America, south to Panama. Nautilus 8(4):39–43.

DALL, W. H. (cont'd.)

1895. New species of shells from the Galapagos Islands. Nautilus 8(11):126–127.

1896. Diagnoses of new species of mollusks from the west coast of America. Proc. U.S. Nat. Mus. 18(1034):7–20.

1908. Reports on the dredging operations off the west coast of Central America to the Galapagos, to the west coast of Mexico, and in the Gulf of California. XIV. The Mollusca and Brachiopoda. Bull. Mus. Comp. Zool., Harvard 43(6):205–487.

1909. Report on a collection of shells from Peru, with a summary of the littoral marine Mollusca of the Peruvian zoological province. Proc. U.S. Nat. Mus. 37(1704):147–294.

1910. New shells from the Gulf of California. Nautilus 24(3):32–34.

1912. New species of fossil shells from Panama and Costa Rica collected by D. F. MacDonald. Smithsonian Inst. Misc. Coll. 59, pt. 2(2077):1–10.

1913. Shells collected at Manzanillo, west of Mexico, Oct. 1910, by C. R. Orcutt, identified by W. H. Dall. Nautilus 26(12):143.

1916. Notes on the West American Columbellidae. Nautilus 30(3): 25–29.

1918a. Descriptions of new species of shells, chiefly from Magdalena Bay, Lower California. Proc. Biol. Soc. Wash. 31:5–8.

1918b. Pleistocene fossils of Magdalena Bay, Lower California, collected by Charles Russell Orcutt. Nautilus 32(1):23–26.

1921a. New fossil invertebrates from San Quentin Bay, Lower California. West Am. Scientist 19(2):17–18.

1921b. New shells from the Pliocene or early Pleistocene of San Quentin Bay, Lower California. West Am. Scientist 19(3):21–23.

D'ASARO, C. N.

1970. Egg capsules of some prosobranchs from the Pacific coast of Panama. Veliger 13(1):37–43.

DESHAYES, G. P.

1839. Nouvelles especes de mollusques, provenant des cótés de la Californie, du Méxique, du Kamtschatka et de la Nouvelle-Zélande. Révue zoologique, par la Société Cuvierienne 2:356–361 (Plates in Magasin de Zool.: 1840).

DURHAM, J. W.

1942. Four new gastropods from the Gulf of California. Nautilus 55(4):120–125.

1950. 1940 E. W. Scripps cruise to the Gulf of California. Part II. Megascopic paleontology and marine atratigraphy. Geol. Soc. Am. Mem. 43:1–216.

DURHAM, J. W., AND E. C. ALLISON

1960. The geologic history of Baja California and its marine faunas. *In* Symposium: The biogeography of Baja California and adjacent seas. Pt. I. Geologic History. Systematic Zool. 9(2): 47–91.

DuShane, Helen

1962. A checklist of mollusks for Puertecitos, Baja California, Mexico. Veliger 5(1):39–50.

1966. A rare Epitonium from the Gulf of California. Veliger 8(4): 311–312.

1969. A new genus and two new species of Typhinae from the Panamic province (Gastropoda: Muricidae). Veliger 11(4):343–344.

1971. The Baja California travels of Charles Russell Orcutt. Dawson's Book Shop, Baja California Travels Series 23. 75 p.

DuShane, Helen, and Twila Bratcher

1965. A new *Scalina* from the Gulf of California. Veliger 8(2): 160–161.

DuShane, Helen, and Ellen Brennan

1969. A preliminary survey of mollusks for Consag Rock and adjacent areas, Gulf of California, Mexico. Veliger 11(4):351–363.

DuShane, Helen, and J. H. McLean

1968. Three new epitoniid gastropods from the Panamic province. Contrib. Sci. Los Angeles Co. Mus. No. 145, 6 p.

DuShane, Helen, and R. Poorman

1967. A checklist of mollusks for Guaymas, Sonora, Mexico. Veliger 9(4):413–441.

DuShane, Helen, and G. G. Sphon

1968. A checklist of intertidal mollusks for Bahia Willard and the southwestern portion of Bahia San Luis Gonzaga, State of Baja California, Mexico. Veliger 10(3):233–246.

Eales, N. B. (Nellie)

1960. Revision of the world species of *Aplysia* (Gastropoda, Opisthobranchia). Bull. British Mus. (Natur. Hist.) Zool. 5(10) 268–404.

Emerson, W. K.

1951. A new scaphopod mollusk, *Cadulus austinclarki,* from the Gulf of California. Journ. Wash. Acad. Sci. 41(1):24–26.

1960. Results of the *Puritan*-American Museum of Natural History expedition to western Mexico. 11. Pleistocene invertebrates from Cerralvo Island. Am. Mus. Novit. 1995. 6 p.

Emerson, W. K., and Anthony D'Attilio

1970. *Aspella myrakeenae,* new species from western Mexico. Nautilus 83(3):88–94.

Emerson, W. K., and L. G. Hertlein

1960. Pliocene and Pleistocene invertebrates from Punta Rosalia, Baja California, Mexico. Am. Mus. Novit. 2004: 1–8.

1964. Invertebrate megafossils of the Belvedere expedition to the Gulf of California. Trans. San Diego Soc. Natur. Hist. 13(17): 333–368.

Emerson, W. K., and E. L. Puffer

1957. Recent mollusks of the 1940 "E. W. Scripps" cruise to the Gulf of California. Am. Mus. Novit. 1825. 57 p.

FARMER, W. F.
1963. Two new opisthobranch mollusks from Baja California. Trans. San Diego Soc. Natur. Hist. 13(6):81–84.
1967. Notes on the Opisthobranchia of Baja California, Mexico, with range extensions-II. Veliger 9(3):340–422.
1968. Tidepool animals from the Gulf of California. Wesword Co., San Diego, California. 70 p.

FARMER, W. F., AND C. L. COLLIER
1963. Notes on the Opisthobranchia of Baja California, Mexico, with range extensions. Veliger 6(2):62–63.

FISCHER-PIETTE, E., AND D. DELMAS
1967. Revision des mollusques lamellibranches du genre *Dosinia* Scopoli. Mem. Mus. Nat. d'Hist. Natur., n.s., Ser. A, Zool. 47(1). 92 p.

GIFFORD, D. S., AND E. W. GIFFORD
1951. Olive shells on opposite shores of the Gulf of California. Nautilus 65(2):55–56.

GOULD, A. A.
1853. Descriptions of shells from the Gulf of California and the Pacific coasts of Mexico and California. Boston Journ. Natur. Hist. 6:374–408.

GOULD, A. A., AND P. P. CARPENTER
1857. Descriptions of shells from the Gulf of California, and the Pacific coasts of Mexico and California. Part II. Proc. Zool. Soc. London, for 1856, part 24:198–208.

GRANT, U. S., IV, AND H. R. GALE
1931. Catalogue of the marine Pliocene and Pleistocene Mollusca of California and adjacent regions. Mem. San Diego Soc. Natur. Hist. 1:1–1036.

GRAU, GILBERT
1959. Pectinidae of the eastern Pacific. Univ. So. Calif. Publ., Allan Hancock Pacific Exped. 23. 308 p.

HANLEY, S. C. T.
1842–1856. An illustrated and descriptive catalogue of recent bivalve shells, London. 392 p.

HANNA, G. D.
1963. West American mollusks of the genus *Conus*. Occas. Papers Calif. Acad. Sci. 35. 103 p.

HANNA, G. D., AND L. G. HERTLEIN
1927. Expedition of the California Academy of Sciences to the Gulf of California in 1921: Geology and palentology. Proc. Calif. Acad. Sci. ser. 4. 16(6):137–157.

HANNA, G. D., AND A. M. STRONG
1949. West American mollusks of the genus *Conus*. Proc. Calif. Acad. Sci. ser. 4. 26(9):247–322.

HERTLEIN, L. G.
1934. Pleistocene mollusks from the Tres Marias Islands, Cedros Island, and San Ignacio Lagoon, Mexico. Bull. So. Calif. Acad. Sci. 33:59–73.
1941. A summary of the knowledge regarding the faunal area of tropical West America, with special reference to mollusks. Proc. dedicatory exercises, Hancock Hall, Univ. So. Calif. 21–24.
1951. Descriptions of two new species of marine pelecypods from west Mexico. Bull. So. Calif. Acad. Sci. 50(2):68–75.
1957. Pliocene and Pleistocene fossils from the southern portion of the Gulf of California. Bull. So. Calif. Acad. Sci. 56(2):57–75.
1958. Descriptions of new species of marine mollusks from west Mexico. Bull. So. Calif. Acad. Sci. 56(3):107–112.

HERTLEIN, L. G., AND W. K. EMERSON
1956. Marine Pleistocene invertebrates from near Puerto Penasco, Sonora, Mexico. Trans. San Diego Soc. Natur. Hist. 12(8): 154–176.
1959. Results of the *Puritan*-American Museum of Natural History expedition to western Mexico. 5. Pliocene and Pleistocene megafossils from the Tres Marias Island. Am. Mus. Novit. 1940. 15 p.

HERTLEIN, L. G., AND A. M. STRONG
1948. Descriptions of a new species of *Trophon* from the Gulf of California. Bull. So. Calif. Acad. Sci. 46(2):79–80.
1951a. Descriptions of three new species of marine gastropods from West Mexico and Guatemala. Bull. So. Calif. Acad. Sci. 50(2):76–80.
1951b. Descriptions of two new species of marine gastropods from West Mexico and Costa Rica. Bull. So. Calif. Acad. Sci. 50(3):152–155.

HIDALGO, J. G.
1893–1898. Obras malacologicas de J. G. Hidalgo. Pt. III. Descripcion de los moluscos recogidos por la comision cientifica envida por el gobierno espanol la America Meridional. Catalogo de los moluscos gasteropods marinos. Mem. Roy. Acad. Sci. Fis. Nat., Madrid, 19:332–432 and 433–608.

HILL, H. R.
1954. Variation in the olive shells of tropical West America. Nautilus 68(2):66–69.
1959. The cone shells of tropical west America. Veliger 2(2):30–32.
1960. Key to the cone shells of western tropical America. Veliger 2(3):51–53.

HOWARD, FAYE B.
1963. Notes on a *Mitrella* (Mollusca: Gastropoda) from the Gulf of California. Veliger 5(4):159–160.

JAECKEL, S.
1927. Zur marinen Molluskenfauna Sudkaliforniens. Zool. Anz. Leipzig 70:45–50.

KEEN, A. MYRA
1958a. New mollusks from tropical West America. Bull. Am. Paleont. 38(172):239–255.
1958b. Sea shells of tropical West America. Stanford Univ. Press. 627 p.
1959. Some side notes on "Seashells of Tropical West America." Veliger 2(1):1–3.
1962. A new west Mexican subgenus and new species of Montacutidae (Mollusca: Pelecypoda), with a list of Mollusca from Bahia de San Quintin. Pacific Natur. 3(9):321–328.
1963. Marine molluscan genera of western North America. Stanford Univ. Press. 126 p.
1971. Sea shells of tropical West America. Stanford Univ. Press. 1064 p.

LAMY, E.
1909. Gasteropodes recueillis par M. L. Diguet dans le Golfe de Californie. Bull. Mus. Hist. Natur., Paris. 15:264–270.
1916. Description d'un lamellibranche nouveau de Golfe de Californie. Bull. Mus. Hist. Natur., Paris. 22(8):443–445.

LANCE, J. R.
1962. A new species of *Armina* (Gastropoda: Nudibranchia) from the Gulf of California. Veliger 5(1):51–54.

LINDSAY, G. E.
1964. Sea of Cortez expedition of the California Academy of Sciences, June 20–July 4, 1964. Proc. Calif. Acad. Sci., ser. 4. 30(11):211–242.

LOWE, H. N.
1913. Shell collecting on the west coast of Baja California. Nautilus 27(3):25–29.
1933. At the head of the Gulf of California. Nautilus 47(2):45–47.
1935. New marine Mollusca from west Mexico, together with a list of shells collected at Punta Penasco, Sonora, Mexico. Trans. San Diego Soc. Natur. Hist. 8(6):15–34.

MACFARLAND, F. M.
1924. Expedition of the California Academy of Sciences to the Gulf of California in 1921. Opisthobranchiate Mollusca. Proc. Calif. Acad. Sci. ser. 4. 13(25):389–420.

MCLEAN, J. H.
1959. A new marine gastropod from west Mexico. Nautilus 73(1): 9–11.
1961. Marine mollusks from Los Angeles Bay, Gulf of California. Trans. San Diego Soc. Natur. Hist. 12(28):449–476.

MARCUS, EVELINE, AND ERNST MARCUS
1967. American Opisthobranch mollusks. Studies in Tropical Oceanography, no. 6: Institute of Marine Sciences, Univ. of Miami, Florida. 256 p. Pt. I: Tropical American Opisthobranchs, p. 1–138. Pt. II: Opisthobranchs from the Gulf of California, p. 141–248.

MARCUS, EVELINE AND ERNST MARCUS (cont'd.)
1970. Some gastropods from Madagascar and West Mexico. Malacologia 10(1):181–223.

MENKE, K. T.
1850–1851. Conchylien von Mazatlan, mit kritischen Anmerkungen. Zeitschr. f. Malakozool, yr. 7, p. 161–173.

MORRIS, PERCY
1966. A field guide to shells of the Pacific coast and Hawaii including shells of the Gulf of California. Houghton Mifflin Co., Boston. Ed. 2, rev. 297 p.

OLDROYD, IDA S.
1918. List of shells from Angel and Tiburon Islands, Gulf of California, with description of a new species. Nautilus 32(1):26–27.

OLSSON, A. A.
1961. Mollusks of the tropical eastern Pacific, particularly from the southern half of the Panamic-Pacific faunal province (Panama to Peru). Panamic-Pacific Pelecypoda. Paleontol. Res. Inst., Ithaca, N.Y. 574 p.

ORCUTT, C .R.
1918. Magdalena Bay shells. West Am. Sci. 22(1):8.

PALMER, KATHERINE E. V. W.
1951. Catalogue of the first duplicate series of the Reigen Collection of Mazatlan shells in the State Museum at Albany, New York. N.Y. State Mus. Bull. 342. 79 p.

PALMER, R. H., AND L. G. HERTLEIN
1936. Marine Pleistocene mollusks from Oaxaca, Mexico. Bull. So. Calif. Acad. Sci. 35:65–81.

PARKER, R. H.
1964. Zoogeography and ecology of some macro-invertebrates, particularly mollusks, in the Gulf of California and the continental slope off Mexico. Vidensk. Medd. fra. Dansk. naturh. Foren. Bd. 126. 178 p.

PERRIER, E., AND A. T. DE ROCHEBRUNE
1894. Sur un *Octopus* nouveaux (*O. digueti*) de la basse Californie, habitant les coquilles des mollusques bivalves. Comptes Rendus, Acad. Sci., Paris 118:770–773.

PILSBRY, H. A.
1910. A new species of *Marinula* from near the head of the Gulf of California. Proc. Acad. Natur. Sci. Philadelphia 62:148.
1932. A misidentified Lower California snail. Nautilus 45(3):105.

ROGERS, M. E.
1962. A new Gulf of California *Periploma*. Bull. So. Calif. Acad. Sci. 61(4):229–231.

RUSSELL, HENRY D.
1963. Notes on methods for the narcotization, killing, fixation, and preservation of marine organisms. Marine Biological Laboratory, Woods Hole, Mass. 70 p.

SHASKY, D. R.
1960. Deep water collecting off Guaymas, Mexico. Veliger 3(1): 22–23.
1961. New deep water mollusks from the Gulf of California. Veliger 4(1):18–21.
1970. New gastropod taxa from tropical western America. Veliger 13(2):188–195.

SHASKY, D. R., AND G. B. CAMPBELL
1964. New and otherwise interesting species of mollusks from Guaymas, Sonora, Mexico. Veliger 7(2):114–120.

SKOGLUND, CAROL
1970. An annotated bibliography of references to marine Mollusca from the northern state of Sonora, Mexico. Veliger 12(4): 427–432.

SMITH, A. G.
1961. Notes on the habitat of *Berthelinia* sp. nov. from the vicinity of La Paz, Baja California, Mexico. Veliger 3(3):81–82.
1963. A revised list of chitons from Guadalupe Island, Mexico. (Mollusca: Polyplacophora). Veliger 5(4):147–149.

SMITH, MAXWELL
1944. Panamic marine shells. Synonymy, nomenclature, range and illustrations. Tropical Photographic Laboratory, Winter Park, Florida. 127 p.
1950. New Mexican and Panamic shells. Nautilus 64(2):60–61.

SORENSON, A.
1942a. Collecting in Mexico. Nautilus 55(4):113–115.
1942b. Traveling and collecting in Mexico. Nautilus 57(1):1–5.

SPHON, G. G.
1961. Notes on the Mitridae of the eastern Pacific. I. *Mitra fultoni* E. A. Smith. Veliger 4(1):32–36.

STEARNS, R. E. C.
1879. Description of new West American land, fresh-water and marine shells, etc. Scientific results of explorations by the U.S. Fish Commission Steamer *Albatross*. Proc. U.S. Nat. Mus. 13:205–225.
1894. The shells of the Tres Marias and other localities along the shores of Lower California and the Gulf of California. Proc. U.S. Nat. Mus. 17:139–204.

STRONG, A. M.
1949. Additional Pyramidellidae from the Gulf of California. Bull. So. Calif. Acad. Sci. 48(2):71–93.

STRONG, A .M., AND G. D. HANNA
1930. (a) Marine Mollusca of the Revillagigedo Island, Mexico; (b) Marine Mollusca of the Tres Marias Islands, Mexico. Proc. Calif. Acad. Sci., ser. 4. 19(2): 7–12 and 19(3): 13–22.

TAYLOR, D. W., AND N. F. SOHL
1962. An outline of gastropod classification. Malacologia 1(1):7–32.

VERRILL, A. E.
1870. Descriptions of shells from the Gulf of California. Am. Journ. Sci., ser. 2. 49(146):217–227.

VIOLETTE, P. E.
1964. Shelling in the Sea of Cortez. King, Tucson, Ariz. 96 p.

VOKES, H. E., AND EMILY H. VOKES
1962. Pelecypods from Barra de Navidad, Mexico. Nautilus 76(2): 61–63.

WILLET, GEORGE
1938. A new *Pseudochama* from Clarion Island, Mexico. Nautilus 52(2):48–49.

WOLFSON, FAY H.
1962. Comparison of two similar species of *Conus* (Gastropoda) from the Gulf of California. Veliger 5(1):23–28.
1968. Spawning notes, I. *Hexaplex erythrostomus*. Veliger 10(3): 292.
1969a. Spawning notes, II. *Mitra dolorosa*. Veliger 11(3):282–283.
1969b. Spawning notes, III. *Strombina maculosa*. Veliger 11(4): 440–441.
1969c. Spawning notes, IV. *Cerithium stercusmuscarum*. Veliger 11(4):441–442.
1970. Spawning notes, V. *Acanthina angelica* I. Oldroyd, 1918, and *Acanthina lugubris* (Sowerby, 1821). Veliger 12(3):375–377.

Sipunculoids

CHAMBERLIN, RALPH V.
1919. Notes on the Sipunculida of Laguna Beach. Pomona College Journ. Ent. and Zool. 12:30–31.

EDMONDS, S. J.
1955. Australian Sipunculoidea, I. Austral. Journ. Mar. Freshwater Research, Vol. 6.
1956. Australian Sipunculoidea II. Austral. Journ. Mar. Freshwater Research, Vol. 7.

FISHER, W. K.
1928. New Sipunculoidea from California. Ann. Mag. Natur. Hist. Ser. 10, 1:194–199.
1950. The sipunculid genus Phascolosoma. Ann. Mag. Natur. Hist. (12)3(30):547–552.
1952. The sipunculid worms of California and Baja California. Proc. U.S. Nat. Mus. 102:371–450.

GEROULD, J. H.
1913. The sipunculids of the eastern coast of North America. Proc. U.S. Nat. Mus. 44:373–437.

HYMAN, LIBBIE H.
1959. The Invertebrates: Vol. V, smaller coelomate groups. McGraw-Hill, New York, p. 610–696.
SELENKA, E., J. DE MAN, AND C. BULOW
1883. Die Sipunculiden. Reis. Arch. Phil. Semper., Teil II, Bd. IV, Abt. 1:XXXII. 131 p.

Echiuroidea

Specific references for the echiuroid fauna of the Gulf are lacking, and few works on the North American spoon worms in general exist. What follows is a list of some of the more significant general papers for this group. The spoon worm *Urechis caupo* has probably undergone more extensive research than any other echiuroid, especially by Dr. G. E. MacGinitie from 1928 to 1938.

CHAMBERLIN, R. V.
1920. The Gephyrea. Report of the Canadian Arctic expedition 1913 to 1918, pt. D.
FISHER, W. K.
1946. Echiuroid worms of the north Pacific Ocean. Proc. U.S. Nat. Mus. 96(3198):215–292.
FISHER, W. K., AND GEORGE E. MACGINITIE
1928. The natural history of an echiuroid worm. Ann. Mag. Natur. Hist., ser. 10. 1:199–204.
GISLEN, TORSTEN
1940. Investigations on the ecology of *Echiurus*. Lunds Univ. Arsskr., new ser. 36(10):1–36.
HYMAN, LIBBIE H.
1940. The Invertebrates: Vol. 1. Protozoa through Ctenophora. McGraw-Hill, New York. 726 p.
SATO, HAYAO
1939. Studies on the Echiuroidea, Sipunculoidea and Priapuloidea of Japan. Sci. Rep. Tohoku Imp. Univ. 4(14):339–460.
SELENKA, EMIL
1885. Report on the Gephyrea collected during the voyage of *H.M.S. Challenger*. Challenger Reports 13:1–24.
WILSON, C. B.
1900. Our North American echiurids. Biol. Bull. 1:163–178.

Bryozoa

CANU, F., AND R. S. BASSLER
1930. The Bryozoan fauna of the Galapagos Islands. Proc. U.S. Nat. Mus. 76:1–78.

HASTINGS, ANNA B.
1930. Cheilostomatous polyzoa from the vicinity of the Panama Canal collected by Dr. C. Crossland on the cruise of the S. Y. "St. George." Proc. Zool. Soc. London 4:697–740.
LEVINSEN, G. M. R.
1909. Morphological and systematic studies on the Cheilostomatous Bryozoa. Copenhagen. 431 p.
OSBURN, RAYMOND C.
1940. Bryozoa of Porto Rico with a resume of the West Indian Bryozoan fauna. New York Acad. Sci., Sci. Surv. Porto Rico and Virgin Is. 16(3):321–468.
1947. Bryozoa of the Allan Hancock Atlantic Expedition, 1939. Report No. 5:1–66.
1950. Bryozoa of the Pacific coast of America. Part I. Cheilostomata-Anasca. Allan Hancock Pac. Exped. 14(1):1–269.
1952. Bryozoa of the Pacific coast of America. Part II. Cheilostomata-Ascophora. Allan Hancock Pac. Exped. 14(2):270–611.
1953. Bryozoa of the Pacific coast of America. Part III. Cyclostomata, Ctenostomata, Entoprocta, and Addenda. Allan Hancock Pac. Exped. 14(3):612–841.
ROBERTSON, ALICE
1900. Studies in Pacific Coast Entoprocta. Proc. Calif. Acad. Sci., Zool. 2. 4:323–348.
1905. Non-incrusting chilostomatous Bryozoa of the west coast of North America. Proc. Calif. Acad. Sci., Zool. 2. 5:235–320.
1908. The incrusting chilostomatous Bryozoa of the west coast of North America. Proc. Calif. Acad. Sci., Zool. 4. 5:253–344.
1910. The cyclostomatous Bryozoa of the west coast of North America. Proc. Calif. Acad. Sci. 6. 12:225–284.

Brachiopods

I have found no literature for the Gulf of California pertaining specifically to brachiopods. For a complete bibliography, see Hyman (1959).

DALL, W. H.
1920. Annotated list of the recent Brachiopoda in the collection of the United States National Museum. Proc. U.S. Nat. Mus. 57:261–377.
ELLIOTT, C.
1951. Geographical distribution of terebratelloid Brachiopoda. Ann. Mag. Natur. Hist., ser. 12, 4.
HYMAN, L. H.
1959. The Invertebrates, vol. V. McGraw-Hill, New York, 516–609.

JONES, G. F., AND J. L. BARNARD
1963. The distribution and abundance of the inarticulate brachiopod *Glottidia albida* (Hinds) on the mainland shelf of southern California. Pac. Natur. 4(2):27–52.

OLDROYD, I. S.
1924. The marine shells of the west coast of North America. Stanford Univ. Publ. Geol. Sci. 1:221–232.

Echinodermata

AGASSIZ, A.
1847–1872. Revision of the Echini (Illus. Cat. Mus. Comp. Zool. 7). Mem. Mus. Comp. Zool. 3(1). 796 p.
1898. Reports on the dredging operations off the west coast of Central America, Mexico and the Gulf of California by the *U.S.S. Albatross* during 1891. Bull. Mus. Comp. Zool. 32(5):71–86.
1904. Panamic deep sea Echini collected by the *U.S.S. Albatross* in 1891. Am. Mus. Comp. Zool. 31.

BOOLOOTIAN, R. A., AND D. LEIGHTON
1965. Brittle stars of southern California. Los Angeles Co. Mus. Publ. Sci. 93. 20 p.

BOONE, LEE
1928. Echinoderms from the Gulf of California and the Perlas Islands. Bull. Bingham Ocean. Coll. 2(6):1–14.
1933. Coelenterata, Echinodermata, and Mollusca. Bull. Vanderbilt Marine Mus. 4:1–217.

CASO, M. E.
1941. Ofivroides de Mexico. Anal. Inst. Biol. Mex. 12.
1946. Equinoideos de Mexico. Anal. Inst. Biol. Mex. 17:247–259.
1948a. Contribucion al conocimiento de los Equiaoideos de Mexico. Anal. Inst. Biol. Mex. 19.
1948b. Contribucion al conocimiento de los Holoturoideos de Mexico, III. Algunas especies de Holoturoideos litorales de al costo Pacifica Mexicana. Anal. Inst. Biol. Mex. 19.

CLARK, H. L.
1912. The Pedinidae and Echinometridae of Hawaiian and other Pacific Echini. Mem. Mus. Comp. Zool. 34(4):1–180.
1913. Echinoderms from Lower California, with descriptions of new species. Bull. Am. Mus. Natur. Hist. 32:185–236.
1914. The Clypeastridae and Scutellidae of Hawaiian and other Pacific Echini. Mem. Mus. Comp. Zool. 46(1):1–80.
1915. Catalogue of recent ophiurans, based on the collections of the Museum of Comparative Zoology. Mem. Mus. Comp. Zool. 25(4):1–252.
1917a. Ophiuroidea. Bull. Mus. Comp. Zool. 61:429–453.
1917b. The Echinoneidae and Spatangidae of Hawaiian and other Pacific Echini. Mem. Mus. Comp. Zool. 46(2):1–204.

CLARK, H. L. (cont'd.)
1920. Holothurioidea. Mem. Mus. Comp. Zool. 39(4):1–40.
1923. Supplementary report on Echinoderms from Lower California. Bull. Am. Mus. Natur. Hist. 48:147–163.
1925. A catalogue of the recent sea urchins (Echinoidea). British Museum. 250 p.
1939. Echinoderms (other than holothurians) collected on the Presidential Cruise of 1938. Smithsonian Misc. Coll. 98(11):1–18.
1940. Notes on echinoderms from the west coast of Central America. Eastern Pacific Exped. of the New York Zoological Society No. 21. Zoologica 25:331–352.
1948. A report on the Echini of the warmer eastern Pacific, based on the collections of the *Velero III*. Allan Hancock Pac. Exped. 8(5):225–351.

DEICHMANN, E.
1937. Holothurians from the Gulf of California, the west coast of Lower California and Clarion Island. Zoologica 22:161–176.
1938. Holothurians from the western coasts of Lower California and Central America, and from the Galapagos Islands. Zoologica 23:361–387.
1941. The Holothurioidea collected by the *Velero III* during the years 1932 to 1938. Part I, Dendrochirota. Allan Hancock Pac. Exped. 8(3):61–195.
1958. The Holothurioidea collected by the *Velero III* and *IV* during the years 1932 to 1954. Part II, Aspidochirota. Allan Hancock Pac. Exped. 11(2):251–358.

DURHAM, J. WYATT
1955. Classification of clypeasteroid echinoids. Univ. Calif. Pub. Geol. Sci. 31(4):73–198.

FELL, H. B.
1963. The phylogeny of sea stars. Phil. Trans. Royal Soc. London, B. 246:381–435.

FISHER, WALTER K.
1928a. Asteroidea of the North Pacific and adjacent waters. Bull. U.S. Nat. Mus. 76(2):1–245.
1928b. Asteroidea of the North Pacific and adjacent waters. Bull. U.S. Nat. Mus. 76(3):1–356.

GLIDDEN, M. L.
1967. Commensal relationships between the sunstar *Heliaster kubinijii* Xantus and an ophelid polychaete. Part II. Univ. Arizona Biol. Studies, Gulf of Calif. IV(2). 11 p.

HARVEY, ETHEL B.
1956. The American Arbacia and other sea urchins. Princeton Univ. Press. 298 p.

HEDING, S. G.
1828. Synapidae. Vidensk. Medd. fra Dansk. Naturh. Foren. 85:105–323.

KOEHLER, RENE
1914. A contribution to the study of Ophiurans of the United States. U.S. Nat. Mus. Bull. 84:1–173.

LEIPOLDT, F.
1895. Asteroidea de *Vettor Pisani* expedition (1882–1885). Zeit. f. wiss. Zool. 59:545–654.

LINDSAY, G. E.
1964. Expedition to the Sea of Cortez. Proc. Calif. Acad. Sci. (4)30:211–242.

LUTKEN, C. F., AND TH. MORTENSEN
1899. The Ophiuridae. Mem. Mus. Comp. Zool. 23(2):1–116.

McCLENDON, J. F.
1909. The ophiurians of the San Diego region. Univ. Calif. Publ. Zool. 6:33–64.

MORTESEN, TH.
1928–1951. A monograph of the Echinoidea. Vols. I–V. Carlsberg Found., Copenhagen.

NEILSEN, EIGEL
1932. Ophiurans from the Gulf of Panama, Gulf of California and the Straits of Georgia. Vidensk. Medd. fra Dansk Naturh. Foren. 91:241–346.

NICHOLS, DAVID
1962. Echinoderms. Hutchinson Co., Ltd., London. 200 p.

SELENKA, E.
1867. Beitrage zur anatomie und systematik der Holothurien. Zeit f. Zool. 17:29–374.

TANZER, E. C.
1967. Commensal relations between the sunstar *Heliaster kubinijii* Xantus and an ophelid polychaete. Part I. Univ. Arizona Biol. Studies, Gulf of Calif., IV(2). 12 p.

VERRILL, A. E.
1867. Notes on the echinoderms of Panama and the west coast of America, with descriptions of new species. Trans. Conn. Acad. 1:251–322.
1868. Notice of a collection of echinoderms from La Paz, Lower California, with descriptions of a new genus. Trans. Conn. Acad. 1:371–376.
1870. Descriptions of echinoderms and corals from the Gulf of California. Am. Journ. Sci. (2), 49:93–100.
1871. The echinoderm fauna of the Gulf of California and Cape San Lucas. Trans. Conn. Acad. 1:593–596.

ZEISENHENNE, FRED C.
1937. Echinoderms from the west coast of Lower California, the Gulf of California, and Clarion Island. Zoologica 22:209–239.
1940. New ophiurians of the Allan Hancock Pacific Expeditions. Allan Hancock Pac. Exped. 8(2):9–42.
1942. New eastern Pacific sea stars. Allan Hancock Pac. Exped. 8(4):197–224.

Enteropneusts

The literature does not deal specifically with the Gulf of California enteropneust fauna. What follows is a list of some of the more significant general papers dealing with this group. Drs. T. H. Bullock, W. E. Ritter, and K. P. Rao are, at the time of this writing, finishing the manuscript for their much anticipated monograph on the Enteropneusta.

BARRINGTON, E. J. W.

1965. The biology of the Hemichordata and Protochordata. W. H. Freeman Co., San Francisco (University Reviews in Biology). 176 p.

BRAMBELL, F. W. R., AND H. A. COLE

1939. The preoral ciliary organ of the Enteropneusta: its occurrence, structure, and possible phylogenetic significance. Proc. Zool. Soc. London, B. 109:181–193.

BULLOCK, T. H.

1940. The functional organization of the nervous system of Enteropneusta. Biol. Bull. 79:91–113.

1944. The giant nerve fiber system in balanoglossids. Journ. Comp. Neurol. 80:355–367.

1945. The anatomical organization of the nervous system of Enteropneusta. Quart. Journ. Micr. Sci. 86:55–111.

BURDON-JONES, C.

1951. Observations on the spawning behavior of *Saccoglossus horsti* Brambell and Goodhart, and other Enteropneusts. Journ. Marine Biol. Assoc. U. K. 29:625–638.

1952. Development and biology of the larva of *Saccoglossus horsti* (Enteropneusta). Phil. Trans. Royal Soc. B 236:553–590.

DAVIS, B. N.

1908. The early life history of *Dolichoglossus pusillus* Ritter. Univ. Calif. Publ. Zool. 4(30):187–226.

DAWYDOFF, C.

1948. Stomocordes. *In* Traete de Zoologie, P. Grasse, ed. Paris: Masson, p. 367–532.

HORST, C. J., VAN DER

1930. Observations on some Enteropneusta. No. 51: Papers from Dr. Th. Mortensen's Pacific Expedition, 1914–1916. Vidensk. Medd. fra Dansk Naturh. Foren. 87:135–200.

1934–1939. Hemichordata. *In* Bronn, Klassen und Ordnungen des Tierreichs 4:4:2:2, p. 1–737.

HYMAN, LIBBIE

1959. The Invertebrates: vol. V. Smaller Coelomate groups. McGraw-Hill, New York, 783 p.

KNIGHT-JONES, E. W.

1952. On the nervous system of *Saccoglossus cambrensis* (Enteropneusta). Phil. Trans. Royal Soc. B 236:315–354.

KNIGHT-JONES, E. W. (cont'd.)
1953. Feeding in *Saccoglossus* (Enteropneusta). Proc. Zool. Soc. London, 123:637–654.
NEWELL, G. E.
1952. The homology of the stomochord of the Enteropneusta. Proc. Zool. Soc. 121:741–746.
RITTER, W. E.
1894. On a new *Balanoglossus* larva from the coast of California, and its possession of an endostyle. Zoologica 1(5):171–210.
1902. The movements of the Enteropneusta and the mechanism by which they are accomplished. Biol. Bull. 3:255–261.
RITTER, W. E., AND B. N. DAVIS
1904. Studies on the ecology, morphology and speciology of young of some Enteropneusta of western North America. Univ. Calif. Publ. Zool. 1:171–210.
TREWAVAS, ETHYLWYNN
1931. Enteropneusta. British Museum, Great Barrier Reef Expedition 4(2):39–67.

Ascidiacea

There are virtually no good references on the ascidians of the Gulf of California. Major references for this group in western North America are as follows:

ARNBACK, C. A.
1929. Chilean tunicates. Ascidians from the Guaitecas Islands. Arkiv f. Zool. 21a(6):1–27.
HERDMAN, W. A.
1898. Descriptions of some simple ascidians collected in Puget Sound, Pacific Coast. Proc. and Trans. Liverpool Biol. Soc. 12:248–267.
RITTER, W. E.
1900. Some ascidians from Puget Sound collections of 1896. Ann. N.Y. Acad. Sci. 12:589–615.
1901. Papers from the Harriman Alaska Expedition. 22, The Ascidians. Proc. Wash. Acad. Sci. 3:225.
1907. The ascidians collected by the U.S. Fisheries steamer *Albatross* on the coast of California during the summer of 1904. Univ. Calif. Publ. Zool. 4(1):1–52.
1913. The simple ascidians from the north-eastern Pacific in the collection of the U.S. National Museum. Proc. U.S. Nat. Mus. 45:427–505.
RITTER, W. E., AND RUTH A. FORSYTH
1917. Ascidians of the littoral zone of southern California. Univ. Calif. Publ. Zool. 16(24):439–512.

STIMPSON, W.
1855. Descriptions of some of the new marine invertebrata from the Chinese and Japanese seas. Proc. Philadelphia Acad. Sci. 7:375–384.

VAN NAME, W. G.
1930. The ascidians of Porto Rico and the Virgin Islands. Sci. Survey of Porto Rico and the Virgin Islands. N.Y. Acad. Sci. 10(4): 403–535.
1945. The North and South American ascidians. Bull. Am. Mus. Natur. Hist. 84:1–476.

ARTHROPODA

Cirripedia

BROCH, HJALMAR
1922. Papers from Dr. Th. Mortensen's Pacific Expedition 1914–1916. X. Studies on Pacific cirripeds. Vidensk. Meddel. Dansk Naturhist. For. 73:215–358.

CORNWALL, IRA E.
1951. The barnacles of California (Cirripedia). Wasmann Journ. Biol. 9(3):311–346.

DARWIN, CHARLES R.
1851. A monograph of the sub-class Cirripedia. The Lepadidae: or, pedunculated cirripedes. London, Ray Society. 400 p.
1854. A monograph on the sub-class Cirripedia. The Balanidae, the Verrucidae, etc. London, Ray Society. 684 p.

HENRY, DORA PRIAULX
1940. The Cirripedia of Puget Sound with a key to the species. Univ. Wash. Publ. Oceanogr. 4(1):1–48.
1941. Notes on some sessile barnacles from Lower California and the west coast of Mexico. Proc. New Eng. Zool. Club 18:99–107.
1942. Studies on the sessile Cirripedia of the Pacific coast of North America. Univ. Wash. Publ. Oceanogr. 4(3):95–134.
1943. Notes on some barnacles from the Gulf of California. Proc. U.S. Nat. Mus. 93(3166):367–373.
1954. Cirripedia: The barnacles of the Gulf of Mexico. *In* Gulf of Mexico, its origin, waters, and marine life, Galtsoff, P. S. (ed.) Fish. Bull. U.S. Fish and Wildlife Service 55(89):443–446.
1958. Intertidal barnacles of Bermuda. Journ. Marine Res. 17:215–234.
1960. Thoracic Cirripedia of the Gulf of California. Univ. Wash. Publ. Oceanogr. 4(4):135–158.

MACDONALD, RODERICK
1929. A report on some cirripeds collected by the S.S. "Albatross" in the eastern Pacific during 1891 and 1904. Bull. Mus. Comp. Zool., Harvard College 69(15):527–538.

NILSSON-CANTELL, C. A.
1957. Thoracic cirripeds from Chile. Reports of the Lund University Chile Expedition 1948–1949. Lunds Univ. Arsskrift 53(9): 1–25.

PILSBRY, HENRY A.
1907a. Notes on some Pacific cirripedes. Proc. Acad. Natur. Sci. Philadelphia 59:360–362.
1907b. Cirripedia from the Pacific coast of North America. Bull. U.S. Bur. Fish. 26:191–204.
1907c. The barnacles (Cirripedia) contained in the collections of the National Museum. Bull. U.S. Nat. Mus. 60:1–122.
1916. The sessile barnacles (Cirripedia) contained in the collections of the U.S. National Museum; including a monograph of the American species. Bull. U.S. Nat. Mus. 93:1–366.
1918. Cirripedia from the Panama Canal Zone. Bull. U.S. Nat. Mus. 103:185–188.
1927. Cirripedia of Curacao. Bijdr. Dierk. 25:36–37.
1928. Littoral barnacles of the Hawaiian Islands and Japan. Proc. Acad. Natur. Sci. Philadelphia 79:305–317.

ROSS, ARNOLD
1962. Results of the *Puritan*-American Museum of Natural History expedition to western Mexico. 15. The littoral balanomorph Cirripedia. Am. Mus. Novit. 2084:1–44.
1968. Bredin-Archbold-Smithsonian biological survey of Dominica. 8. The intertidal Balanomorph Cirripedia. Proc. U.S. Nat. Mus. 125(3663):1–23.
1969. Studies on the Tetraclitidae (Cirripedia: Thoracica): Revision of *Tetraclita*. Trans. San Diego Soc. Natur. Hist. 15(15): 237–251.
1971. A new genus of Chthamalidae (Cirripedia) from the southeastern Pacific island of San Ambrosio. Trans. San Diego Soc. Natur. Hist. 16(11):265–278.

WERNER, WILLIAM E.
1967. The distribution and ecology of the barnacle *Balanus trigonus*. Bull. Marine Sci. 17(1):64–84.

ZULLO, VICTOR A.
1961. A new subgenus and species of coral-inhabiting barnacle from the Gulf of California. The Veliger 4(2):71–75.

Amphipoda

(Note: for general reference and comprehensive bibliographies see Barnard, 1969a and 1969b.)

BARNARD, J. LAURENS

1952a. A new amphipod of the genus *Ceradocus (Denticeradocus)* from Lower California. Bull. So. Calif. Acad. Sci. 51:55–59.

1952b. A new species of amphipod from Lower California (genus *Eriopisa*). Pacific Sci. 6:295–299.

1954. Amphipoda of the family Ampeliscidae collected in the eastern Pacific Ocean by the *Velero III* and *Velero IV*. Allan Hancock Pac. Exped. 18:1–37.

1960. The amphipod family Phoxocephalidae in the eastern Pacific Ocean, with analyses of other species and notes for a revision of the family. Allan Hancock Pac. Exped. 18:175–368.

1964. Marine Amphipoda of Bahia de San Quintin, Baja California. Pacific Natur. 4:55–139.

1968. Gammaridean Amphipoda of Bahia de Los Angeles, Baja California. Trans. San Diego Soc. Natur. Hist.

1969a. Gammaridean Amphipoda of the rocky intertidal of California: Monterey Bay to La Jolla. Bull. U.S. Nat. Mus. 258:1–230.

1969b. The families and genera of marine Gammaridean Amphipoda. Bull. U.S. Nat. Mus. 271:1–535.

SHOEMAKER, C. R.

1925. The Amphipoda collected by the United States Fisheries Steamer *Albatross* in 1911, chiefly in the Gulf of California. Bull. Am. Mus. Natur. Hist. 52:21–61.

1942. Amphipod crustaceans collected on the presidential cruise of 1938. Smithsonian Misc. Coll. 101(11):1–52.

Isopoda

(Note: there is a vast amount of published literature dealing with the isopod Crustacea, but the following references are for papers dealing specifically with the Panamic region or are important monographic works, such as Richardson [1905], Van Name [1936], and Schultz [1969].)

HANSEN, H. J.

1897. Reports on the dredging operations off the west coast of Central America to the Galapagos . . . by the U.S. Fish Comm. Steamer *Albatross* . . . part XXII. The Isopoda. Bull. Mus. Comp. Zool. 31(5):95–130.

MENZIES, R. J.

1962. The marine isopod fauna of Bahia de San Quintin, Baja California, Mexico. Pac. Natur. 3(11):338–348.

MENZIES, R. J., AND D. FRANKENBERG

1966. Handbook of the common marine isopod crustacea of Georgia. Univ. Ga. Press, Athens. 93 p.

MULAIK, STANLEY B.
1960. Contribución al conocimiento de los isopodos terrestres de México (Isopoda, Oniscoidea). Revista de la Sociedad Mexicana, de Historia Natural 21(1): 79–294.

OHLINC, AXEL
1901. Isopoda from Tierra del Fuego and Patagonia, 1. Valvifera. Svenska Exp. Till Magellanslanderna Bd. II (11):261–306.

OHMART, ROBERT D.
1964. Ecology and taxonomy of some isopoda in the northern Gulf of California. Univ. Ariz. Biol. Studies in the Gulf of California, II: 11 p.

RICHARDSON, HARRIET
1901. Papers from the Hopkins-Stanford Galapagos Expedition, 1898–1899, VI. The Isopods. Proc. Wash. Acad. Sci. III: 565–568.
1904. Contributions to the natural history of the Isopoda. Proc. U.S. Nat. Mus. 27:1–31.
1905a. Description of a new species of *Livoneca* from the coast of Panama. Proc. U.S. Nat. Mus. 29:445–446.
1905b. A monograph on the isopods of North America. Bull. U.S. Nat. Mus. 54:1–727.
1910. Report on isopods from Peru collected by Dr. R. E. Coker. Proc. U.S. Nat. Mus. 38:79–85.
1914. Reports on the scientific results of the expedition to the tropical Pacific . . . on the U.S. Fish Comm. Steamer *Albatross* . . . XVII . . . and 1904–1905. Bull. Mus. Comp. Zool., Harvard 58(8):361–372.

SCHULTZ, GEORGE A.
1969. The marine Isopod Crustacea (How to know . . .). William C. Brown, Pub., Dubuque, Iowa. 359 p.

SHINO, S.
1964. On two species of bopyrid isopods parasitic on *Callianassa uncinata* Milne-Edwards, from Chile. Rept. Fac. Fisheries, Prefectural Univ. of Mie 5:27–32.

VAN NAME, W. G.
1924. Isopods from the Williams Galapagos Expedition. Zoologica 5(18): 181–210.
1936. The American land and fresh-water Isopod Crustacea. Bull. Am. Mus. Natur. Hist. 71:1–535.

Stomatopoda

BIGELOW, R. P.
1894. Report upon the crustacea of the order Stomatopoda collected by the Steamer *Albatross* between 1885 and 1891, and on other specimens in the United States National Museum. Proc. U.S. Nat. Mus. 17:489–550.

BOONE, P. L.
1930. Crustacea: Stomatopoda and Brachyura. Scientific results of cruises of the yachts *Eagle* and *Ara,* 1921–1928, William K. Vanderbilt, commanding. Bull. Vanderbilt Marine Mus. 2:1–228.
1934. Crustacea: Stomatopoda and Brachyura. Bull. Vanderbilt Marine Mus. 5:1–210.
LUNZ, G. R.
1937. Stomatopoda of the Bingham Oceanographic Collection. Bull. Bingham Ocean. Coll. 6(50): 1–17.
RATHBUN, M. J.
1910. The stalk-eyed crustacea of Peru and the adjacent coast. Proc. U.S. Nat. Mus. 38:531–620.
SCHMITT, W. L.
1940. The stomatopods of the west coast of America. Allan Hancock Pac. Exped. 5(4):129–225.

Natantia

ANDERSON, W. W., AND M. LINDNER
1965. Clave provisional para camarones de la familia Penaeidae con referencia especial a las especies Americanas. Contrib. del Inst. Nacional de Investigaciones Biológico-Pesqueras, Serie Trabajos de Divulgación IX (90): 1–61.
BARNARD, K. H.
1950. Descriptive catalogue of South African decapod Crustacea. Ann. So. Afr. Mus. 38: 1–837.
BURKENROAD, M. D.
1934. Littoral Penaeidae chiefly from the Bingham Oceanographic collection. Bull. Bingham Ocean. Coll. 4(7): 1–109.
1937. Sergestidae (Crustacea Decapoda) from the Lower California region, with descriptions of two new species, and some remarks on the organs of Pesta in *Sergestes*. Zool. 22:315–329.
1938. Penaeidae from the region of Lower California and Clarion Island, with descriptions of four new species. Zool. 23:55–91.
CASTRO, PETER
1966. The shrimps (Crustacea Decapoda Natantia) of the Sonoran coast of the Gulf of California. Univ. of Ariz. Biol. studies in the Gulf of Calif. IV (2). 36 p.
CHACE, FENNER A.
1937. Caridean decapod Crustacea from the Gulf of California and the west coast of Lower California. Zoologica 22:109–138.
1951. The grass shrimps of the genus *Hippolyte* from the west coast of North America. Journ. Wash. Acad. Sci. 41(1):35–39.
COUTIERE, HENRI
1909. The American species of snapping shrimps of the genus *Synalpheus*. Proc. U.S. Nat. Mus. 36:1–93.

DEMAN, J. G.
1911. Alpheidae. Résultats des explorations zoologiques, botaniques, océanographiques et géologiques entreprises aux Indes Orientales en 1899–1900 á bord du Siboga. Siboga Exped. 39a(60): 133–466.

GLASSELL, S. A.
1936. New and obscure decapod Crustacea from the west American coasts. Trans. San Diego Soc. Natur. Hist. 8:411–454.

HART, JOSEPHINE F. L.
1964. Shrimps of the genus *Betaeus* on the Pacific coast of North America, with descriptions of three new species. Proc. U.S. Nat. Mus. 115(3490):431–466.

HOLMES, S. J.
1900. Synopsis of the California stalk-eyed Crustacea. Occ. Pap. Calif. Acad. Sci. 7:1–262.

HOLTHUIS, L. B.
1947. The Hippolytidae and Rhynchocinetidae collected by the Siboga and Snellius expeditions with remarks on other species. The Decapoda of the Siboga Expedition, part IX. Siboga Exped. 39a:1–100.
1950. The Palaemonidae collected by the Siboga and Snellius expeditions with remarks on other species. I. Subfamily Palaemonidae. The Decapoda of the Siboga expedition, part X. Siboga Exped. 39a:1–268.
1951. A general revision of the Palaemonidae (Crustacea Decapoda Natantia) of the Americas. I. Subfamilies Euryrhynchinae and Pontoniinae. Occ. Pap. Allan Hancock Found. 11:1–332.
1952. A general revision of the Palaemonidae (Crustacea Decapoda Natantia) of the Americas. II. The Subfamily Palaemoninae. Occ. Pap. Allan Hancock Found. 12:1–396.
1955. The recent genera of the caridean and stenopodidean shrimps (Crustacea Decapoda Natantia) with keys for their determination. Zool. Verh. 26:1–157.

KINGSLEY, J. S.
1878. Notes on the North American Caridea in the Museum of the Peabody Academy of Science at Salem, Mass. Proc. Acad. Natur. Sci. Philadelphia 30:89–98.

LOCKINGTON, WILLIAM N.
1878. Remarks on some new Alphei, with a synopsis of the North American species. Ann. Mag. Natur. Hist. (5) 1:465–480.
1879. Notes on Pacific coast Crustacea. Bull. Essex Inst. 10:159–165.

RATHBUN, M. J.
1910. The stalk-eyed crustaceans of Peru and the adjacent coast. Proc. U.S. Nat. Mus. 38:531–620.

RIOJA, E.
1948. Estudios carcinológicos. XX. Valor diagnóstico del tercer maxilípedo y de sus cerdas en algunas especies del género *Penaeus* de las costas de Mexico. An. Inst. Biol. 19:453–457.

SCHMITT, W. L.
1921. The marine decapod Crustacea of California. Univ. Calif. Publ. Zool. 23:1–470.
1924. Crustacea (Macrura and Anomura). XXIV. Expedition of the California Academy of Sciences to the Gulf of California in 1921. Proc. Calif. Acad. Sci. (4) 13 (24):381–388.
1935. The west American species of shrimp of the genus *Penaeus*. Proc. Biol. Soc. Wash. 35:15–24.

STIMPSON, WILLIAM
1860. Notes on North American crustacea in the museum of the Smithsonian Institution, No. II. Ann. Lyc. Natur. Hist. N.Y. 7:176–246.
1871. Notes on North American crustacea in the museum of the Smithsonian Institution, No. III. Ann. Lyc. Natur. Hist. N.Y. 10:92–136.

STREETS, T. HALE
1871. Description of five new species of Crustacea from Mexico. Proc. Acad. Natur. Sci. Philadelphia 23:225–227.

Macrura

ALLEN, B. J.
1916. Notes on the spiny lobster. Univ. Calif. Publ. Zool. 16:139–152.

BACKUS, JOHN
1960. Observations on the growth rate of the spiny lobster. Calif. Fish and Game 46(2):177–181.

BOONE, LEE
1931. Anomuran, Macruran Crustacea from Panama and the Canal Zone. Bull. Am. Mus. Natur. Hist. 63:137–189.

DEES, L. T.
1968. Spiny lobsters. U.S. Dept. Int., Fisheries Leaflet 523. 7 p.

HERRNKIND, W. F.
1970. The migration of the spiny lobster. Nat. Geographic 137(5): 36–43.

LINDBERG, ROBERT G.
1955. Growth, population dynamics, and field behavior in the spiny lobster, *Panuliris interruptus* (Randall). Univ. Calif. Publ. Zool. 59(6):157–248.

SCHMITT, W. L.
1935. Crustacea, Macrura and Anomura of Porto Rico and the Virgin Islands. N.Y. Acad. Sci., Sci. Survey of Porto Rico and the Virgin Islands 15(2):125–227.

Thalassinoidea

BOUVIER, E. L.
1895. Sur une collection de Crustaces Decapodes recueillis en Basse-Californie par M. Diguet Bull. Mus. Hist. Natur. Paris 1:6–8.

DEMAN, J. G.
1925a. Sur deux especes encore imparfaitement connues du genre *Axius* Leach. Bull. de la Societe Zoologique de France 50(2–3): 50–61.
1925b. The Decapoda of the Siboga Expedition, Pt. VI. The Axiidae collected by the Siboga Expedition. Siboga Expeditie 39a 5.

LOCKINGTON, WILLIAM N.
1878. Remarks upon the Thalassinidea and Astaeidea of the Pacific coast of North America. Ann. Mag. Natur. Hist. 2(5):299–304.

MACGINITIE, G. E.
1930. The natural history of the mud shrimp *Upogebia pugettensis* (Dana). Ann. Mag. Natur. Hist. 10(6):36–44.
1934. The natural history of *Callianassa californiensis* Dana. Am. Midland Natur. 15(2):166–177.

WESTERVELT, C. A., JR.
1967. The littoral anomuran decapod crustacean fauna of the Punta Penasco-Bahia La Cholla area in Sonora, Mexico. Ph.D. dissertation, Univ. Ariz., Tucson. Also in Dissertation Abstracts, 27B: 4183-B.

Porcellanidae

BENEDICT, JAMES E.
1902. Description of a new genus and 46 new species of crustaceans of the Family Galatheidae, with a list of the known marine species. Proc. U.S. Nat. Mus. 15:1–26.

BOONE, PEARL L.
1931a. The littoral crustacean fauna of the Galapagos Islands, Part II. Anomura. Zoologica 14:1–62.
1931b. Anomuran, Macruran Crustacea from Panama and the Canal Zone. Bull. Am. Mus. Natur. Hist. 63:137–189.

BOUVIER, E. L.
1895. Sur une collection de Crustaces decapodes recueillis en Basse-Californie par M. Diguet. Bull. Mus. Hist. Natur. Paris 1:6–8.
1898. Sur quelques crustaces anomures et brachyures recueillis par M. Diguet en Basse-Californie. Bull. Mus. Hist. Natur. Paris 4:371–384.

BRUSCA, RICHARD C., AND JANET HAIG
1972. Range extensions of porcelain and hermit crabs in the Gulf of California. Bull. So. Calif. Acad. Sci. 71(1):56.

CHACE, FENNER A., JR.
1962. The non-brachyuran decapod crustaceans of Clipperton Island. Proc. U.S. Nat. Mus. 113:605-635.

DEL SOLAR, E. M., F. BLANCAS, AND R. MAYTA
1970. Catalogo de Crustaceos del Peru. Lima. 53 p.

GLASSELL, STEVE A.
1936. New porcellanids and pinnotherids from tropical North American waters. Trans. San Diego Soc. Natur. Hist. 8:277–304.
1937. Porcellanid crabs from the Gulf of California. Zoologica 22:79–88.
1938a. Three new anomuran crabs from the Gulf of California. Allan Hancock Pac. Exped. 5(1):1–6.
1938b. New and obscure decapod crustacea from the west American coast. Trans. San Diego Soc. Natur. Hist. 8:411–454.
1945. Four new species of North American crabs of the genus Petrolisthes. Journ. Wash. Acad. Sci. 35(7):223–229.

HAIG, JANET
1956. Notes on two anomuran crustaceans new to California waters. Bull. So. Cal. Acad. Sci. 55(2):79–81.
1960. The Porcellanidae (crustacea Anomura) of the eastern Pacific. Allan Hancock Pac. Exped. 25:1–226.
1962. Papers from Dr. Th. Mortensen's Pacific Expedition, 1914–1916, 79. Porcellanid crabs from eastern and western America. Vidensk. Medd. fra Dansk naturh. Foren., bd. 124:171-192.
1968. Eastern Pacific expeditions of the New York Zoological Society. Porcellanid crabs (Crustacea: Anomura) from the west coast of tropical America. Zoologica 53(2):57–74.

HAIG, JANET, T. S. HOPKINS, AND T. B. SCANLAND
1970. The shallow water anomuran crab fauna of southwestern Baja California, Mexico. Trans. San Diego Soc. Natur. Hist. 16(2): 13–32.

HENDERSON, J. R.
1888. Report on the Anomura collected by *H.M.S. Challenger* during the years 1873–1876. Rpt. Zool. Challenger Exped., XI, 221 p.

SCHMITT, WALDO L.
1921. The marine decapod crustacea of California. Univ. Calif. Publ. Zool. 23:1–470.
1924a. Crustacea (Macrura and Anomura). No. 36 in: "Expedition of the California Academy of Sciences to the Gulf of California in 1921." Proc. Calif. Acad. Sci. (4), 13(24):381–388.
1924b. The Macrura and Anomura collected by the Williams Galapagos Expedition, 1923. Zoologica 5(15): 161–171.

STREETS, T. HALE
1871. Catalogue of Crustacea from the Isthmus of Panama, collected by J. A. McNeil. Proc. Acad. Natur. Sci. Philadelphia 23: 238–243.

VOGEL, B. R.
1966. A report on a collection of crabs from the Gulf of California. Southwestern Naturalist 11:139–140.
WESTERVELT, C. A.
1967. The littoral anomuran decapod crustacean fauna of the Punta Penasco-Bahia la Cholla area in Sonora, Mexico. Dissertation Abstracts, 27B:4183–B, Univ. Ariz.

Paguroidea

BENEDICT, JAMES E.
1892. Preliminary description of 37 new species of hermit crabs of the genus *Eupagarus* in the United States National Museum. Proc. U.S. Nat. Mus. 15:1–26.
1902. Description of a new genus and 46 new species of crustaceans of the family Galatheidae, with a list of the known marine species. Proc. U.S. Nat. Mus. 26:243–334.
BOONE, PEARL LEE
1931. A collection of anomuran and macruran crustacea from the Bay of Panama and the fresh waters of the Canal Zone. Bull. Am. Mus. Natur. Hist. 63:137–189.
1932. The littoral crustacean fauna of the Galapagos Islands. Part II. Anomura. Zoologica 14(1):1–62.
GLASSELL, STEVE A.
1937. Hermit crabs from the Gulf of California and the west coast of Lower California. Zoologica 22:241–263.
GORDON, J. A.
1956. A bibliography of pagurid crabs, exclusive of Alcock, 1905. Bull. Am. Mus. Natur. Hist. 108(3):257–352.

Brachyura

BAKER, C. F.
1912. Notes on the Crustacea of Laguna Beach. 1st Ann. Rpt. Laguna Marine Lab.: 100–117.
BEHRSTOCK, ROBERT A.
1966. A guide to some of the brachyuran crabs of the northern Gulf of California. Biol. Studies in the Gulf of Calif. IV, 17 p. Mimeographed reports of student projects, Univ. Ariz.
BOONE, P. L.
1927. The littoral crustacean fauna of the Galapagos islands. Part 1: Brachyura. Zoologica 8:127–288.
1929. A collection of brachyuran Crustacea from the Bay of Panama and the fresh waters of the Canal Zone. Bull. Am. Mus. Natur. Hist. 58:561–583.
1930. Scientific results of the cruises of the yachts 'Eagle' and 'Ara,' 1921–1928, William K. Vanderbilt, Commanding. Crustacea: Stomatopoda and Brachyura. Vanderbilt Marine Mus. Bull. 2:1–228.

BOTT, R.
1955. Dekapoden (Crustacea) aus El Salvador. 2. Litorale Decapoden, ausser *Uca*. Senckenbergiana, Biologia 36:45–70.

BOUVIER, E. L.
1895. Sur une collection de Crustaces decapodes recueillis en Basse-Californie par M Diguet. Bull. Mus. Natur. Hist. Paris. 1:6–8.

BUITENDIJK, ALIDA M.
1950. Note on a collection of Decapoda Brachyura from the coasts of Mexico, including the description of a new genus and species. Leiden, Rijksmus. Natur. Hist., Zool. Mededel. 30:269–282.

CONTRERAS, F.
1930. Contribución al conocimiento de las jaibas de México. México. Univ. Nac. Inst. Biol. An. 1A 227–241.

COVENTRY, G. A.
1944. Results of the fifth George Vanderbilt expeditions (1941). (Bahamas, Caribbean Sea, Panama, Galapagos Archipelago and Mexican Pacific Islands). The Crustacea. Acad. Natur. Sci. Philadelphia, Monogr. 6:531–544.

CRANE, JOCELYN
1937a. The Templeton Crocker Expedition. III. Brachygnathous crabs from the Gulf of California and the west coast of Lower California. Zoologica 22(3):47–78.
1937b. The Templeton Crocker Expedition. III. Oxystomatous and Dromiaceous crabs from the Gulf of California and the west coast of Lower California. Zoologica 22(3):97–108.
1940. On the post-embryonic development of brachyuran crabs of the genus *Ocypode*. No. 18 in Eastern Pacific Expeditions of the New York Zoological Society. Zoologica 25:65–82.
1941. Crabs of the genus *Uca* from the west coast of central America. Zoologica 26(3):145–208.
1947. Intertidal brachygnathous crabs from the west coast of tropical America with special reference to ecology. Zoologica 32:69–95.

DANA, J. D.
1852. Crustacea, pt. 1. *In* The United States Exploring Expedition during the years 1838, 1839, 1840, 1841, and 1842, under the command of Charles Wilkes, U.S.N. Philadelphia. 685 p.

DOLEZAL, R. J.
1964. The brachyuran crabs of Puerto Penasco, Sonora, Mexico, with an annotated key. Biol. studies in the Gulf of California II. 10 p. Mimeographed reports of student projects, Univ. Ariz.

FAXON, W.
1893. Reports of the dredging operations off the west coast of Central America to the Galapagos, to the west coast of Mexico, and in the Gulf of California . . . by the U.S. Fish Commission steamer 'Albatross,' during 1891 . . . VI. Preliminary descriptions of new species of Crustacea. Harvard Mus. Comp. Zool. Bull. 24:149–220.

FAXON, W. (cont'd.)

1895. Reports on an exploration off the west coasts of Mexico, Central and South America, and off the Galapagos Islands . . . by the U.S. Fish Commission steamer 'Albatross,' during 1891 . . . XV. The stalk-eyed crustacea. Harvard, Mus. Comp. Zool. Mem. 18, 292 p.

FINNEGAN, SUSAN

1931. Report on the Brachyura collected in Central America, the Gorgona and Galapagos Islands, by Dr. Crossland on the 'St. George' Expedition to the Pacific, 1924–1925. Linn. Soc. London, Journ. Zool. 37:607–673.

FRICKE, D.

1968. Crustacea of the northeastern coast of the Gulf of California. Biol. Studies of the Gulf of Calif. IV, 21 p. Mimeographed reports of student projects, Univ. Ariz.

GARTH, J. S.

1939. New brachyuran crabs from the Galapagos Islands. Allan Hancock Pac. Exped. 5(2):9–48.

1940. Some new species of brachyuran crabs from Mexico and the Central and South American mainland. Allan Hancock Pac. Exped. 5:53–127.

1946. Littoral brachyuran fauna of the Galapagos Archipelago. Allan Hancock Pac. Exped. 5:341–601.

1948. The Brachyura of the 'Askoy' Expedition with remarks on carcinological collecting in the Panama Bight. Am. Mus. Natur. Hist. Bull. 92(1):1–66.

1957. Reports of the Lund University Chile Expedition 1948–1949. 29. The Crustacea Decapoda Brachyura of Chile. Lunds Univ. Arsskr., n.f., Avd. 2, 53(7):1–128.

1958. Brachyura of the Pacific Coast of America. Oxyrhyncha. Allan Hancock Pac. Exped. 21, in 2 vols.: 1–858, 1–55.

1960. Distribution and affinities of the brachyuran Crustacea (Symposium: The biogeography of Baja California and adjacent seas. Part II. Marine Biotas) Systematic Zool. 9:105–123.

1961. Eastern Pacific Expeditions of the New York Zoological Society. XLV. Non-intertidal brachygnathous crabs from the west coast of tropical America. Part 2: Brachygnatha Brachyrhyncha. Zoologica 46:133–159.

1965. The brachyuran decapod crustaceans of Clipperton Island. Calif. Acad. Sci. Proc. (4)33:1–46.

GARTH, J. S., AND W. STEPHENSON

1966. Brachyura of the Pacific coast of America. Brachyrhyncha: Portunidae. Allan Hancock Monog. in Marine Biol. 1:1–154.

GLASSELL, STEVE A.

1933. Description of five new species of Brachyura collected on the west coast of Mexico. Trans. San Diego Soc. Natur. Hist. 7:331–344.

GLASSELL, STEVE A. (cont'd.)

1934a. Affinities of the brachyuran fauna of the Gulf of California. Journ. Wash. Acad. Sci. 24(7):296–302.

1934b. Some corrections needed in recent carcinological literature. San Diego Soc. Natur. Hist. Trans. 7:453–454.

1935. New or little known crabs from the Pacific coast of northern Mexico. San Diego Soc. Natur. Hist. Trans. 8:93–105.

1936. Six new brachyuran crabs from the Gulf of California. Zoologica 21:213–218.

HILTON, W. A.

1916. Crustacea from Laguna Beach. Journ. Ent. and Zool., Pomona College 8:65–73.

HOLMES, S. J.

1900. Synopsis of the California stalk-eyed Crustacea. Calif. Acad. Sci., Occas. Pap. 7:1–262.

HOLTHUIS, L. B.

1954a. On a collection of decapod Crustacea from the republic of El Salvador (Central America). Leiden, Rijksmus. Natur. Hist., Zool. Verhandel, 23:1–43.

1954b. Observaciones sobre los Crustaceos decapodos de la Republica de El Salvador. El Salvador, Univ., Inst. Trop. Invest. Cien., Comm. 3:159–166.

HULT, J.

1938. Crustacea Decapoda from the Galapagos Islands collected by Mr. Rolf Blomberg. Arkiv. for Zool. 30A(5):1–18.

LEENE, JENTINA E.

1938. The Decapoda Brachyura of the Siboga Expedition. VII. Brachygnatha: Portunidae. *In* Siboga-Expeditie Mon. 39C, Liv. 131:1–156.

LOCKINGTON, W. N.

1877a. Description of seventeen new species of Crustacea. Calif. Acad. Sci. Proc. 7:41–48.

1877b. Remarks on the Crustacea of the west coast of North America, with a catalogue of the species in the museum of the California Academy of Sciences. Cancroidea. Calif. Acad. Sci., Proc. 7:94–108.

MENZIES, ROBERT J.

1948. A revision of the brachyuran genus *Lophopanopeus*. Occ. Paps. Allan Hancock Found. 4. 44 p.

MILNE-EDWARDS, A.

1879. Etudes sur les Xiphosures et les Crustaces de la region mexicaine. 1873–1880. *In* Mission Scientifique au Mexique et dans l'Amerique centrale, Recherches Zoologiques. Paris, 5. 368 p.

NININGER, H. H.

1918. Crabs taken at Laguna Beach in the summer of 1916. Journ. Ent. and Zool., Pomona College 10:36–42.

NOBILI, G.
1897. Decapodi e Stomatopodi raccolti dal Dr. Enrico Festa nel Darien, a Curacao, La Guayra, Porto Cabello, Colon, Panama, ecc. Turin Univ., Mus. Zool. Anat. Compar., Bol. 12(280), 8 p.
1901. Viaggio del Dr. Enrico Festa nella Repubblica dell'Ecuadore e regione vicine. XXIII. Decapodi e Stomatopodi. Turin Univ., Mus. Zool. Anat. Compar. Bol. 16(415). 58p.
1902. Decapodi raccolti dal Dr. F. Silvestri nell' Chile. Rev. Chilena Hist. Natur. 6:233–238.

ORDWAY, A.
1863. Monograph of the genus *Callinectes*. Boston Journ. Natur. Hist. 7:567–583.

RATHBUN, MARY J.
1893. Scientific results of explorations by the U.S. Fish Commission steamer *Albatross*. XXIV. Descriptions of new genera and species of crabs from the west coast of North America and the Sandwich Islands. U.S. Nat. Mus. Proc. 16:223–260.
1894. Notes on the crabs of the family Inachidae in the United States National Museum. Proc. U.S. Nat. Mus. 17:43–75.
1896. The genus *Callinectes*. U.S. Nat. Mus. Proc. 18 (for 1895): 349–375.
1897. A revision of the nomenclature of the Brachyura. Biol. Soc. Wash. Proc. 11:153–167.
1898. The Brachyura collected by the U.S. Fish Commission steamer *Albatross* on the voyage from Norfolk, Virginia, to San Francisco, California, 1887–1888. U.S. Nat. Mus. Proc. 21: 567–616.
1900. Synopses of North-American invertebrates. VII. The cyclometopous or cancroid crabs of North America. Am. Natur. 34:131–143.
1902a. Papers from the Hopkins-Stanford Galapagos Expedition, 1898–1899. VIII. Brachyura and Macrura. Washington Acad. Sci. Proc. 4:275–292.
1902b. Descriptions of new decapod crustaceans from the west coast of North America. Proc. U.S. Nat. Mus. 24:885–905.
1904. Decapod crustaceans of the northwest coast of North America. *In* Harriman Alaska Exped. Washington. 10, Crustaceans. 210 p.
1907. Reports on the scientific results of the expedition to the tropical Pacific . . . by the U.S. Fish Commission steamer 'Albatross,' from August, 1899 to March, 1900 . . . IX. Reports on the scientific results of the expedition to the eastern tropical Pacific . . . by the U.S. Fish Commission steamer 'Albatross' from October, 1904 to March, 1905 . . . X. The Brachyura. Harvard, Mus. Comp. Zool. Mem. 35:23–74.

RATHBUN, MARY J. (cont'd.)
1908. Descriptions of fossil crabs from California. Proc. U.S. Nat. Mus. 35:341–349.
1910. The stalk-eyed Crustacea of Peru and the adjacent coast. U.S. Nat. Mus. Proc. 38:531–620.
1918. The grapsoid crabs of America. Bull. U.S. Nat. Mus. 97:1–461.
1923. Scientific results of the expedition to the Gulf of California . . . by the U.S. Fisheries steamship 'Albatross,' in 1911 . . . XIII. The Brachyuran crabs collected by the U.S. Fisheries steamer 'Albatross,' in 1911, chiefly on the west coast of Mexico. Am. Mus. Natur. Hist. Bull. 48(20):619–637.
1924a. Expedition of the California Academy of Sciences to the Gulf of California in 1921, Crustacea (Brachyura). Calif. Acad. Sci. Proc. (4)13:373–379.
1924b. Brachyuran crabs collected by the Williams Galapagos Expedition, 1923. Zoologica 5:153–159.
1925. The spider crabs of North America. Bull. U.S. Nat. Mus. 129:1–613.
1926. The fossil stalk-eyed Crustacea of the Pacific slope of North America. U.S. Nat. Mus. Bull. 138. 155 p.
1930 The cancroid crabs of America. Bull. U.S. Nat. Mus. 152:1–609.
1933. Description of new species of crabs from the Gulf of California. Proc. Biol. Soc. Wash. 46:147–150.
1937. The oxystomatous and allied crabs of America. Bull. U.S. Nat. Mus. 166:1–278.

SCHMITT, W.L.
1921. The marine decapod Crustacea of California. Univ. Calif. Publ. Zool. 23:1–470.
1924. Expedition of the California Academy of Sciences to the Gulf of California in 1921. Crustacea (Brachyura). Proc. Calif. Acad. Sci. ser. 4, 13(24):373–379.
1933. Crustacea. *In* G. H. Banning, Hancock Expedition of 1933 to the Galapagos Islands. General Report. Zool. Soc. San Diego, Bull. 10:21–26.
1939. Decapod and other Crustacea collected on the Presidential Cruise of 1938. Smithsonian Misc. Coll. 98(6):1–29.

SIVERTSEN, E.
1933. The Norwegian Zoological Expedition to the Galapagos Islands 1925, conducted by Alf Wollebaek . . . VII. Littoral Crustacea Decapoda from the Galapagos Islands. Oslo Mus., Meddel. Zool. 38:1–23.

SMITH, S. I.
1869. Notes on new or little known species of American cancroid Crustacea. Boston Soc. Natur. Hist. Proc. 12:274–289.
1871. List of the Crustacea collected by J. A. McNiel in Central America. Peabody Acad. Sci., Ann. Rpt. 2–3, 1869–1870: 87–98.

STEPHENSON, W.
1965. A morphometric analysis of certain western American swimming crabs of the genus *Portunus* Weber 1795. *In* symposium on Crustacea, Marine Biol. Assoc. India. Abstracts of Papers 4.

STIMPSON, W.
1857. On the Crustacea and Echinodermata of the Pacific shores of North America. Part I. Crustacea. Journ. Boston Soc. Natur. Hist. 6:444–532.
1859. Notes on North American Crustacea, No. 1. New York, Lyc. Natur. Hist. Ann. 7:49–93.
1860. Notes on North American Crustacea, in the Museum of the Smithsonian Institution. No. II. New York, Lyc. Natur. Hist. Ann. 7:176–246.
1871. Notes on North American Crustacea in the Museum of the Smithsonian Institution. No. III. New York, Lyc. Natur. Hist. Ann. 10:92–136.

STREETS, T. H.
1871. Catalogue of Crustacea from the Isthmus of Panama, collected by J. A. McNeil. Proc. Acad. Natur. Sci. Philadelphia 23:238–243.

Index